Cell Analysis

Volume 1

Cell Analysis

Volume 1

Edited by
NICHOLAS CATSIMPOOLAS
Boston University School of Medicine
Boston, Massachusetts

PLENUM PRESS • NEW YORK AND LONDON

Library of Congress Cataloging in Publication Data

Main entry under title:

Cell analysis.

Bibliography: p.
Includes index.
1. Cytology—Technique. 2. Cells—Analysis. I. Catsimpoolas, Nicholas. [DNLM: 1.
Cells—Analysis—Periodical. W1 CE126H]

QH585.C44	574.87′028	82-5289
ISBN 978-1-4684-4099-7	ISBN 978-1-4684-4097-3 (eBook)	AACR2
DOI 10.1007/978-1-4684-4097-3		

© 1982 Plenum Press, New York
Softcover reprint of the hardcover 1st edition 1982
A Division of Plenum Publishing Corporation
233 Spring Street, New York, N.Y. 10013

Contributors

R. B. Allan, *Department of Cell Biology, University of Glasgow, Glasgow G12 8QQ, U.K.*

James W. Bacus, *Medical Automation Research Unit, Rush–Presbyterian–St. Luke's Medical Center, Chicago, Illinois 60612*

S. A. Ben-Sasson, *The Hubert H. Humphrey Centre for Experimental Medicine and Cancer Research, The Hebrew University–Hadassah Medical School, Jerusalem, Israel*

Michael W. Berns, *Department of Developmental and Cell Biology, University of California, Irvine, Irvine, California 92717*

Francis Dumont, *Unité de Cancérologie Expérimentale et de Radiobiologie, INSERM U95, 54500 Vandoeuvre-les-Nancy, France. Present address: Department of Immunology, Merck, Sharp and Dohme Research Laboratories, Rahway, New Jersey 07065*

N. B. Grover, *The Hubert H. Humphrey Centre for Experimental Medicine and Cancer Research, The Hebrew University–Hadassah Medical School, Jerusalem, Israel*

Volker Kachel, *Max-Planck-Institut für Biochemie, D-8033 Martinsried, West Germany*

J. M. Lackie, *Department of Cell Biology, University of Glasgow, Glasgow G12 8QQ, U.K.*

J. Naaman, *The Hubert H. Humphrey Centre for Experimental Medicine and Cancer Research, The Hebrew University–Hadassah Medical School, Jerusalem, Israel*

Gary C. Salzman, *Life Sciences Division, Los Alamos National Laboratory, University of California, Los Alamos, New Mexico 87545*

Robert J. Walter, *Department of Developmental and Cell Biology, University of California, Irvine, Irvine, California 92717*

P. C. Wilkinson, *Department of Bacteriology and Immunology, Western Infirmary, University of Glasgow, Glasgow G11 6NT, U.K.*

Preface

The selective combination of physical, biochemical, and immunological principles, along with new knowledge concerning the biology of cells and advancements in engineering and computer sciences, has made possible the emergence of highly sophisticated and powerful methods for the analysis of cells and their constituents. This series on *Cell Analysis* is, therefore, aiming at providing the theoretical and practical background on how these methods work and what kind of information can be obtained. *Cell Analysis* will cover techniques on cell separation, cell identification and classification, characterization of organized cellular components, functional properties of cells, and cell interactions. Applications in cell biology, immunology, genetics, toxicology, specific diseases, diagnostics and therapeutics, and other areas will be covered whenever relevant results exist.

Nicholas Catsimpoolas
Boston, Massachusetts

Contents

Chapter 3

Combination of Two Physical Parameters for the Identification and Separation of Lymphocyte Subsets

Francis Dumont

Chapter 4

Electrical Sizing of Cells in Suspension

N. B. Grover, S. A. Ben-Sasson, and J. Naaman

Chapter 5

Light Scattering Analysis of Single Cells

Gary C. Salzman

Quantification of Red Blood Cell Morphology

JAMES W. BACUS

I. HISTORY

The description of red blood cells extends back to the earliest days of the development of the microscope. In 1658 Swammerdam described the oval red blood cells of the frog, and in 1673 Leeuwenhoek discovered the red cells in human blood and described them as circular in contrast to amphibian cells. In 1770 Hewson published detailed descriptions of structure, form, and dimensions of red blood cells from different animals, including man. In 1846 Gulliver published measurements of red cell size (diameter) on 485 species of vertebrates.

Until this time, descriptions and measurements concentrated on the average diameter of cells from specimens and not on size distributions of cells in individuals. However, by the 1880s various workers had reported red cell diameter sample distributions. During this same period, there was concomitant growth in knowledge and understanding of hemoglobin in red cells. In 1747 Menghini demonstrated the presence of iron in red cells; Funke isolated hemoglobin crystals in 1851; and Hoppe-Seyler had indicated the functional characteristics of oxygen uptake and discharge by 1867. Beginning with the observations of Soret (1878), it was understood that the primary hemoglobin absorption band was in the violet near the visible limit, around 415 nm (Jope, 1949). However, this was largely ignored because of the development of techniques for staining blood by Ehrlich in 1877 (Conn, 1948). The development

JAMES W. BACUS • Medical Automation Research Unit, Rush–Presbyterian–St. Luke's Medical Center, Chicago, Illinois 60612.

of these staining techniques stimulated the morphologic study of blood and the relationship between the size and hemoglobin content of the erythrocyte, even though there was no precise stoichiometric relationship between the amount of hemoglobin and the perceived color. In summary, then, by about 1900 the importance of both the size and hemoglobin content of red cells was becoming clear.

From 1880 to 1930, a gradual definition of hypochromic anemia and iron deficiency developed. Lange had first described the disease chlorosis in 1554; by the 1930s this was gradually understood to be iron deficiency anemia as we understand it today (Witts, 1969). Price-Jones (1933) published an exhaustive study of red cell diameter sample distributions for blood from normal individuals and for blood from some individuals with anemia, including iron deficiency. He indicated size changes in both the mean and variances of sample population distributions associated with anemia. Gradually, from the 1920s to the 1940s, the use of the mean red cell size (measured by the hematocrit, divided by the cell count) as proposed by Wintrobe (1930, 1931, 1932, 1934) and others (Jorgensen and Warburg, 1927) became a common measure used to differentiate microcytic anemias from normal. Cooley and Lee (1925) first described thalassemia in the homozygous state. This was later reported in the heterozygous form by Wintrobe et al. (1940). Also, Cartwright and Wintrobe (1952, 1966) described the anemia of infection and other chronic disorders in the 1940s and 1950s. Each of these three types of anemia resulted in the production of small hypochromic red cells and thus were potentially confused with each other during differential diagnosis based on changes in mean cell size. The variation in cell size, demonstrated to be important by Price-Jones, was not easily measured and thus was largely ignored as a routine quantitative parameter.

Sorenson (1876) recognized that cells larger than normal were characteristic of pernicious anemias, and Addison (1855) provided the first clinical description of megaloblastic anemia. The description of other macrocytic anemias, e.g., from liver disease and macrocytosis caused by stimulated erythropoiesis, followed this early work. Similarly, other anemias were gradually described by their characteristic red cell morphology. For example, Haden (1934) described measurement relationships between the thickness and diameter of red cells from normal blood and from the blood of patients with spherocytosis. The first morphologic descriptions of the red cells in sickle cell anemia were provided by Herrick (1910).

In many ways, our ability to describe red cells has depended upon the sophistication of our observational and measurement apparatus. The above described development of measurement and quantification of red cells depended upon the development of the microscope, simple counting chambers, hematocrit tubes, and, later, upon advances in chemistry and absorption spectroscopy. Ultimately, our current understanding of the disease process and its description have resulted from these measurement capabilities.

In the mid-1900s further advances in microscopy occurred in the area of quantitative microdensitometry. This is probably best characterized by the work of Casperson (1936, 1940), who performed precise microphotometric absorption studies of the nucleic acid of tissue smears. Although this work was limited by current standards with regard to detailed absorption measurements, since the measurement apertures usually encompassed the entire cell or the entire cell nucleus, they provided a clear direction for the potential of quantitative and more automated density and distributional measurements from previously subjective impressions and tedious manual measurement methods. In the 1950s advances in electronics enabled faster scanning with flying spot, television scanning, automatically controlled stage movement and focus, and very small apertures of about 0.15 μm.

In the 1970s several investigators studied red blood cells with variations of these faster, high-resolution scanning methods (Bentley and Lewis, 1975, 1976; Eden, 1973; Green, 1970; James and Goldstein, 1974; Hammarsten *et al.*, 1953). This chapter details studies in our laboratories in this regard (Westerman *et al.*, 1980; Navarro, 1979; Bacus, 1972, 1980; Bacus *et al.*, 1976; Bacus and Weens, 1977). The methods of red cell measurement are described in considerable detail. Then, although the techniques enable, in general, an advanced discriminatory potential when applied to the differential diagnosis of all of the above-mentioned anemias, the hypochromic microcytic anemias are discussed in detail. Through this example it should become clear that just as our current description and understanding of the disease process and our diagnostic ability have resulted from former measurement methods, these new techniques provide additional capabilities for the future.

II. DETAILS OF RED CELL MEASUREMENTS

The specific instrument that we have used in our studies was developed and constructed in our laboratory (Bacus, 1980b). A monolayer blood film preparation is used (Bacus, 1974), and the cells are analyzed microdensitometrically at 415 nm. The cells are not stained prior to analysis. The instrument uses microscope stage motors which move the slide and focus the microscope objective. Conventional microscope optics project cell images on a television scanning imaging sensor. As the microscope stage is moved in a scanning pattern, different red blood cells are located for analysis. Each time the microscope stage stops, all the cells in a given field are measured until a large number of cells has been analyzed.

These techniques have allowed us to make precise detailed measurements of individual cells. A large number of different measurements has been and can be made since the digitized image of each cell is available for analysis. However, experience has indicated that measurements of hemoglobin content, shape, pallor, and size can be done in a very robust fashion, and these measurements will be discussed in detail below since they form a basis for the quantification of red blood cell morphology by this technology. Other measurements will surely evolve, and, in fact, we have also used additional measurements, but those detailed below form a standard subset.

For the purposes of description, each image may be described as an $n \times n$ measurement matrix, $h(x,y)$, where each value of $h(x,y)$ is a measured point of absorbance obeying the Beer–Lambert absorption law:

$$h(x,y) = \log \frac{I_0}{I_t} = \frac{km}{a} \qquad (1)$$

where I_0 is the incident light, I_t is the transmitted light, k is the specific absorptivity of hemoglobin under these conditions, m is the mass of hemoglobin for that picture element measurement point, and a is the measurement spot size.

A convenient way to locate objects, i.e., red blood cells, for measurement is to set a clipping level at a low absorbance and to make measurements on objects that are above this clipping level. Thus for a single red blood cell in a field of view,

$$H = \frac{a}{k} \sum_x \sum_y h(x,y) \qquad (2)$$

results in a measurement of the mass of hemoglobin (H) for that cell.

The cell area (A) and the boundary perimeter (P) are computed from an octal chain code representation of the outer cell boundary. The octal chain code is determined by searching around the clipping threshold level to outline the cell of interest. Generalized techniques of octal chain code analysis have been described by Freeman (1974), and some of the included descriptions follow his notation. Each chain code element is indicated by a_i, $i = 1,2,3, \ldots , n$. Each a_i is an integer from 0 to 7. Also, each element has a length of 1 for even a_i and $\sqrt{2}$ for odd a_i. Thus the perimeter (P) of an object, with N_e the total number of even valued elements and N_o the total number of odd elements, is:

$$P = N_e + N_o \times \sqrt{2} \qquad (3)$$

The following relationship exists between the x,y coordinate system, or the digitized grid, and the elements a_i of the chain.

a_i	a_{ix}	a_{iy}	
0	1	0	
1	1	1	
2	0	1	
3	-1	1	(4)
4	-1	0	
5	-1	-1	
6	0	-1	
7	1	-1	

Given the initial y value, Y_0, this relationship is used according to equation (3) to find the area (A) enclosed by the chain code.

$$A = \left| \sum_{i=1}^{n} a_{ix}(Y_{i-1} + \tfrac{1}{2}a_{iy}) \right| \tag{5}$$

A robust measurement of shape, which is particularly applicable to red cells, comes from a theorem in the real plane known as the isoperimetric inequality (Courant and Hilbert, 1937; Polya and Szego, 1951). This states that the perimeter squared divided by the area is greater than or equal to 4π, with equality holding if and only if the perimeter is in the shape of a circle. If two shapes are geometrically similar, i.e., they have the same shape but different sizes, they will have the same p^2/A since the perimeter increases linearly with the size, while the area increases with the square of the size. However, a multiplicity of different shapes may result in the same p^2/A greater than 4π. Thus, for digital red blood cell images, p^2/A is a convenient measure of circularity; however, because of the digital nature of the image, the boundaries of the image are represented as a polygon. Examples of area and circularity measurement results are shown for three octal chain coded line drawings of the closed contours in Fig. 1. Table I details the feature calculation for the contour of Figure 1a. Table I is arranged such that the calculated value of equation (5), for each summation step, is in the left column. The information relevant to the calculation is in the next five columns, in the same row. For example, the octal chain code for each contour is in column 3.

Of course, a circle cannot be represented exactly on a course grid. Figure 2 indicates circular polygons of the best fitting circles to a grid with different radius values. It turns out that, depending on the placement of an object on the sampling grid, an error will result in the p^2/A measure which is a function of the fineness of the digital resolution. Figure 3 is a graph of p^2/A as a function of the radius of digitally generated round objects. For circular polygons the

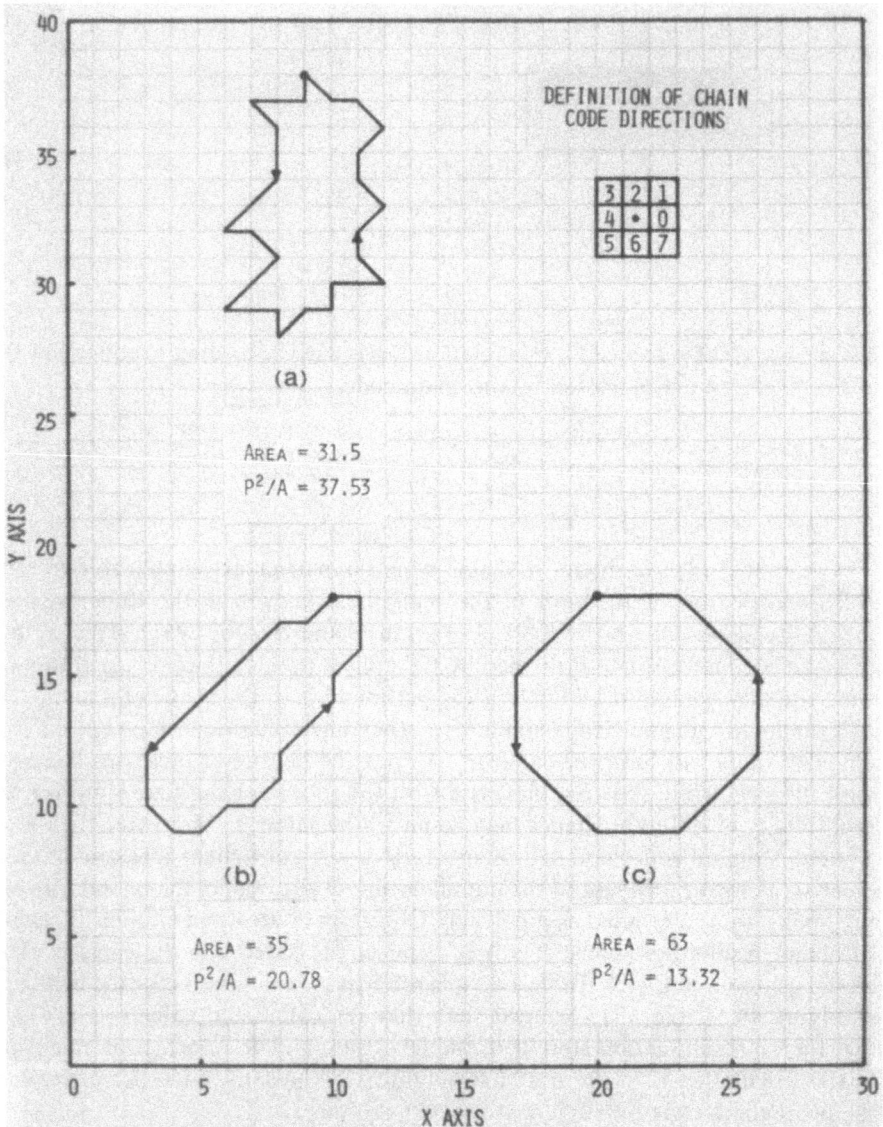

FIGURE 1. Three closed contour octal chain coded boundary shapes, similar to red cell outlines. Area and p^2/A have been computed for each of the boundaries. Detailed computations for (a) are in Table I. The arrow indicates the direction of boundary tracing to form the chain code from ●, the initial x,y coordinate starting point.

TABLE I
Method of Calculating Area and p^2/A from Octal Chain Code of Closed Contour of Figure 1a

$a_{ix}(Y_{i-1} + \frac{1}{2}a_{iy})$	i	a_i	Y_{i-1}	a_{ix}	a_{iy}
0	1	6	38	0	-1
-37	2	4	37	-1	0
-37	3	4	37	-1	0
37	4	7	37	1	-1
0	5	6	36	0	-1
0	6	6	35	0	-1
-33.5	7	5	34	-1	-1
-32.5	8	5	33	-1	-1
32	9	0	32	1	0
31.5	10	7	32	1	-1
-30.5	11	5	31	-1	-1
-29.5	12	5	30	-1	-1
29	13	0	29	1	0
29	14	0	29	1	0
0	15	6	29	0	-1
28.5	16	1	28	1	1
29	17	0	29	1	0
0	18	2	29	0	1
30	19	0	30	1	0
30	20	0	30	1	0
-30.5	21	3	30	-1	1
0	22	2	31	0	1
32.5	23	1	32	1	1
-33.5	24	3	33	-1	1
0	25	2	34	0	1
35.5	26	1	35	1	1
-36.5	27	3	36	-1	1
-37	28	4	37	-1	0
-37.5	29	3	37	-1	1
$|\Sigma| = 31.5$		Area = 31.5	$P = 16 + 13\sqrt{2} = 34.38$		
		$p^2/A = 37.53$			

limiting value is approximately 13.9810 (Navarro, 1979). It can also be proven that p^2/A has a lower bound of 13.25484, which is the lowest digitized value that can possibly be achieved (Navarro, 1979).

One of the more important morphologic features of mammalian red blood cells is their "central pallor." Typically, this is a round, central, biconcave area of decreased thickness, resulting in a decreased central absorbance. The hypochromic red cell is described as a cell with an exaggeration of normal central pallor (Bacus, 1980a). Our quantitative definition of central pallor uses the

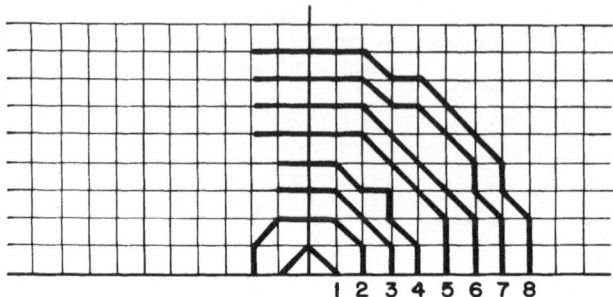

FIGURE 2. Right upper quadrants of circular polygons from the best-fitting circles on a grid. Each circle (polygon) has a different radius value in grid units, and each was generated by starting at one of the labeled radius values and proceeding upward toward the next node point on the grid with the closest euclidian distance to the actual radius value.

concept of red cell "absorbance profiles" as illustrated in Fig. 4. Each profile is composed of a series of absorbance measurements through the center of the cell, and two profiles are orthogonal to each other. Figure 5 illustrates the concept of the pallor measurement, i.e., the percentage by volume not occupied by the round biconcave red cell when compared to a disk of the same area with uniform absorbance equal to the average maximum values of the four-edge peak values from the X and Y profiles. This is developed in the following manner.

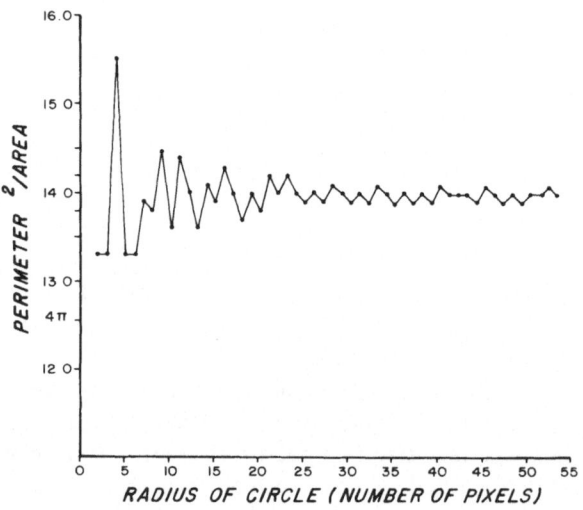

FIGURE 3. Graph of p^2/A as a function of the radius for the digitally generated circles described and illustrated in Fig. 2.

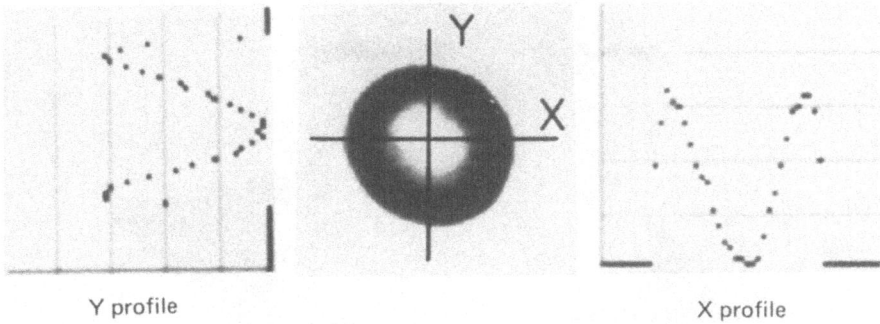

Y profile   X profile

FIGURE 4. Generation of absorbance profiles from two orthogonal traces, x and y, through the center of the red cell. Each profile indicates a series of absorbance measurements along the lines superimposed on the cell image.

If the area (A) of the cell and the average maximum absorbance (T) of the absorbance profiles are known, then the calculated mass (D) of the circumscribing disk is given by:

$$D = AT \qquad (6)$$

Given the measured integrated absorbance (H) for the entire digitized cell from equation (2), then the pallor measurement (ϕ) is:

$$\phi = D - H \qquad (7)$$

Since the areas of D and H are equal, and if it is assumed that the chromophore, hemoglobin, is homogeneously distributed, then ϕ reflects the accumulated differences in thickness between the numerous discretely sampled points of the digitized image and the disk model. These accumulated thickness differences over the given area result in the volume measurement of pallor. As an absolute measurement, this result also, of course, reflects the factors of hemoglobin concentration and specific absorptivity, i.e.,

$$\phi = ck\phi' \qquad (8)$$

where c is the concentration of hemoglobin in the specific cell being measured, k is the specific absorptivity constant of hemoglobin for the wavelengths used according to the bandpass of the measuring equipment and under the conditions of fixation and mounting employed, and ϕ' is the unknown true volume of pallor.

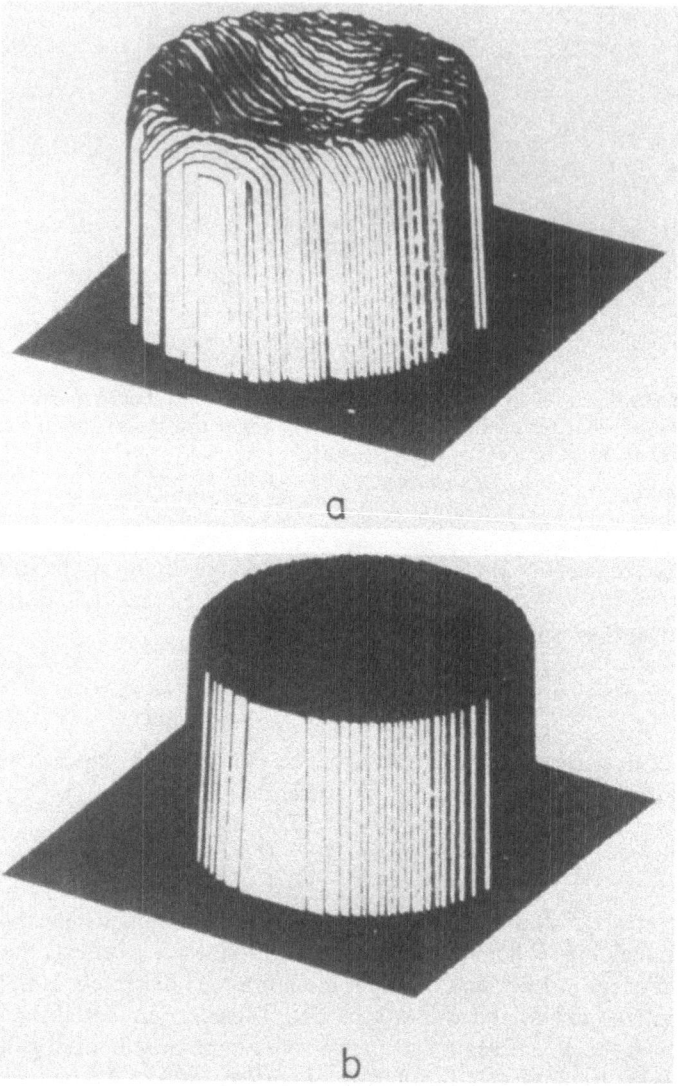

FIGURE 5. Comparison of a digitized red cell in (a) to the uniform absorbance disk model for that cell in (b). The central pallor measurement expresses the differences between the cell and the disk as the percentage volume of indentation.

A relative measure of central pallor (Φ) that factors out these constants is:

$$\Phi = \frac{D - H}{D} \times 100 \qquad (9)$$

This expresses the central pallor as the percentage volume of indentation, comparing the red cell to a disk of uniform absorbance equal to the maximum found at the cell edge.

The foregoing discussion concerning size and shape measurements, as well as an examination of Fig. 3, would indicate that high digital resolution is important for the accurate measurement of red blood cells. Digital resolution on the order of 25 to 30 pixels per object radius would seem appropriate. This implies 2000 to 3000 picture elements per object. Unfortunately, other practical design constraints—including the size of image-sensing arrays, the size of core memory for image analysis, light source shading and contrast, stepping motors to move the object into the field of view, centering objects, and the natural biological variation in the size of objects—often limit the aperture size of measuring equipment. We use approximately 1000 picture elements per red cell in our work.

III. CELL SAMPLE POPULATION DISTRIBUTIONS

Given the discussions in the two previous sections, it should be evident that a quantitative morphologic description of individual red blood cells is demonstrably feasible. The sample population distributions of these measurements from the red cells of a given specimen therefore form the natural framework for a quantitative description of the population of red cells in a given blood specimen. In this regard, typical sample distributions for each of the four parameters, size, hemoglobin, pallor, and circularity, are indicated at the top left in Figs. 6–9 for blood from a normal individual. For cell size, hemoglobin content, and pallor, these distributions are unimodal and symmetric. In the case of circularity (Fig. 9), the distribution is unimodal, but not symmetric. As indicated previously, there is a mathematically defined lower limit to the distribution and *any* deviations in shape from circularity result in a higher number. Thus the distributions tend to be skewed toward the right even for blood from normal individuals because of slightly non-ideally-circular red cells.

Some of the important considerations when new measurement or measurement techniques are proposed are the considerations of the analytic or instrument variations compared to the biological components of variation within individuals and the biological components of variation among individuals (Harris, 1975, 1976). Thus Figs. 6–9 include an evaluation of the reproducibility of the individual measurements. This was accomplished by repeatedly measuring the same cell for 1000 times rather than acquiring 1000 different cells. Also, some idea of the normal biological component of variation is given, since in each figure the results from several different normal persons have been overplotted. For cell size, hemoglobin, and pallor, it is evident that

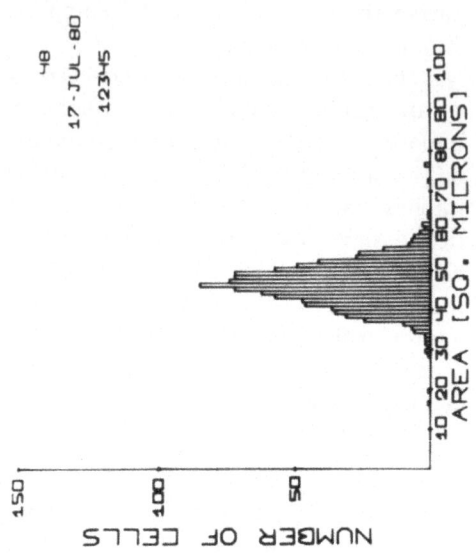

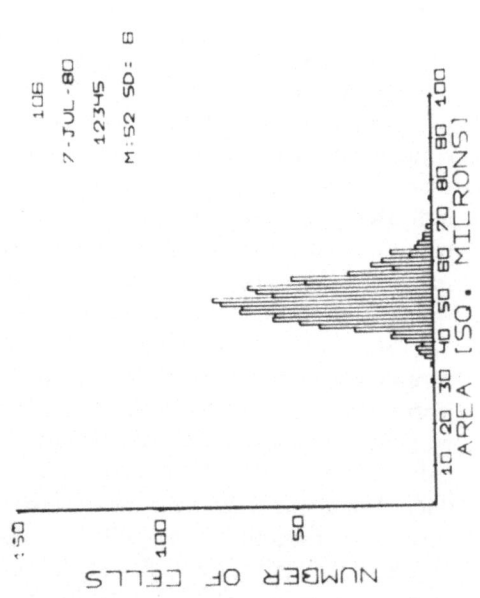

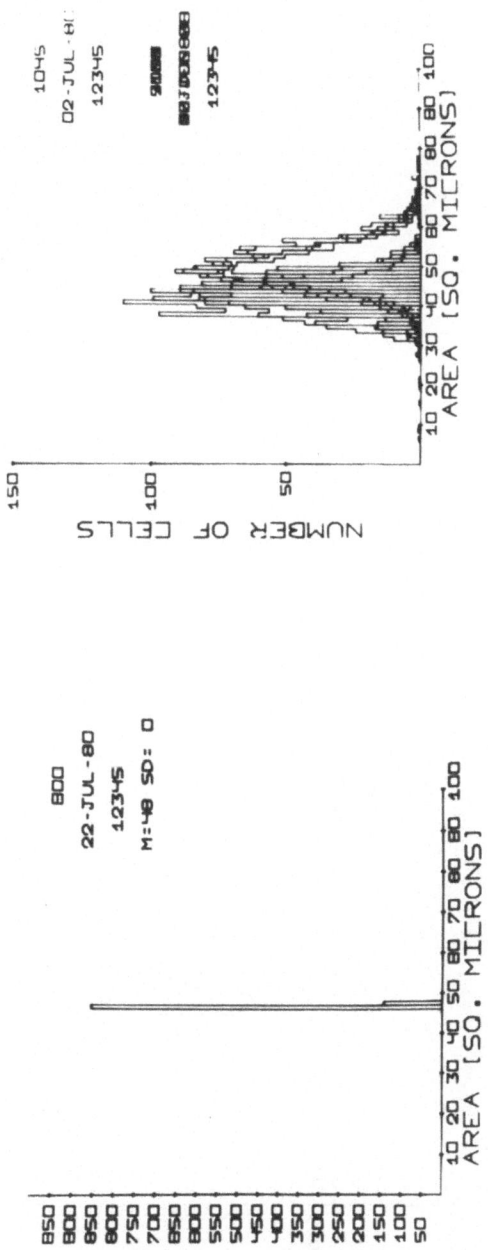

FIGURE 6. Red cell area sample population distributions. The upper left distribution is from blood from a normal individual. The upper right distribution is from the blood of a patient with iron-deficiency anemia, indicating the shift in the distribution toward smaller cells. The lower left distribution indicates analytical variation and was obtained by measuring the same individual cell 1000 times successively, rather than measuring 1000 different cells from a sample. The lower right distribution indicates the variation between different individual normal persons, since several distributions have been overplotted for comparison.

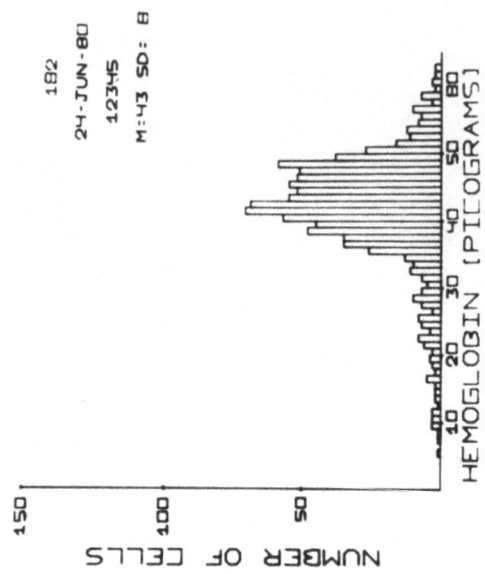

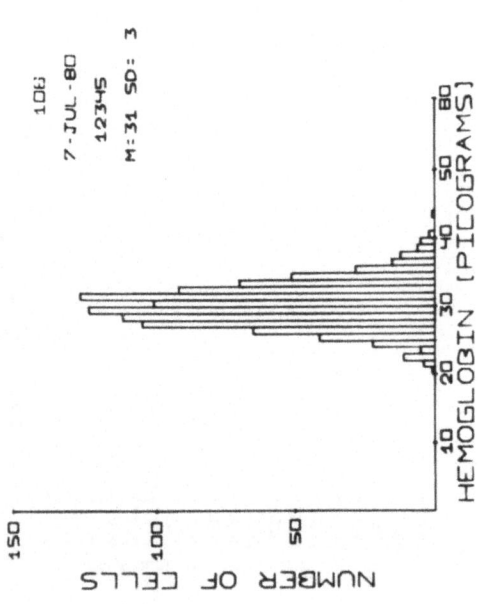

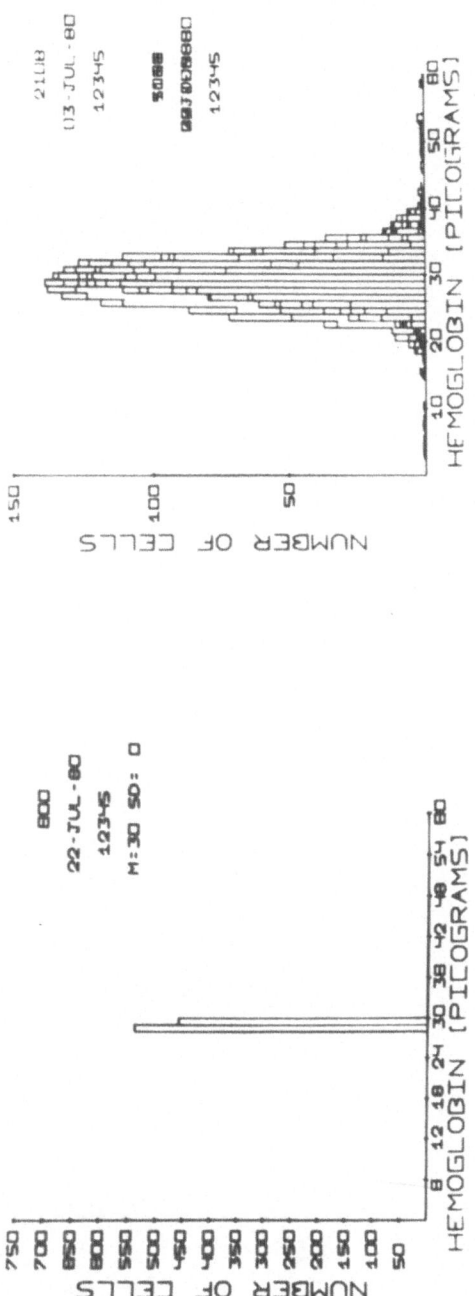

FIGURE 7. Red cell hemoglobin sample population distributions. The upper left distribution is from blood from a normal individual. The upper right distribution is from blood of a patient with megaloblastic anemia, indicating the shift in the distribution toward larger cells. The lower left distribution indicates analytical variation and was obtained by measuring the same individual cell 1000 times successively, rather than measuring 1000 different cells from a sample. The lower right distribution indicates the variation between different individual normal persons, since several individual distributions have been overplotted for comparison.

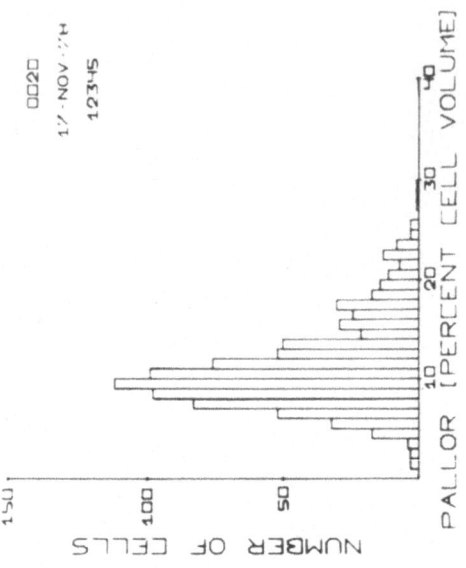

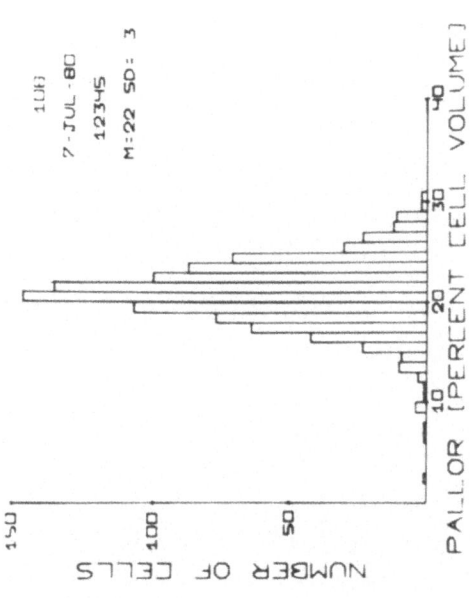

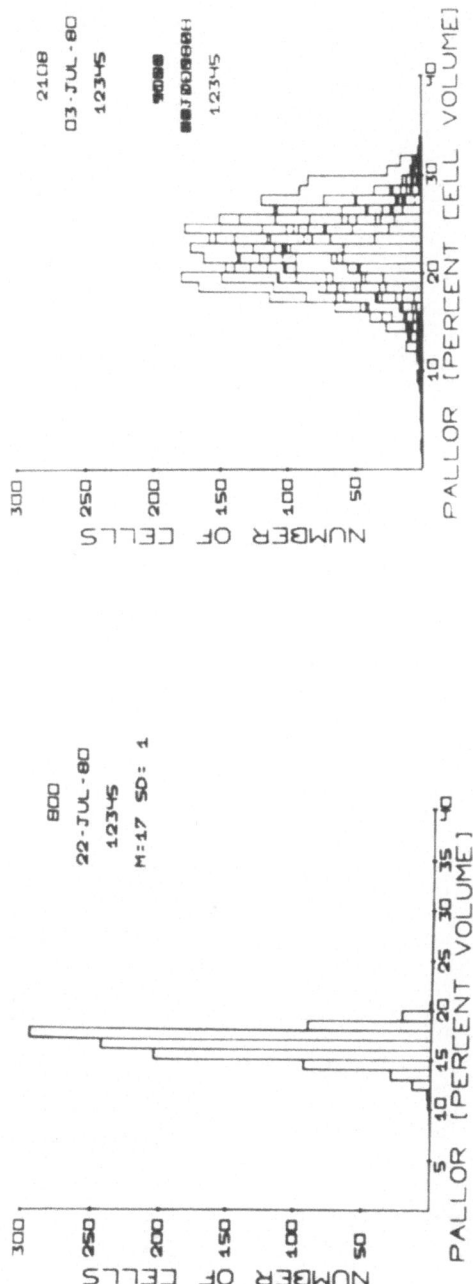

FIGURE 8. Red cell central pallor sample population distributions. The upper left distribution is from blood from a normal individual. The upper right distribution is from the blood of a patient with spherocytic anemia, indicating the shift in the distribution toward cells with less central pallor. The lower left distribution indicates analytical variation and was obtained by measuring the same individual cell 1000 times successively. The lower right distribution indicates the variation between different individual normal persons, since several individual distributions have been overplotted for comparison.

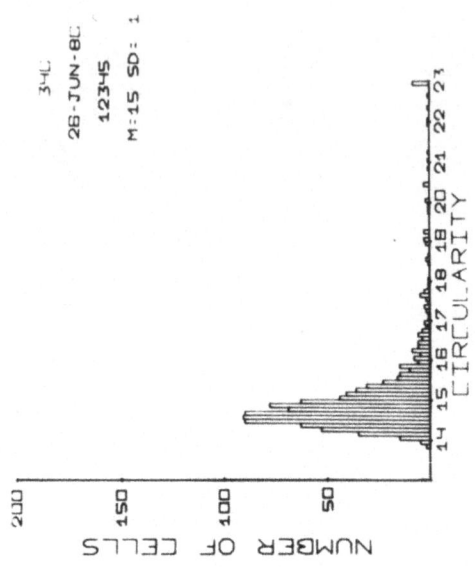

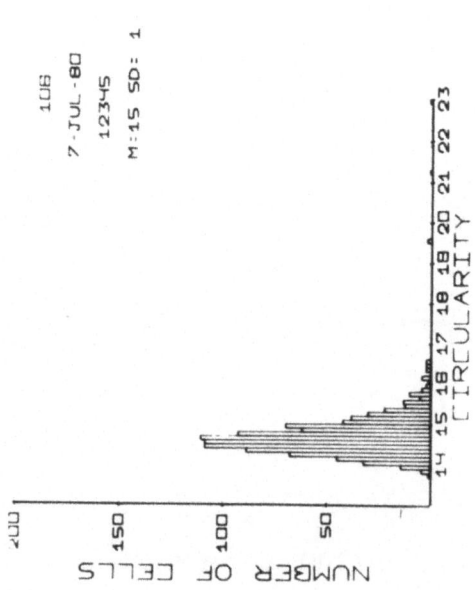

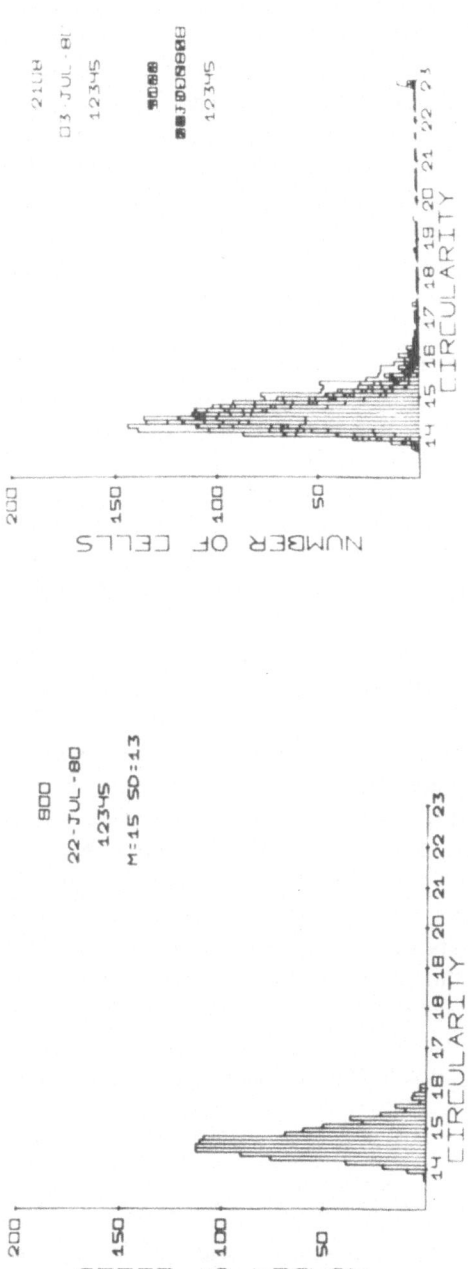

FIGURE 9. Red cell circularity, i.e., perimeter squared divided by area, sample population distributions. The upper left distribution is from blood from a normal individual. The upper right distribution is from blood of a patient with sickle cell anemia, indicating the skewing of the distribution to the right for more irregularly shaped cells. The lower left distribution indicates analytical variation and was obtained by measuring the same individual cell 1000 times successively. The lower right distribution indicates the variation between different individual normal persons, since several individual distributions have been overplotted for comparison.

the individual component of variation for these normal individuals is similar in each case but that the mean of the population distribution has shifted slightly.

Finally, it is important also to realize what happens to these sample population distributions in different abnormal or disease conditions. In this regard, each top right portion of Figs. 6–9 illustrates typical distributions for various clinical conditions that have major and specific results for the measurements under consideration. For example, the cell sample population distribution of area is shown for iron deficiency in Fig. 6, and the distribution of pallor is shown for spherocytic anemia in Fig. 8. Thus, the component of variation caused by the disease state can be compared to these other components of variation.

Since cell area and hemoglobin content are two important parameters which characterize red blood cells, an adequate description of their sample distributions is essential to a red cell analysis. In addition to their univariate distributions, as illustrated above, we have discovered that additional information can be obtained by considering their bivariate distributions as indicated in Figs. 10 and 11. Each bivariate distribution can be described by four parameters. The first two parameters describe the central tendency of the distribution, i.e., the mean cell area (MCA) in square micrometers and the mean cell hemoglobin (MCH) in picograms, and the second two describe the dispersion of the distribution by the scatter along the two principal orthogonal axes, i.e., by the eigenvalues (EV1 and EV2). This is illustrated in Fig. 10. EV1 and EV2 can be thought of as the corresponding standard deviations of the distribution if it were rotated by approximately 45° and projected toward the rotated axes of the coordinate system. They are analogous to the standard deviation in the sense that larger values indicate more dispersion along the corresponding major and minor axes of the distribution. The reason that the distributions change in their dispersion characteristics, of course, relates to the 120-day life-span of the red blood cell and the fact that as an anemic condition is either developing or changing toward normal the various cells that are characteristic of that anemia in their size and hemoglobin content are existing in concert with other cells that are produced under different conditions. The distributions illustrated in Fig. 11 are typical in that they are oblong, showing a high correlation between hemoglobin and size. This relates to the fact that smaller cells have

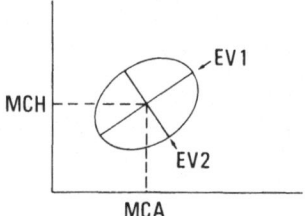

FIGURE 10. Diagrammatic representation of the bivariate distribution of cell hemoglobin vs. cell size, illustrating the relationship of the mean cell area (MCA), mean cell hemoglobin (MCH), and the two eigenvalues of the distribution (EV1 and EV2).

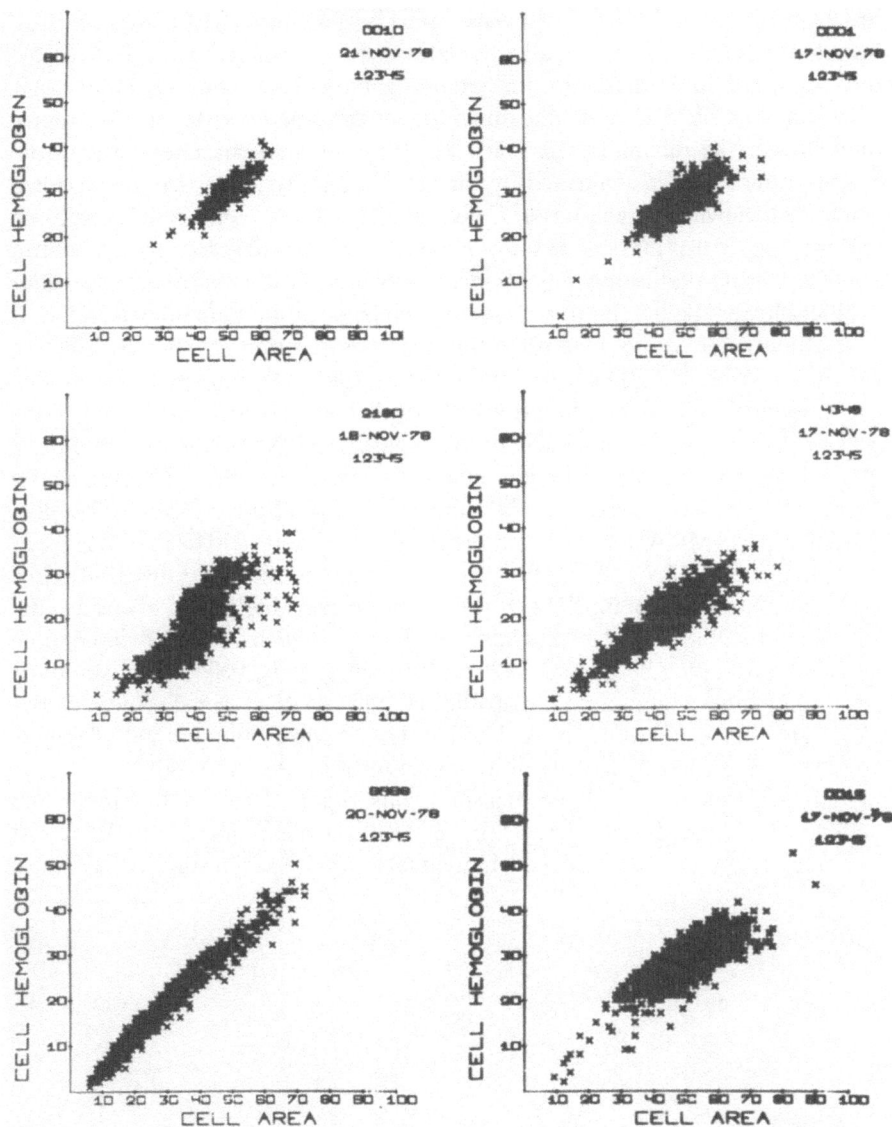

FIGURE 11. Bivariate cell sample population distributions of cell area and cell hemoglobin. Each plotted point represents a single cell with its corresponding measurement of cell area and cell hemoglobin. The two upper distributions are from the blood of two different normal persons. The distribution in the middle left is from the blood of a patient with iron deficiency after transfusion with normal blood. The sample distribution in the middle right is from a patient with mechanical hemolysis with iron deficiency. The distribution in the lower left is from a patient with hereditary elliptocytosis with a membrane disorder, and the distribution on the lower right is from a patient with sickle cell anemia.

less hemoglobin and larger cells have more hemoglobin. If this rule were followed precisely, with no biological variation, all the data points in these distributions would lie on a straight line extending to the right with a constant slope. The fact that they do not indicates inherent biological variation. The reproducibility of the technique does not contribute to the variation shown in the scattergrams. This is confirmed by the reproducibility of measurements for a single cell shown in Figs. 6 and 7, i.e., area is within 1 μm^2 and hemoglobin within 1 pg. Variations in certain anemias from the diagonal line of central tendency for normal is one indication of hypochromasia. Some anemias exhibit distinct clusters below the line of central tendency and others do not.

Much is left to be done regarding the consideration of what parameters can be extracted from these distributions, e.g., Figs. 6–9 and Fig. 11, to provide useful clinical reports for quantitative morphologic descriptions of red blood cells and for diagnostic purposes. In this regard, we have published some work indicating directions for important values that are relevant to different anemic conditions. The basic report form is indicated in Fig. 12 for a normal specimen. In that report, MCA refers to the mean cell area (μm^2); MCH to mean cell hemoglobin (pg); EV1 to the first eigenvalue of the bivariate distribution of cell area and hemoglobin; EV2 to the second eigenvalue of the bivariate distribution of cell area and hemoglobin; PAL to the mean value for percentage by volume occupied by central pallor in the round cells; PSD to the standard deviation of the pallor sample distribution; and SKW to the skewness of the circularity shape distribution. A detailed description of the method of generating this report has been fully described elsewhere (Bacus, 1980b).

In addition to the above measurements reported for each sample and obtained from the distributions (Figs. 6–9), the bottom portion of the report lists numerical values for normal individuals and for seven different types of

DATE:	24-Jan-78	SAMPLE:	0179					
97.0%	BICONCAVE						MCA	MCH
	MCA	49		1.2%	SPHEROCYTES		50	32
	MCH	30		0.2%	ELONGATED		50	32
	EV1	35		1.4%	IRREGULAR		48	30
	EV2	3		0.2%	TARGETS		69	35
AVERAGE		49 MCA	30 MCH		19 PAL	3 PSD	10 SKW	
	0.961	NORMAL	4.251	MEGALOBLASTIC				
	4.026	IRON DEFICIENT	6.352	HEMOGLOBIN SS				
	2.291	CHRONIC DISEASE	4.829	HEMOGLOBIN SC				
	3.800	β-THALASSEMIA	4.636	SPHEROCYTIC				

FIGURE 12. Printout report for a normal specimen.

anemias. These numerical values are measures of the similarity or closeness of the patient blood specimen under analysis to typical examples of each of the anemias listed, i.e., it is a distance measure in a multidimensional space. The anemia with the lowest numerical value is the one that most closely matches the patient blood sample analyzed; the next lowest number indicates the next closest match, etc.

The following details an example of the use of these new techniques in a recent clinical study (Westerman *et al.,* 1980). The diagnostic profile was used to compare the assessment of the anemia of chronic disease to serum ferritin levels and serum iron and iron-binding capacity, diagnostic tests which are commonly used to distinguish the anemia of chronic disease from iron deficiency and β-thalassemia. Traditionally, this has been a very difficult distinction to make using standard measurements. In that study, the primary diagnosis according to the image-processing method was obtained by choosing the anemia with the lowest numerical value from the report, described and illustrated in Fig. 12. The results showed that analysis obtained by image processing were diagnostic for 83% of the patients with chronic disease. Serum ferritin levels were supportive of diagnosis for 33% of the patients, and serum iron levels were useful for approximately 25% of the patients.

To examine this example in more detail, Figs. 13, 14, and 15 illustrate the use of some parameters of the population distributions described in distinguishing the four conditions of normal (1), iron deficiency (2), chronic disease (3), and β-thalassemia (4). In these figures, each plotted point, i.e., 1, 2, 3, or 4, relates to one of the above types of clinical specimens, with the measured values. For example, in Fig. 13 (top), measurements of EV1 are plotted against measurements of EV2 to show the clustering separability of normal from chronic disease.*

An examination of Figs. 13 and 15 illustrates the distinct separability of these four clinical conditions, even in the various two-dimensional scattergram representations. It is clear from this data that chronic disease is characterized by having a high value of EV2, e.g., approximately 4–8 when compared to either normal or β-thalassemia. Additionally, the condition is characterized by a low value of PSD—approximately 3.5—and an MCH higher than that of iron deficiency, as shown in Fig. 15.

In order to relate these findings to the profile numbers illustrated in the report printout of Fig. 12, it should be considered that the profile numbers are actually computed from a sixteen-dimensional space using all the values in the report printout. Each dimension is normalized to zero mean and unit standard deviation over all the conditions in the profile, so that one dimension is not

*None of these cases were the same as those in the previously reported study, which was an independent assessment of the technique at the Mount Sinai Hospital Medical Center in Chicago.

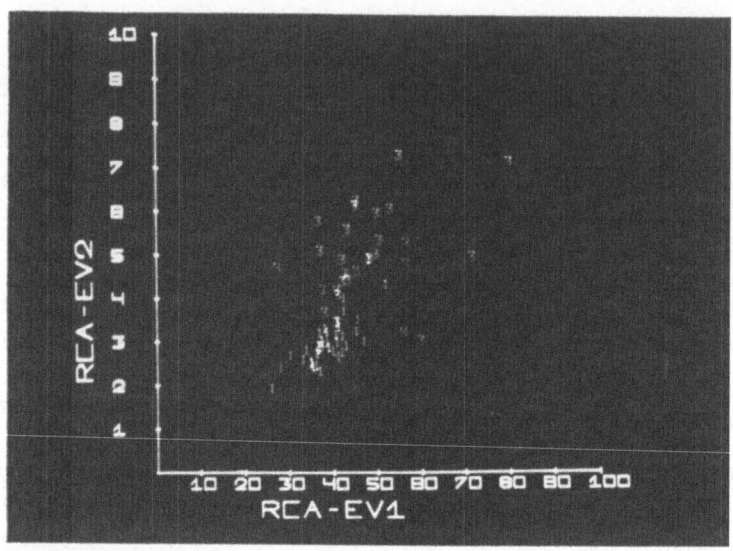

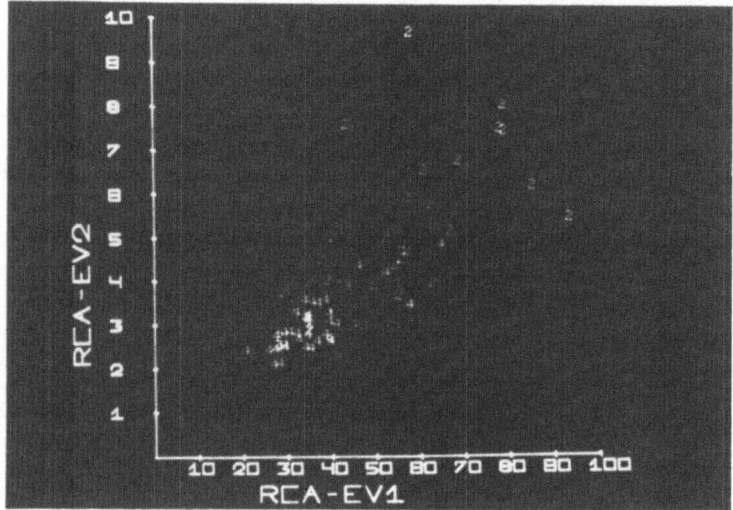

FIGURE 13. Scatterplots, using two parameters from the report printouts, EV1 and EV2, as described in Fig. 10 and illustrated in Fig. 11. Each plotted point indicates a different blood specimen with the corresponding values of EV1 and EV2 computed from 1000 measured red cells. The clustering of the different scattergram points for the conditions of normal (1), iron deficiency (2), chronic disease (3), and β-thalassemia (4), indicate good separability in this parameter space.

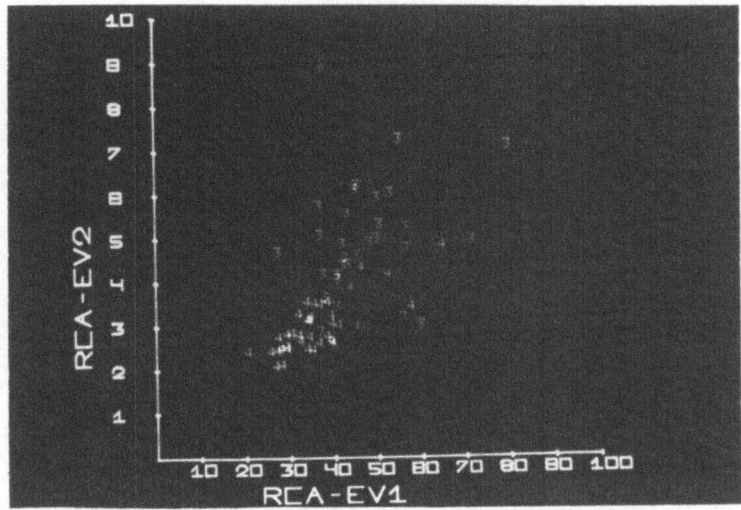

FIGURE 13. (Cont.)

unduly weighted in the calculations. This is not convenient to illustrate dia-
grammatically. However, Fig. 16 illustrates the concept of the computations of
the report printout in the two-dimensional space of MCH and EV2. In this
figure each clinical condition is indicated by one point which is the mean value
of the scattergram values indicated in Fig. 14. Given a new sample, which is
represented by point X in Fig. 16, a distance is computed to each of the means.
These distances are simply reported as the profile numbers. Thus a very low
distance value, e.g., 0.961, close to zero, would indicate a specimen which is
nearly identical to the prototype mean for that clinical condition, according to
the quantitative morphological descriptions discussed.

IV. DISCUSSION AND SUMMARY

The foregoing has reviewed the past efforts at quantitation of red cell measurements and has indicated the current status of new measurement possibili-
ties that have evolved during the last 10 years. These new techniques are based
on measurements from high-resolution microdensitometry. Our research has
indicated that size, hemoglobin content, shape, and central pallor seem to be
the most robust measurements that can be extracted at the moment with this
new technology. Sample population distributions of these measurements can
be obtained in a rapid and reliable fashion. Examples of some population dis-

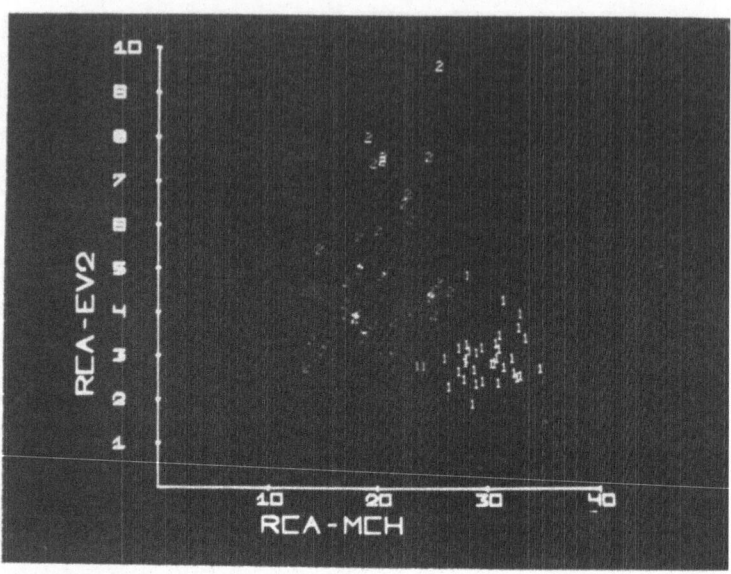

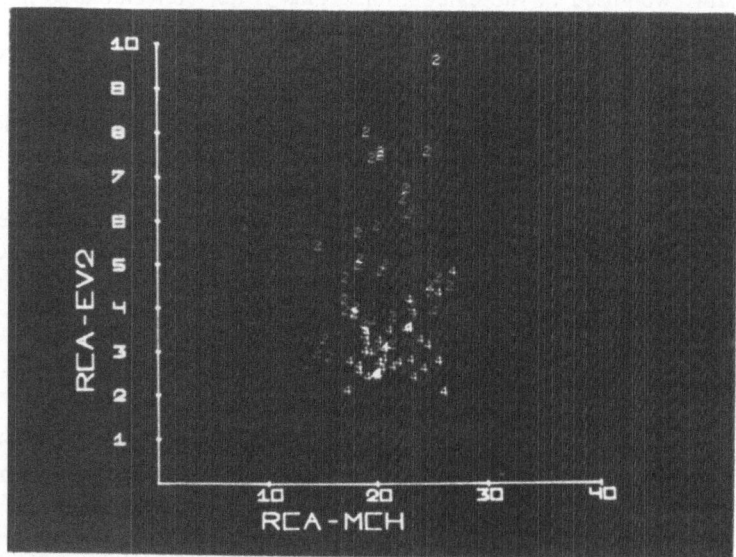

FIGURE 14. Scattergrams of EV2 and MCH as taken from the report printouts, indicating the further enhancement of separability of chronic disease, iron deficiency and β-thalassemia from each other and from normal when MCH is used with EV2.

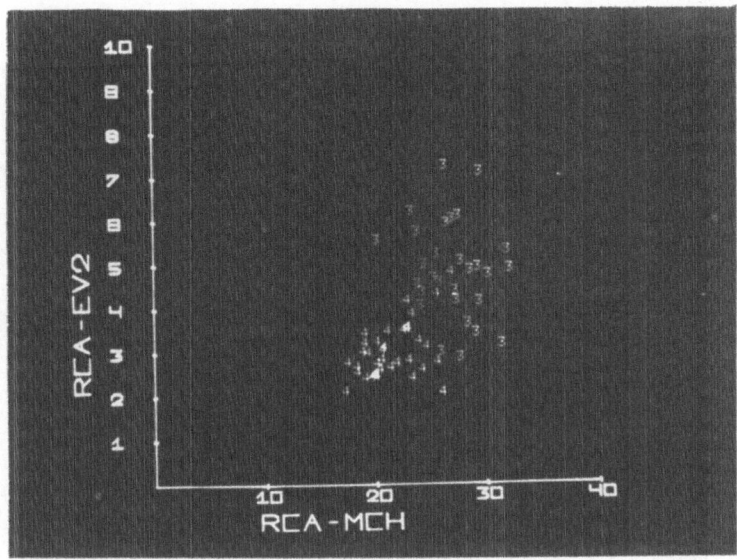

FIGURE 14. (Cont.)

tributions were presented, along with some indication of the analytic intraindividual and interindividual variation, and the variations in disease conditions which can be expected with these measurements.

As indicated in the beginning of this chapter, our ability to describe red blood cells has depended upon the sophistication of our observational and measurement apparatus. Until recently cell size was the only parameter routinely available as a quantitative measure from individual red cells. Although, as indicated, historically it had been recognized that there was significant diagnostic information in the morphology of red blood cells, these characteristics were only subjectively evaluated during a hematological examination. Also, important from the perspective of these studies, changes in the sample distribution parameters of individual cell measurements, e.g., the standard deviation of size demonstrated to be diagnostically significant by Price-Jones over 50 years ago, have not been easily measured and have been largely ignored as routine quantitative parameters. When considering the differential diagnosis of hypochromic microcytic anemias, the limit of previous techniques was an assessment of mean cell size, mean cell hemoglobin, and red cell counts. Thus differentiation from the normal was possible on the basis of cell analysis, but the discrimination between subcategories, e.g., iron deficiency, β-thalassemia, and chronic disease, was not possible. However, as indicated by the studies described in this chapter, we now have available more discriminating individual

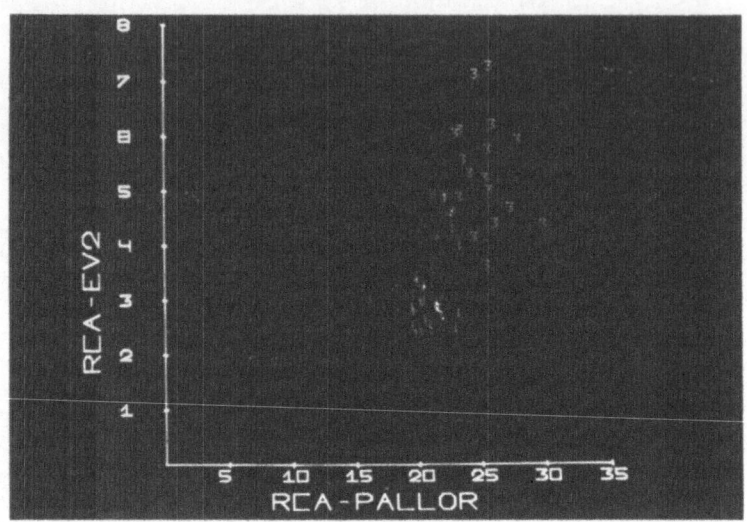

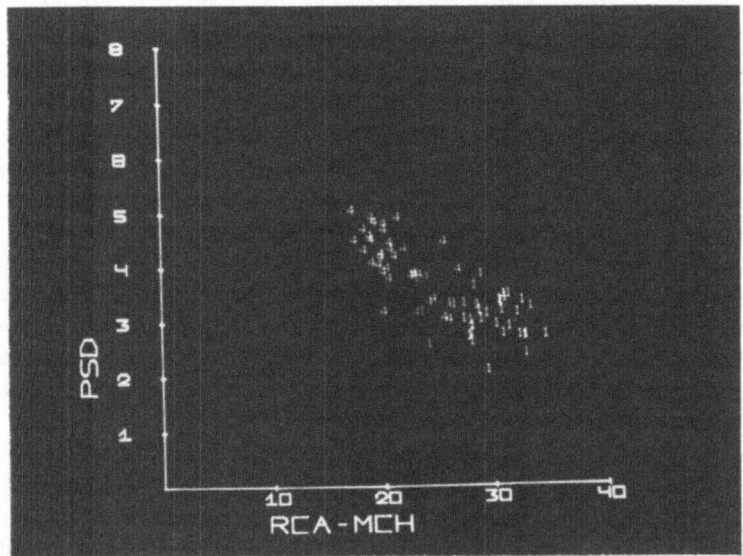

FIGURE 15. The further effect of pallor and pallor standard deviation in clustering the specimen sample points illustrated in Figs. 13 and 14.

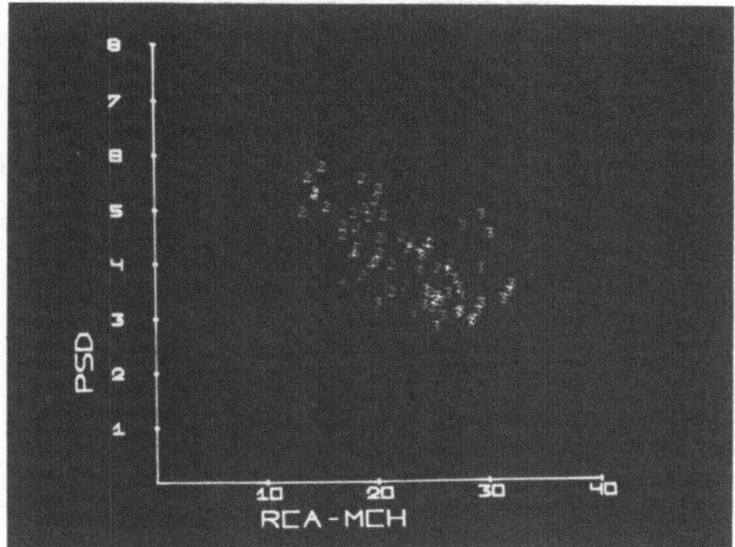

FIGURE 15. (Cont.)

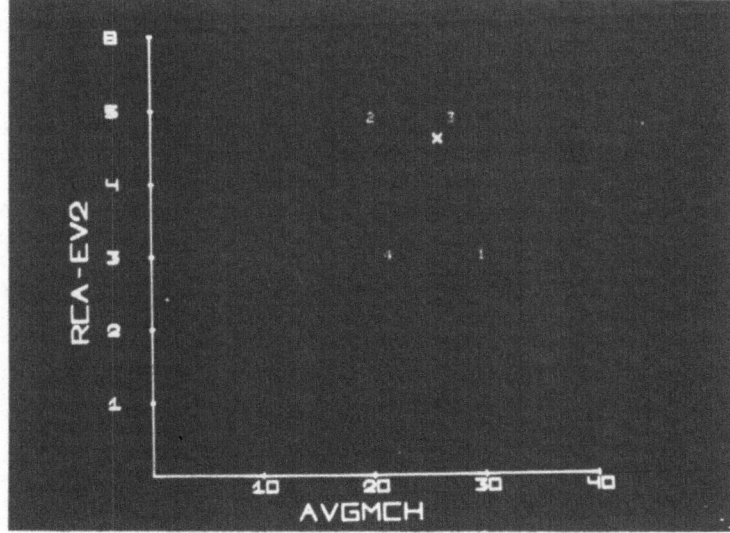

FIGURE 16. A two-space scattergram representation of the concept of the computation of the profile distance numbers of Fig. 12 and also relating to Figs. 13, 14, and 15. Each of the numbers 1, 2, 3, and 4 represents the mean of normal, iron deficiency, chronic disease, and β-thalassemia, respectively, in the two-space of MCH and EV2. The point x represents an unknown blood with the given MCH and EV2. The distance in this space is computed to the reference means, resulting in four numbers. These are analogous to the profile numbers in the report printout. In the actual report the profile numbers are computed from all the report parameters except for the minor subpopulation MCA and MCH values.

cell measurements and analytical parameters of their sample population distributions.

Obviously, we are entering a very new and exciting area of quantitation in the measurement of red cells in clinical hematology. These techniques should change and extend our definitions of anemias and red cell disorders. The results presented here, relating to the hypochromic microcytic anemias, indicate the potential of these methods.

REFERENCES

Addison, T. (1855) *On the Constitutional and Local Effects of Disease of the Suprarenal Capsules,* S. Highly, London.

Bacus, J. W. (1974) Erythrocyte morphology and centrifugal "spinner" blood film preparation, *J. Hist. Cytol.* **22**:506.

Bacus, J. W. (1980a) Quantitative measurement of red blood cell central pallor and hypochromasia, *Anal. Quant. Cytol.* **2**:123.

Bacus, J. W. (1980b) Quantitative morphological analysis of red blood cells, *Blood Cells* **6**:295.

Bacus, J. W., and Weens, J. H. (1977) An automated method of differential red blood cell classification with application to the diagnosis of anemia, *J. Hist. Cytol.* **25**:614.

Bacus, J. W., Belanger, M. G., Aggarwal, R. K., and Trobaugh, F. E. (1976) Image processing for automated erythrocyte classification, *J. Hist. Cytol.* **24**:195.

Bentley, S. A., and Lewis, S. M. (1975) The use of an image analysing computer for the quantitation of red cell morphological characteristics, *Br. J. Haematol.* **29**:81.

Bentley, S. A., and Lewis, S. M. (1976) The morphological classification of red cells using an image analysing computer, *Br. J. Haematol.* **32**:205.

Cartwright, G. E., (1966) The anemia of chronic disorders, *Sem. Hem.* **3**:351.

Cartwright, G. E., and Wintrobe, M. M. (1952) The anemia of infection. XVII. A review, in: *Advances in Internal Medicine,* Volume 5 (W. Dock and I. Snapper, eds.), The Year Book Publishers, Chicago, p. 165.

Casperson, T. (1936) Über den chemischen Aufbau der Strukturen des Zellkernes, *Skand. Arch. Physiol.* **73**:1.

Casperson, T. (1940) Methods for the determination of the absorption spectra of cell structures, *J. Roy. Microscop. Soc.* **60**:8.

Conn, H. J. (1948) *The History of Staining,* Biotech Publication, Geneva, N.Y., pp. 103–109.

Cooley, T. B., and Lee, P. (1925) Series of cases of splenomegaly in children with anemia and peculiar bone change, *Trans. Am. Pediatr. Soc.* **37**:29.

Courant, R., and Hilbert, D. (1937) *Methods of Mathematical Physics I,* Interscience, New York, pp. 97–98.

Eden, M. (1973) Image Processing Techniques in Relation to Studies of Red Cell Shape, in *Red Cell Shape.* (M. Bessis, R. I. Weed, and P. F. Leblond, eds.), Springer-Verlag, New York, p. 141.

Freeman, H. (1974) Computer processing of off line-drawing images, *Comput. Surv.* **6**:57.

Green, J. E. (1970) *Computer Methods for Erythrocyte Analysis,* Proceedings of the Symposium on Feature Extraction and Selection in Pattern Recognition, IEEE Catalog No. 70CS1C, Argonne, Illinois, October, p. 100.

Haden, R. L. (1934) The mechanism of the increased fragility of the erythrocytes in congenital hemolytic jaundice, *Am. J. Med. Sci.* **188**:441.

Hammarsten, E., Thorell, B., Aquist, S., Eliasson, N., and Akerman, L. (1953) Studies on the hemoglobin formation during regenerative erythropoiesis, *Exp. Cell Res.* **5**:404.

Harris, E. K. (1975) Some theory of reference values. I. Stratified (categorized) normal ranges and a method for following an individual's clinical laboratory values, *Clin. Chem.* **21**:1457.

Harris, E. K. (1976) Some theory of reference values. II. Comparison of some statistical models of intraindividual variation in blood constituents, *Clin. Chem.* **22**:1343.

Herrick, J. B. (1910) Peculiar elongated and sickle-shaped red blood corpuscles in a case of severe anemia, *Arch. Inter. Med.* **6**:517.

James, V., and Goldstein, D. J. (1974) Haemoglobin content of individual erythrocytes in normal and abnormal blood, *Br. J. Haemotol.* **28**:89.

Jope, E. M. (1949) The ultraviolet spectral absorption of haemoglobins inside and outside the red cell, in: *Haemoglobin* (F. I. W. Roughton and J. C. Kendrew, eds.), Butterworths, London, pp. 205–219.

Jorgensen, S., and Warburg, E. J. (1927) The indices and diameters of the erythrocytes and the best haematological criterion of pernicious anaemia. I. Historical notes and normal values, *Acta Med. Scand.* **66**:109.

Navarro, E. F. (1979) The P2A function circularity and ellipticity of digital images and polygonal curves, M.S. thesis, Cornell University.

Polya, G., and Szego, G. (1951) *Isoperimetric Inequalities in Mathematical Physics,* Princeton University Press, Princeton, pp. 8–9.

Price-Jones, C. (1933) *Red Blood Cell Diameters,* Oxford Medical, London.

Sorensen, S. T. (1876) *Undersogelser over Antallet af rode og hvide Blodlegemer under physiologiske og pathologiske Tilstande,* (Disp.) Kjobenhavn, p. 236.

Soret, J. L. (1878) Recherche sur l'absorption des rayons ultra-violets par diverses substances, *Arch. Sci. Phys. Nat.* **61**:322.

Westerman, M. P., O'Donnell, J., and Bacus, J. W. (1980) Assessment of the anemia of chronic disease by digital image processing of erythrocytes, *Am. J. Clin. Pathol.* **74**:163.

Wintrobe, M. M. (1930) The erythrocyte in man, *Medicine* **9**:195.

Wintrobe, M. M. (1931) The direct calculation of the volume and hemoglobin content of the erythrocyte, *Am. J. Clin. Pathol.* **1**:147.

Wintrobe, M. M. (1932) The size and hemoglobin content of the erythrocyte, *J. Lab. Clin. Med.* **17**:899.

Wintrobe, M. M. (1934) Anemia, *Arch. Int. Med.* **54**:256.

Wintrobe, M. M., Matthews, E., Pollack, R., and Dobyns, B. (1940) A familial hemopoietic disorder in Italian adolescents and adults, *J. Am. Med. Assoc.* **114**:1530.

Witts, L. J. (1969) *Hypochromic Anaemia,* F. A. Davis, Philadelphia, pp. 3–5.

Laser Microirradiation and Computer Video Optical Microscopy in Cell Analysis

MICHAEL W. BERNS AND ROBERT J. WALTER

I. INTRODUCTION

Laser light is intense, coherent, monochromatic electromagnetic radiation. Because of these properties it can be a unique probe of cellular structure and function. The damage produced by a focused laser beam may be caused by classical absorption by natural or applied chromophores and the subsequent generation of heat (Berns and Salet, 1972), or it may be caused by a photochemical process. An example of such a process would be the production of monoadducts or diadduct cross-linking in the case of laser light's stimulated binding of psoralens to nucleic acids (Peterson and Berns, 1978a). However, a third possibility is the generation of damage by an uncommon physical effect that occurs when ultra-high photon densities are achieved in very short periods of time (a few nanoseconds or picoseconds). The resulting nonlinear optical effects such as multiphoton absorption occur when the classic law of reciprocity does not hold. These effects may be responsible for some of the disruption observed in biological material (Berns, 1976). Whichever of the above damage-producing mechanisms is operating, be it "classical" or "uncommon," the damage often can be confined to a specific cellular or subcellular target in a consistent and controllable way. In addition, once the biophysical mechanism of laser interaction with the molecules is ascertained, the investigator has a method for precise disruption of a specific class of molecules within a strictly delimited region of the living cell. The size of this region may be considerably

MICHAEL W. BERNS AND ROBERT J. WALTER • Department of Developmental and Cell Biology, University of California, Irvine, Irvine, California 92717.

smaller than the size of the focused laser beam because of the distribution of the target molecules in the target zone. However, the size of the focused laser spot also is of paramount importance because it defines the maximum volume of biological material that will be available for direct interaction with the laser photons. Though the diameter of the focused laser spot is a direct function of the wavelength, the magnification of the focusing objective, and the numerical aperture of the objective, the actual diameter of the "effective" lesion area may be considerably less than the theoretical limit of the focused laser beam. This is because a high-quality laser beam can be generated in the TEM_{oo} mode, which results in a beam with a gaussian energy profile across it. The profile is carried over to the focused spot, which results in a "hot spot" of energy in the center. It has been demonstrated consistently (Berns, 1974a) that by careful attenuation of the raw laser beam, the damage-producing portion in the focused spot can be confined to the central hot spot (e.g., that is the only region within the focused spot that is above the threshold for damage production). As a result, lesions can be routinely produced less than 0.25 μm in diameter, and frequently down to 0.1 μm in diameter.

II. LASER MICROBEAMS

Just over 12 years ago, the blue-green argon ion laser microbeam was introduced as a potential tool for subcellular microsurgery (Berns et al., 1969a; Berns and Rounds, 1970). Previous to that, there had been limited success with the red ruby laser (Bessis et al., 1962) and the classical ultraviolet microbeam (Moreno et al., 1969). The earlier work with the blue-green argon laser led to subsequent development of a tunable wavelength flashlamp pumped dye laser microbeam (Berns, 1972), followed by a dye laser that was pumped by the green wavelength of a low-power neodymium YAG laser (Berns, 1975). Recently, we have constructed a completely tunable dye laser microbeam (from 217 to 800 nm) by employing the second (532-nm), third (355-nm), and fourth (265-nm) harmonic wavelengths of a high-power 10-nsec pulsed neodymium YAG laser (Fig. 1). In addition, a separate high-power 25-psec neodymium YAG laser has been integrated into the system to permit exposure of living cells to ultrashort pulses of light. This dual-laser system is interfaced with an inverted Zeiss axiomat microscope and an image-array processing computer. The development of this system now permits the exposure of groups of cells, single cells, or individual organelles within single cells to a wide variety of wavelengths at various power densities, with time exposures as short as 25 psec or longer. In addition, the use of the sophisticated Zeiss axiomat microscope and the image-processing computer permits a state-of-the-art optical and photometric examination of the biological material.

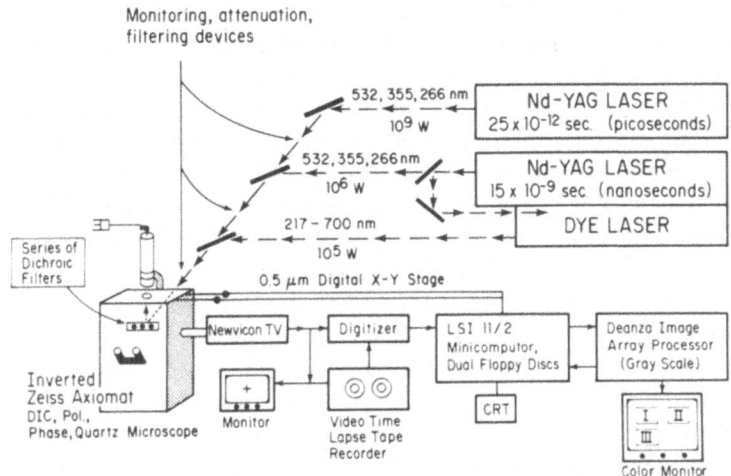

FIGURE 1. Laser microbeam system. The three basic components of the system are the lasers (Quantel YAG #400 and 481/TDL III), the microscope (Zeiss inverted AXIOMAT equipped for phase contrast, bright field, polarization, and differential interference contrast), and the T.V. computer system [De Anza IP 5000 image-array processor, Sierra #LST-1, television camera, and GYYR #DA5300 MKIII (video tape system)]. An LSI-11 minicomputer is used to drive the image array processor. In addition, the image processor LSI combination is interfaced to the X–Y digital microscope stage in order to provide cell-tracking capabilities.

III. COMPUTER-ENHANCED VIDEO MICROSCOPY FOR LASER MICROSURGERY

While microbeam irradiation can be performed using almost any laboratory microscope, the use of the highest-quality optical components is preferred. This requirement stems from the fact that many microbeam experiments utilize diffraction-limited laser spots focused onto cellular structures whose sizes are at the theoretical limit of resolution of the light microscope. Since both the size of the laser spot and the image resolution are determined by the quality of the microscope optics, the need for superior components is readily apparent.

In addition to this resolution requirement, there is also a need for images with very high contrast. As the purpose of most microbeam experiments is to produce some alteration in the cellular target, the ultimate success of the experiment is limited by the ability to see and record these alterations. In many cases the laser-induced alteration will only be seen as a local darkening or paling in the target region, which results in subtle changes in contrast after the irradiation. The ability to distinguish these subtle changes in contrast is therefore of critical importance.

Certain optical techniques such as phase contrast, polarization, and differential interference contrast (DIC) microscopy can be used to produce images with some degree of enhanced local contrast by taking advantage of specific optical properties of the specimen. These techniques can be used to perform microbeam experiments; however, their usefulness is limited to only certain types of specimens, and they still may not provide the required resolution or contrast. Finally, the experiment may also be limited by the inability of the human eye to perceive images that are either too light or too dark but otherwise have sufficient contrast and resolution.

These problems can be overcome by coupling the microscope to a high resolution video camera or, as in our system, to a video camera and an image-processing computer. The use of a video imaging system has several advantages. First, video cameras can be selected that are sensitive to light intensities beyond the range of the unaided human eye. Second, the sophisticated video cameras have manual controls for manipulating the brightness and contrast of the image. With the use of such a camera, detailed high-contrast images can be obtained from microscope specimens that otherwise cannot be distinguished by the human eye.

We first incorporated a silicon intensified-target (SIT) video camera into a low-light-level microbeam system in 1972 in order to view the specimen while the irradiation was taking place and also to allow us to minimize undesirable effects caused by the illuminating light on the cells. In our most recent system, we can use either a silicon diode tube or a newvicon tube to produce the video signal. The newvicon tube is not as light-sensitive as the silicon tubes; however, it has other advantages, such as increased response time, which makes it preferable to a SIT camera when high light sensitivity is not needed. Recent advances in microscopy, such as the AVEC polarization technique, the single side-band technique, and the video-enhanced polarization method, all utilize the versatility of the video camera to produce superior microscope images (Allen *et al.*, 1981a,b). These techniques, while still new, give every indication of being the start of a real revolution in optical microscopy.

Computer image processing can provide much more sophisticated enhancement and manipulation of microscope images than is possible with video techniques alone. While the field of digital image processing has been advancing for more than 20 years, there has been little application of these techniques to the field of cell biology. This is due in part to the cost and complexity of the equipment, and also to the fact that the processed image is not produced in real time, but only after an extended processing period. During this lag period, the image must be stored on some medium such as photographic film or magnetic tape while awaiting processing. A combination of these problems makes computer processing unattractive to the investigator, who can use less complicated photographic or optical processing techniques instead.

Recent advances in computer technology have led to the development of small, relatively inexpensive image-array processors that are capable of performing sophisticated image-processing routines on video images in real time. We have incorporated such a processor into our AXIOMAT microscope and video system in order to produce a microbeam system of extreme versatility. Real-time processing allows sophisticated routines such as edge detection, background subtraction, and pseudocolor enhancement to be performed on the microscope image during the time of the actual experiment. In addition, the processor can be used for more analytical tasks such as calculation of object areas, boundary lengths, or intracellular distances. Some of the unique features of this system will now be described.

A. Contrast Enhancement

A photograph of a PTK_2 tissue-culture cell, taken using phase-contrast optics, is shown in Fig. 2a. This photograph was taken directly from a video monitor before any computer enhancement had been performed. Analysis of this image showed that most of the picture elements (called "pixels") in this image had gray tones that were tightly distributed around a middle gray value. When such a tight distribution of gray values occurs, the probability of any pixel being surrounded by other pixels with gray values equal or similar to its own value is high; consequently, the local contrast around that pixel is low. The contrast within such an image can be enhanced by using digital processing techniques that can physically reassign the gray value of a pixel to any other gray value.

An intensity transformation function was derived, using the image in Fig. 2a, that maximized the local contrast around each pixel in the image by directly reassigning new gray values to those pixels with low contrast. The gray-value reassignments defined by this function were loaded into the appropriate registers of the image processor, and each subsequent video frame read into the processor had its contrast enhanced in the same manner. Thus the contrast of the microscope image is maximized and displayed as quickly as the video signal is generated. A photograph of the contrast enhanced image, taken directly from the video monitor, is shown in Fig. 2b.

B. Edge Enhancement

Edge enhancement is a frequently used processing technique for detecting objects and suppressing background. Several different enhancement strategies can be used. Figure 2c is a photograph of the same field of tissue-culture cells shown in Fig. 2a after the image has been processed to enhance image boundaries. As with the previous technique, this contrast enhancement was accomplished by directly reassigning gray values within the range where cellular

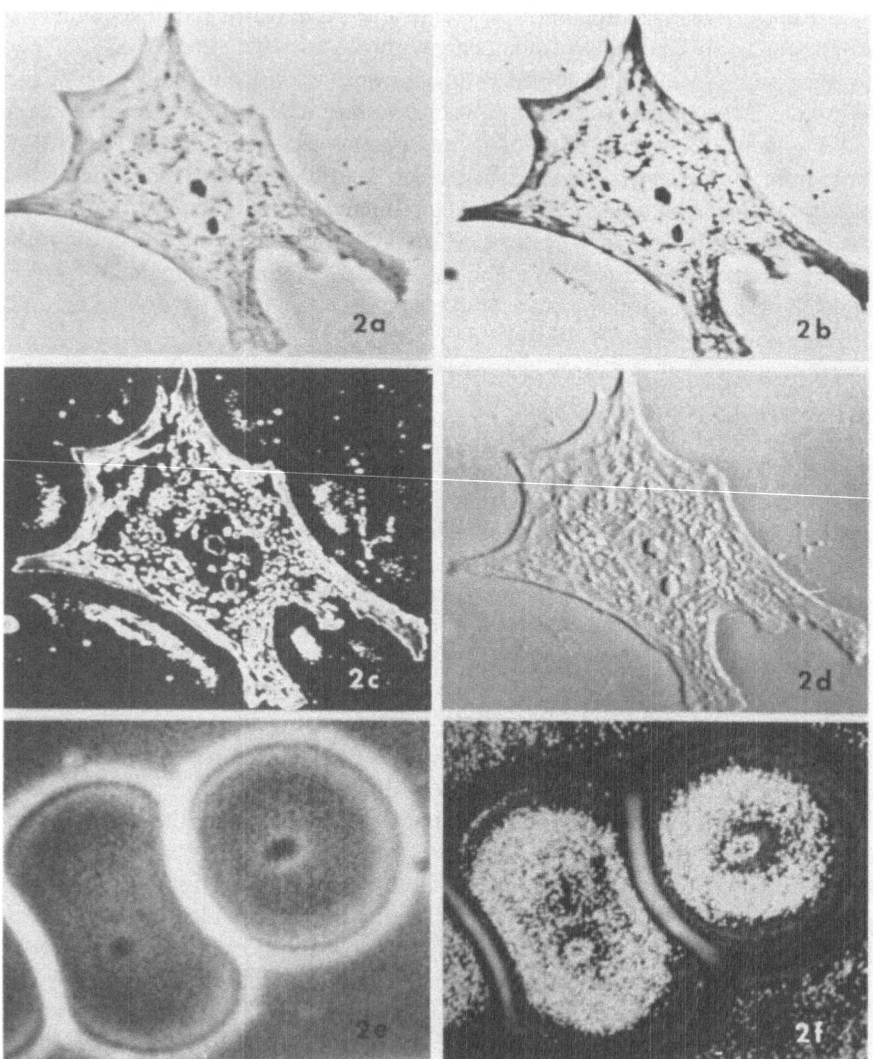

FIGURE 2. (a) Live phase contrast image of PTK$_2$ kangaroo kidney cell; image photographed directly from television monitor. (b) Same cell after computer contrast enhancement involving an intensity transformation resulting in the reassignment of gray values, and redisplay of the image in real time. (c) Same cell redisplayed following contrast enhancement of image boundaries. (d) Real time image of same cell following boundary enhancement by subtracting a slightly offset image from an original unshifted image. (e) Phase contrast image of air-dried red blood cells following the placement of a laser lesion 1–2 μm in diameter in each (dark spots). (f) Black and white photos of same cells enhanced by pseudocolor. Note that many of the various shades of gray in (e) are more easily delineated from each other in (f); in particular, note details around the lesions, and specifically the detection of a more severely damaged region within the lesion itself (arrows). Most of the colors in (f) are reds, purples, and pinks.

boundaries are most likely to be found. However, unlike the enhanced image of Fig. 2b, contrast in other parts of the image that do not contain edges has been suppressed by assigning these regions to a uniform black intensity value. This ability to enhance contrast in some parts of an image while suppressing contrast in others is a type of manipulation that can be quickly and routinely performed with an image processor but is nearly impossible with photographic or optical techniques alone. This type of manipulation accentuates regions of interest within an image while suppressing uninteresting details and background. Like the previous technique, this enhancement can be performed on the video signal in real time.

Figure 2d shows the same field of cells processed to enhance edges and boundaries in another manner. This image shows sharp contrast around the boundaries of the plasma membrane and the intracellular organelles, while the uneven background shading seen in Fig. 2a, b, and c has been eliminated. This enhancement was accomplished using the image processor by slightly offsetting the image in the X direction and then subtracting the offset image from the original unshifted image. For small offset values, this operation is a close approximation of the partial derivative of the image-intensity function in the X direction. Areas within the image where the contrast is changing rapidly in the X direction will have a large nonzero derivative, while areas where the contrast is changing slowly will have derivative near zero. Consequently, the cell boundaries appear prominently in this enhanced image because the contrast was changing rapidly in the neighborhood of these structures in the original image. As with the other techniques, this type of processing can be performed on our system with any video signal in real time.

This technique has some similarities to the optical technique of DIC microscopy. In this technique, two offset images are created using orthogonally polarized light. One of these images is optically retarded in relation to the other; consequently, the images will optically interfere and be "subtracted" when they are recombined. Since the images are formed with polarized light, any unique interaction the specimen might have with polarized light (such as birefringence or optical rotation) will contribute to the final image of the specimen.

Our computer-processing edge-enhancement technique has several fundamental differences when compared to DIC images, even though many specimens may appear nearly identical using either technique. While the DIC image is made with polarized light, and consequently any unique interactions the specimen may have with polarized light may contribute to the final image, our technique can be performed on any microscope image, regardless of whether it was formed with bright-field, phase-contrast, or fluorescence optics. This freedom to select the optical setup used for forming the edge-enhanced image allows us to utilize some of the advantages these other techniques may have over the DIC technique. We have found that when using phase-contrast optics, images can be obtained that are virtually identical to normal DIC images but that also have increased depth of field and can be obtained at substantially lower specimen illumination. We have chosen to call the computer edge-enhancement technique *differential phase-contrast* (DPC) microscopy in order to avoid confusion when discussing images made by these two techniques.

C. Pseudocolor Enhancement

Contrast enhancement utilizing the direct reassignment of gray values is best applied to images that do not contain the full range of all possible gray values. When an image contains many pixels at each possible gray value, the contrast cannot be enhanced over any given range of values without decreasing the contrast over another range. This problem can be overcome, however, by using the technique of pseudocolor contrast enhancement. The basis for this technique is the fact that the human eye can distinguish many more shades of color than it can shades of gray. Contrast between nearly identical gray tones in an image can be enhanced by replacing each gray tone with a different color. Since there are many more colors to choose from than there are shades of gray, different colors can be selected that are both of higher contrast than the original gray tones and are also pleasing to the eye.

Pseudocolor enhancement can be implemented in our system in much the same manner as the previous method. A pseudocolor contrast-enhancement function is derived that maximizes the local contrast in an image; then the gray value to color reassignments defined by this function are loaded into the appropriate registers of the image processor. Once the reassignment values have been loaded, each subsequent image generated by the video camera will be converted to pseudocolor as it is displayed on the video monitor, that is, the conversion is done in real time. In Fig. 2e, an image of blood cells with a laser lesion is depicted. Figure 2e is the original black and white, and Fig. 2f is a black and white photo made of the pseudocolor transformation. Note the easy delineation between nearly identical gray tones. The net result is the resolution of detail beyond that detected in the unmanipulated image.

IV. CHROMOSOME MICROSURGERY

In 1969 a low-power pulsed argon ion laser was focused onto chromosomes of living mitotic salamander cells that had been photosensitized with the vital dye acridine orange. The result was the production of a 0.5 μm lesion in the irradiated region of the chromosome (Fig. 3). Subsequent studies on the salamander and rat kangaroo cells (PTK_1 and PTK_2) demonstrated that the laser microbeam could be used to selectively inactivate the nucleolar genes (Berns *et al.*, 1969a,b). These studies employed three different laser microbeam systems: the low-power argon laser with acridine orange sensitization, a high-power argon laser without dye photosensitization (most likely a multiphoton process mechanism), and, recently, the fourth harmonic (265 nm) of a neodymium YAG laser. Not only can the nucleolar genes be selectively deleted, thus resulting in a loss of nucleoli in the subsequent cell generations, but a corresponding lack of one light-staining Giemsa band in the nucleolar organizer region of the chromosome can be demonstrated in cells cloned from the single irradiated cell (Fig. 4) (Berns *et al.*, 1979). Recent experiments employing *in situ* hybridization with [³H]-RNA and selective silver staining for the nucleolar organizer have demonstrated the loss of one group of ribosomal genes in the clonal population of cells (Figs. 5–8). The use of the laser to direct the loss of selected chromosome regions with the subsequent maintenance of this genetic loss is clearly feasible. In addition, it is now relatively easy to manipulate the ribosomal genes *in vitro* in order to study their regulation and function. This is a problem of considerable interest in light of the classic genetic studies on the bobbed mutant in *Drosophila* (Tartof, 1974) and gene amplification in amphibian oocytes (Davidson, 1968). Our preliminary evidence with silver staining and *in situ* hybridization suggest the possibility of ribosomal gene "magnification" *in vitro* following the laser deletion of one group of ribosomal genes (see Figs. 5–8). This suggestion is based on the finding that the one remaining nucleolar organizer region (following deletion of one) appears to silver-stain twice as deeply as those in cells with the normal two nucleolar organizers. Similarly, there appears to be roughly twice the normal amount of *in situ* hybridization to the one nucleolar organizer region in the cells cloned from the irradiated cell.

Other studies have involved removal of entire chromosomes from mitotic cells (Berns, 1974b). This can be accomplished by irradiation of the centromere region at mitotic metaphase. When a centromere with its microtubule attachment site (the kinetochore) is destroyed, the chromatid no longer remains attached to the mitotic spindle. Frequently the chromatid remains behind at the metaphase plate and is caught within the constriction body at cytokinesis. The chromosome may be incorporated into the cytoplasm of one of the daughter cells. The genetic result is the frequent production of one daughter cell that is missing an entire chromosome and one daughter cell that has an extra chromosome enclosed within a micronucleus. These daughter cells have been followed through subsequent mitosis (Brenner *et al.*, 1980), and the irradiated chromosome has been shown to duplicate itself without a functional kinetochore. At the next mitosis, the duplicated irradiated chromosome cannot attach to the spindle, and once again a micronucleus is formed. The capability of directed whole chromosome removal permits a class of cytogenetic studies in which investigators could selectively delete chromosomes and thus have a

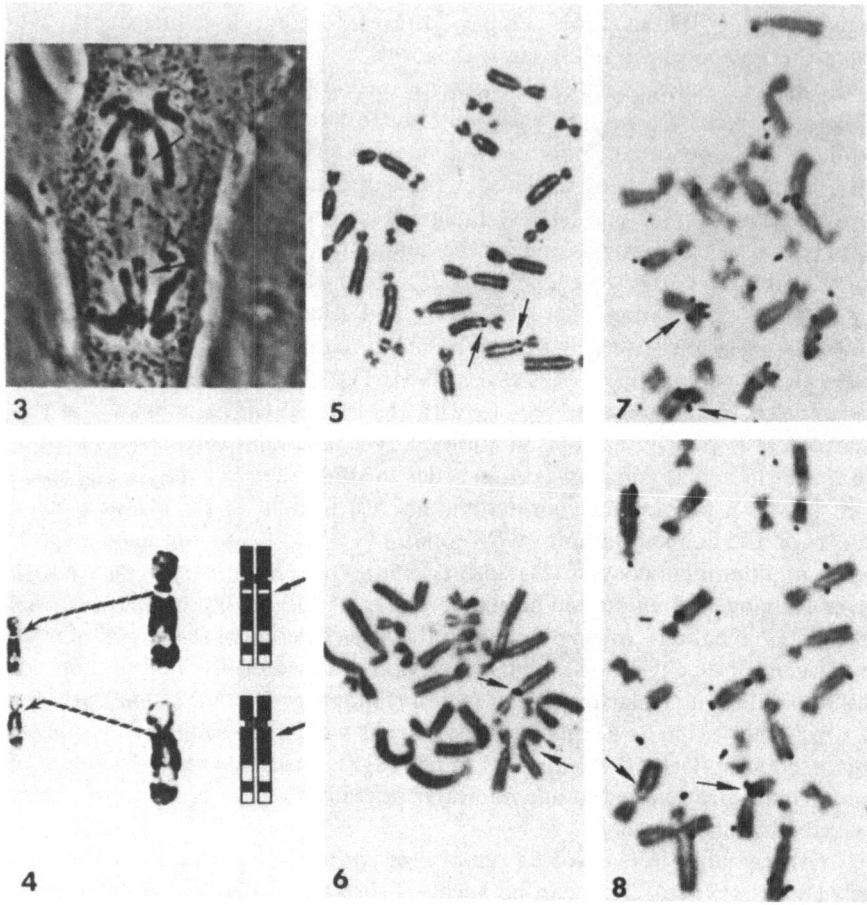

FIGURE 3. Phase contrast photomicrographs of anaphase PTK$_2$ chromosomes after placement of a two 1-m-diameter lesions on chromosome arms (arrows). Lesions were produced by irradiation with the 514-nm beam of an argon laser with an energy density of 1000 μJ/μm^2 without dye sensitization.

FIGURE 4. Giemsa–trypsin banded chromosomes from PTK$_2$ clone in which originating cell had one nucleolar-organizer secondary constriction irradiated with a 265-nm beam of a YAG laser. Note the deletion of one light-staining chromosome region (arrows).

FIGURE 5. Silver-stained chromosomes from control nonirradiated PTK$_2$ cells. Note two clearly stained nucleolar organizer regions (arrows).

FIGURE 6. Silver-stained chromosomes from irradiated clone. Note one heavily stained nucleolar organizer.

FIGURE 7. *In situ* hybridization of [^{3}H]-rRNA to control nonirradiated PTK$_2$ cell. Note two chromosomes with selective hybridization to the nucleolar organizer.

FIGURE 8. *In situ* hybridization to cell cloned from the irradiated cell. Note only one chromosome with hybridization to the nucleolar organizer.

method to compliment the already well-developed methods of somatic cell fusion. In addition, the ability to damage a restricted chromosome region (such as the centromere) and follow the cell through its cell cycle to a subsequent mitosis permits studies on chromosome damage and repair from a different perspective.

A. Mitotic Organelles

1. The Centriolar Zone

Extensive work has been devoted to using the laser to selectively disrupt three mitotic structures (centrioles, kinetochores, and microtubules) in order to elucidate their function in the process of cell division. These studies have been the most demanding in terms of understanding and applying the "selective damage" principles discussed earlier in this chapter.

Centrioles are just within the resolution of the light microscope. In the PTK$_2$ cell line the centriolar duplex is frequently visible in prophase as a phase dark dot 0.25 μm in diameter within a perinuclear clear zone (Rattner and Berns, 1976a). Ultrastructurally, the centriolar complex is composed of the centriole proper and a surrounding cloud of material called the "pericentriolar cloud." Treatment of prophase cells with nontoxic levels of acridine orange selectively sensitized the pericentriolar cloud to the green beam of either of the green lasers (Berns *et al.,* 1977). Irradiation of the centriolar complex following acridine orange treatment resulted in selective disruption of the cloud without apparently affecting the centriole. The cells progressed toward metaphase, but no anaphase movement of chromosomes occurred even though the cells went through cytokinesis (Figs. 9–13). Since acridine orange binds selectively to nucleic acid, this study supports earlier investigations (Smith-Sonneborn and Plaut, 1967) suggesting that some nucleic acid is located in pellicle of *Paramecium.* However, in our studies the high degree of sensitivity of the pericentriolar cloud implicated this region as a major site of nucleic acid localization. In addition, the lack of microtubule organization following disruption of the cloud suggests that this region is a microtubular organizing center *in vivo.* This fact was simultaneously confirmed by Gould and Borisy (1977) using isolated pericentriolar material. In later centriolar-zone laser microbeam experiments, a psoralen compound that is photochemically bound to DNA did not inhibit mitosis following exposure of the centriolar region to the appropriate cross-linking wavelength (365 nm) of laser light (Peterson and Berns, 1978c). However, another psoralen compound, which upon exposure to long-wavelength uv binds to both DNA and RNA, very effectively inhibited mitosis after laser microirradiation of the centriolar region. These results suggest that an RNA in the

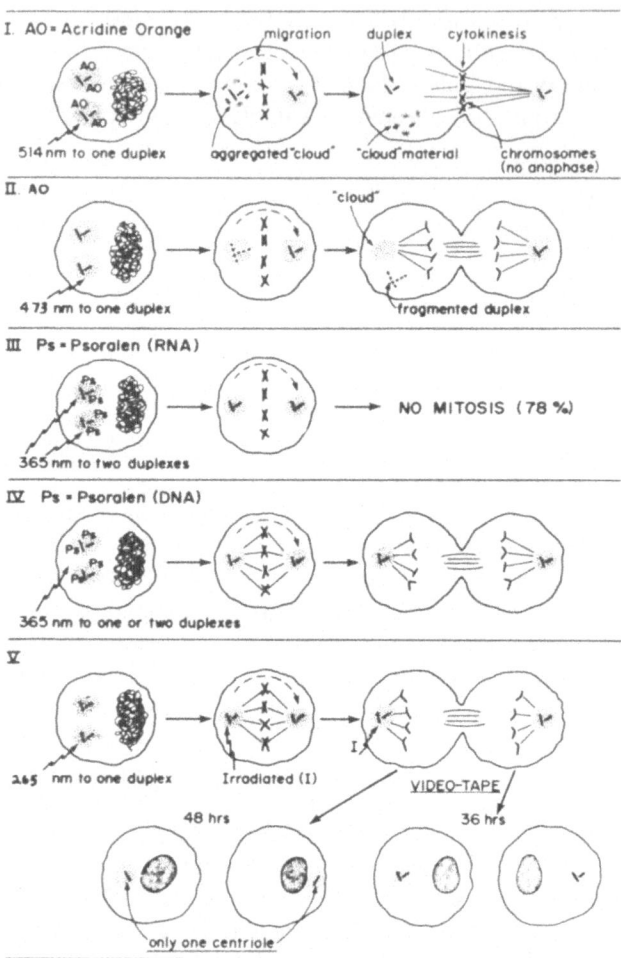

FIGURE 9. Summary of experiments in which the centriolar region was irradiated under conditions to produce selective disruption to specific components. (I) Acridine orange sensitized the pericentriolar cloud to 488 and 514 nm. (II) The centriole proper was selectively disrupted and the pericentriolar cloud unaffected by 473 nm; acridine orange was used as a photosensitizer. (III) Psoralen (AMT) selectively sensitized the pericentriolar cloud by binding to RNA. (IV) Psoralen (HMT, MMT, TMP) specific for DNA had no effect on the immediate process of cell division. (V) Irradiation with 265-nm laser light was effective in preventing centriole replication but did not inhibit the immediate cell-division process. Furthermore, the fact that the irradiated cell went through a subsequent division without duplicating centrioles demonstrates that centriole replication is not needed for mitosis to occur. This result also implicates the nucleic acid in the process of centriole duplication.

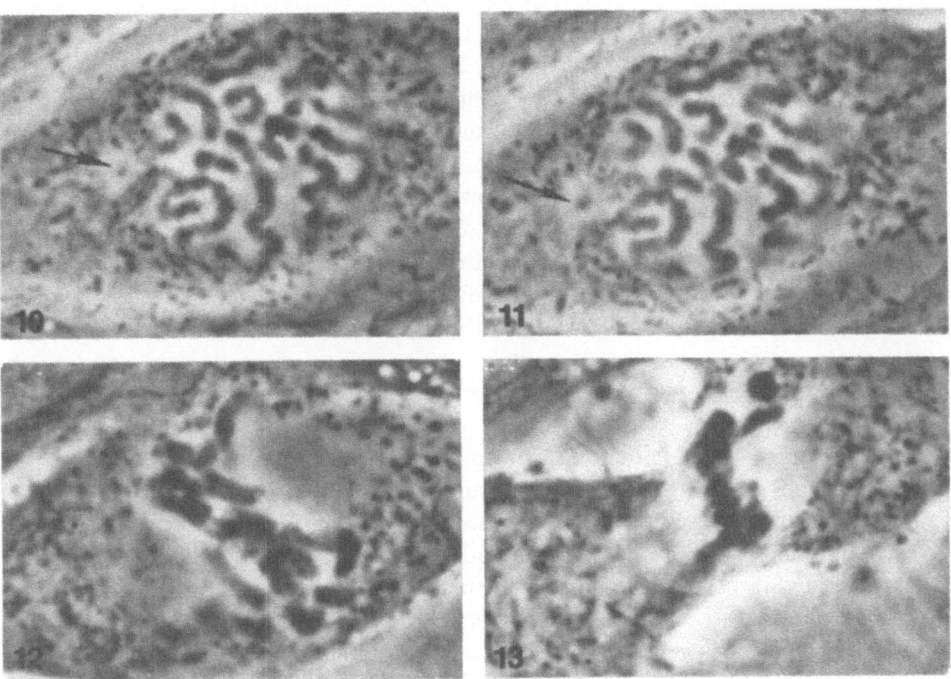

FIGURE 10. Prophase PTK_2 cell treated with acridine orange to sensitize the centriolar region to the argon ion laser beam of 514 nm. Arrow indicates the centriolar region (dark spot) in the perinuclear clear zone. This is a preirradiation picture.

FIGURE 11. The centriolar region immediately following irradiation. Note the slight increase in the extent of darkening (arrow).

FIGURE 12. The irradiated cell about 15 min following irradiation. Note that the chromosomes have continued to condense and align in a metaphase-like configuration.

FIGURE 13. At 30 min postirradiation the cell undergoes cytokinesis without any anaphase movement of chromosomes. Ultrastructural examination of this and similarly irradiated cells demonstrated that the pericentriolar material had been selectively damaged.

pericentriolar region has a major role in the organization and function of the mitotic spindle. Recent RNase digestion studies by Brinkley and Pepper (1980) support this finding.

A final series of laser microbeam studies on the centriolar region involved selective destruction of the centriole proper without damage to the pericentriolar cloud (Berns and Richardson, 1977). These studies employed the blue second-harmonic wavelength (473 nm) of the YAG laser with acridine orange as a sensitizing agent. The biophysical mechanism of damage production was most likely an "uncommon" physical effect because of the high-power density and short-pulse duration (180 nsec). It was demonstrated that cells with destroyed centrioles—but intact pericentriolar material—were capable of proceeding through mitosis in a normal fashion. Just what the role of the centriole is in mitosis is presently a highly debatable issue. In collaboration with R. W. Tucker, Department of Oncology, Johns Hopkins University, studies are currently under way at the LAMP facility to alter selective centriolar region components and follow the cells by computerized tracking through the cell cycle. This approach should elucidate centriolar replication and its relationship to progression through the cell cycle and the subsequent control of mitosis.

2. Kinetochores

The other major structure involved in the organization of microtubules is the kinetochore. Using a very finely focused green laser beam, it has been possible to destroy this region of the chromosome and then investigate the dynamics of chromosome movement. When both kinetochores of a metaphase double-chromatid chromosome are destroyed, the chromosome drifts about in the cell and the chromatids separate slightly from each other at the exact time that the rest of the chromosomes initiate their anaphase movements. This observation illustrates that the initial separation of chromatids at anaphase is not a microtubule force-mediated event (Brenner *et al.,* 1980).

Further studies have been conducted in which only one kinetochore is destroyed, and the chromosome with both chromatids, but only one functional kinetochore, is tracked (McNeill and Berns, 1980) (see Fig. 14). The results are quite dramatic and demonstrate the following features of mitosis:

1. Two functional kinetochores are necessary for the alignment of a chromosome on the metaphase plate and for normal anaphase movement.
2. Bipolar tension on the kinetochore is necessary to stabilize the orientation of the chromosome on the metaphase plate.
3. Irradiation and inactivation of one kinetochore lead to nondisjunction of the irradiated chromosome.
4. Chromatids with irradiated kinetochores retain their ability to replicate but are unable to repair the damaged kinetochore region.
5. Within limits, the velocity with which a kinetochore moves is independent of the mass associated with it.

3. Microtubules

The microtubules were one of the first mitotic structures irradiated fruitfully with the classic uv-microbeam instruments (Forer, 1966). In our labora-

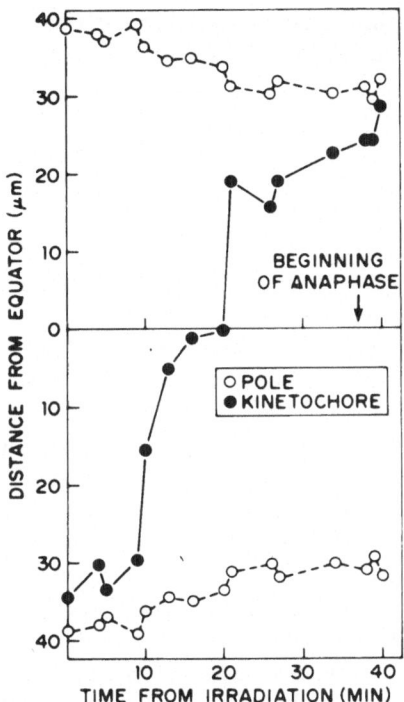

FIGURE 14. Graphic depiction of the movement of a double-chromatid chromosome following laser irradiation of one kinetochore. The black circle represents the centromere with *two* kineto-chores, one on each side. The kinetochore closest to the bottom pole was irradiated at time 0. The entire double-chromatid chromosome with only one functional kinetochore subsequently went through the movements depicted by the black circle in the figure. The rate of movement was equivalent to the normal rate of anaphase movement even though the chromosome mass was twice that of a single chromatid.

tory, we have initiated microtubule studies with the laser microbeam on the highly visible dense band of microtubules in dividing fungal cells.

Of particular interest is the function of the dense microtubular band that extends between the two separating nuclei at the end of fungal mitosis (Howard and Aist, 1977). Earlier observations led to the hypothesis that this band of microtubules served to "push" the two nuclei apart. However, laser disruption of the band resulted in a threefold increase in rate of nuclear separation. In addition, damage produced to the outside of the nucleus (distal to the band) resulted in a significant decrease in the rate of nuclear separation. These experiments indicate that the internuclear band of microtubules is rate-limiting (i.e., slowing down the movement of the nuclei) rather than force-producing. Furthermore, it appears that the forces for nuclear separation may be coming from the other side of the nuclei where EM has revealed a substantial network of microtubules.

B. Cytoplasm

1. Mitochondria

In addition to the mitotic and other nuclear organelles (chromosomes and nucleoli), the laser microbeam has been extensively applied to the subcellular disruption of single mitochondria (Berns *et al.,* 1970b). Much of this work has been conducted in contracting mammalian cardiac cells in culture and has had as its major aim elucidation of the factors regulating cardiac cell contractility. Morphologically distinct lesions can be placed in individual mitochondria and the subsequent contractile, electrical, and morphologic responses of the cell carefully analyzed. Salet *et al.* (1979) appear to have demonstrated that the laser light energy can be trapped and converted directly to ATP by the irradiated organelle. The irradiated cells also undergo a transient increase in beat rate. In other studies, cells have been impaled with microelectrodes prior to selective irradiation, and a distinct depolarization of the cell membrane has been demonstrated following irradiation of one mitochondrion. Furthermore, only those cells with the classic "pacemaker" action potential (Kitzes *et al.,* 1977) can be shown to enter a fibrillatory state following irradiation. The non-pacemaker cells still exhibit a laser-induced depolarization, but they maintain normal electrical and contractile activity. In all the irradiated cells, the cell membrane eventually returns to its normal resting potential, thus suggesting that the laser effect on the cell membrane is transient, probably resulting in a temporary alteration of membrane permeability to specific ions. These types of investigations are permitting precise alteration in cardiac cell contractility by producing a defined lesion at a predetermined subcellular site. Subsequent repair, recovery, and pharmacological control of beat arrhythmia can then be studied.

2. Myofilaments, Stress Fibers, and 100-Å Filaments

The laser microbeam can be used to study other motility-related cytoplasmic cell structures. For example, individual myofibers can be microirradiated at specific subfilament points. It has been possible to selectively damage a single Z line, or A band, in an actively contracting cell (Strahs *et al.,* 1978), and then analyze the changes in both contractile pattern and myofilament structure.

The cytoplasmic stress fibers of cultured endothelial cells are very amenable to selective microirradiation (Strahs and Berns, 1979). It has been possible to sever a single stress fiber and then follow its repair and regeneration. Studies are currently in progress to elucidate the repair phenomenon by testing various metabolic and microtubule inhibitors. In addition, selective alteration of a specific number of stress fibers and at specific locations within the cytoplasm permits detailed studies on the role of these cytoskeletal elements in cell migration and cell shape changes.

Intracellular motility patterns have been studied by placing multiple 0.25-μm lesions in preselected regions of bands of 100-Å filaments and then examining the relative movement of the lesion sites with respect to each other (Strahs and Berns, 1979).

V. PLANT CELL DEVELOPMENT (CHLOROPLAST IRRADIATION)

Though no detailed microbeam studies have been conducted on the chloroplasts in plant cells, the potential for such studies is great. Cells with large chloroplasts, or multiple distinct chloroplasts, would be particularly amenable to study. Entire chloroplasts, parts of a chloroplast, or specific ultrastructural elements of a chloroplast could be selectively damaged by appropriate matching of laser wavelength and chloroplast pigment.

The efficient absorbency of the argon laser wavelengths (488 and 514 nm) by chloroplasts has been demonstrated in the green alga Coleochaete (McBride et al., 1974). In this alga single cells in the developing multicellular thallus were destroyed by selective irradiation of the one large chloroplast in the cell.

A series of developmental studies was conducted in which a specific number of non-seta (flagellum)-bearing cells in the thallus was destroyed, and the subsequent mitotic and differentiative pattern of the thallus was studied (McBride et al., 1974). These studies revealed that mitosis could be stimulated in the thallus by merely reducing the number of cells in a given region. Mitosis apparently was stimulated when thallus cells were no longer contacted on all sides by other cells. In addition, the selective destruction of seta-bearing cells consistently resulted in new seta-bearing cells' being differentiated from non-seta cells so that the number of these cells was always maintained. These studies demonstrated built-in self regulatory developmental mechanisms for both seta cell differentiation and vegetative cell growth. In addition, it was possible to induce a differentiative process by selective removal of a specialized cell type (the seta cells).

VI. DEVELOPMENTAL CELLULAR NEUROBIOLOGY

The laser microbeam is very useful in developmental studies when it is necessary to precisely destroy specific cells or groups of cells in the embryo or larva. One of the first studies involved the use of the ruby laser microbeam to destroy the supraesophageal ganglion in spiders and then analyze the altered web-building behavior (Witt, 1969). However, most extensive microbeam work on the nervous system has been on the nematode *Caenorhabditis elegans* [personal communication with R. Russel (University of Pittsburgh) and J. G. White (University of Colorado)]. In these studies, specific cells in the embryonic or juvenile nervous systems are destroyed by laser microirradiation, and subsequent nervous system development and behavior of the organism are analyzed. The approach is being used to examine the developmental fate of specific cells in development of the nervous system and the genetic control of behavior.

In other studies by the group of Lohs-Schardin *et al.* (1979), the 257-nm wavelength of a frequency doubled argon laser has been used to destroy selected regions of the developing *Drosophila* germ band and blastoderm. Using a 10 to 30-μm focused laser beam, the group destroyed 0–45 nuclei, and the subsequent defects were used to derive what the authors termed "defect maps." These authors feel that the laser microbeam approach provides for a more detailed and accurate developmental fate map than the earlier methods employing lesion production. This is perhaps because the ability to selectively destroy a smaller group of cells in a specific target area exists with laser microirradiation. The groups of Cremer *et al.* (1978) and Zorn *et al.* (1979) have also used their laser system to perform extensive studies on chromatin damage and repair.

More recently, we have used the 265-nm fourth harmonic of the YAG laser and the 280-nm second harmonic of the YAG–pumped dye laser to study the development of a neurosensory system in the cricket (Edwards *et al.*, 1980). The hypothesis that pioneer fibers, which develop relatively early in the differentiation of insect appendages, serve to organize the peripheral sensory nerves was tested by ablating apical regions of the cercal rudiments in embryos of *Acheta domesticus*. Multiple nerve bundles rather than the normal mid-dorsal and mid-ventral pair of nerves were formed within the cercus following laser ablation of the cercal tips before pioneer fiber differentiation, but the cercal nerve was normal when lesions were made after formation of the pioneer fiber tracts and associated glia. These results indicate a necessary morphogenetic role for the pioneer fibers.

VII. PATTERN FORMATION

Another area of considerable importance is pattern formation. One of the most fruitful systems for study has been the imaginal disks of *Drosophila melanogaster*. Because of the wealth of genetic information and the ease of handling the larvae and embryos, *Drosophila* has become one of the most useful eukaryotic systems for the study of developmental genetics.

The laser microbeam has just recently been used to induce specific pattern abnormalities by the production of small areas of localized cell death in individual imaginal disks of *Drosophila* larvae. Specific regions of dissected disks were treated with the 265-nm fourth harmonic wavelength of a YAG laser in order to induce cell death. The effects of the cell death were analyzed using *in vivo* culture and induced metamorphosis to detect pattern duplications and triplications. The key feature of this system was the ability to confine effects of the laser treatment to preselected regions of the disks. The irradiated disks were incubated *in vitro* for a short time after irradiation and then transplanted into the abdomens of host larvae that were followed through metamorphosis. Using this method, it has been possible to determine the potential for pattern regulation of a small group of cells *in situ*, with a resolution much greater than in previous studies.

VIII. CONCLUSIONS

The purpose of this article has been to demonstrate the wide range of application of laser microbeam irradiation in cell biology. Studies in which this approach has already contributed to the resolution of specific problems have been discussed. Studies have also been discussed in which the application of laser microsurgery is just beginning to be applied, and the ultimate contribution of the approach has yet to be realized. In addition, a new approach to optical microscopy has been described which employs a high-sensitivity television system combined with an image-array processing computer. The resulting capabilities in image enhancement and image manipulation appear to extend greatly the capabilities of optical microscopy.

ACKNOWLEDGMENTS. This research has been supported by the following grants: NIH HL15740, NIH GM 23445, NIH RRO 1192, USAF OSR 80-0062, and NIH NB 07778. Large segments of this chapter come from an article published in *Science* magazine: 213:505, 1981.

REFERENCES

Adkisson, K. P., Baic, D., Burgott, S., Cheng, W. K., and Berns, M. W. (1973) Argon laser microirradiation of mitochondria in rat myocardial cells in tissue culture. IV. Ultrastructural and cytochemical analysis of minimal lesions, *J. Mol. Cell. Cardiol.* 5:5598.

Allen, R. D., Allen, N. S., and Travis, J. L. (1981a) Video-enhanced contrast, differential interference contrast (AVEC-DIC) microscopy: New methods capable of analyzing microtubule-related motility in the reticulopodial network of *Allogromia iaticollaris, Cell Motility* 1:291–302.

Allen, R. D., Travis, J. L., Allen, N. S., and Yilmaz, H. (1981b) Video-enhanced contrast polarization (AVEC-POL) microscopy: A new method applied to the detection of birefringence in the motile reticulopodial network of *Allogromia iaticollaris, Cell Motility* 1:275–289.

Berns, M. W. (1972) Partial cell irradiation with a tunable organic dye laser, *Nature (London)* **240**:483.

Berns, M. W. (1974a) *Biological Microirradiation* (Biological Techniques Series), Prentice-Hall, New York.

Berns, M. W. (1974b) Directed chromosome loss by laser microirradiation, *Science* **186**:700.

Berns, M. W. (1975) Dissecting the Cell with a Laser Microbeam, in, *Lasers in Physical Chemistry and Biophysics* (J. Joussot-Dubien, eds.), Elsevier, New York, pp. 389–401.

Berns, M. W. (1976) A possible two-photon effect *in vitro* using a focused laser beam, *Biophys. J.* **16**:973.

Berns, M. W., and Cheng, W. K. (1971) Are chromosome secondary constrictions nucleolar organizers? A re-evaluation using a laser microbeam, *Exp. Cell Res.* **69**:185.

Berns, M. W., and Floyd, A. D. (1971) Chromosome microdissection by laser: A functional cytochemical analysis, *Exp. Cell Res.* **67**:305.

Berns, M. W., and Richardson, S. M. (1977) Continuation of mitosis after selective laser microbeam destruction of the centriolar region, *J. Cell Biol.* **75**:977.

Berns, M. W., and Rounds, D. E. (1970) Cell surgery by laser, *Sci. Am.* **22**:98.

Berns, M. W., and Salet, C. (1972) Laser microbeams for partial cell irradiation, *Int. Rev. Cytol.* **33**:131.

Berns, M. W., Olson, R. S., and Rounds, D. E. (1969a) *In vitro* production of chromosomal lesions using an argon laser microbeam, *Nature (London)* **221**:74.

Berns, M. W., Rounds, D. E., and Olson, R. S. (1969b) Effects of laser microirradiation on chromosomes, *Exp. Cell Res.* **56**:292.

Berns, M. W., Ohnuki, Y., Rounds, D. E., and Olson, R. S. (1970a) Modification of nucleolar expression following laser microirradiation of chromosomes, *Exp. Cell Res.* **60**:133.

Berns, M. W., Gamaleja, N., Duffy, C., Olson, R., and Rounds, D. E. (1970b) Argon laser microirradiation of mitochondria in ray myocardial cells in tissue culture, *J. Cell Physiol.* **76**:207.

Berns, M. W., Cheng, W. K., Floyd, A. D., and Ohnuki, Y. (1971) Chromosome lesions produced with an argon laser microbeam without dye sensitization, *Science* **171**:903.

Berns, M. W., Rattner, J. B., Brenner, S., and Meredith, S. (1977) The role of the centriolar region in animal cell mitosis: A laser microbeam study, *J. Cell Biol.* **72**:351.

Berns, M. W., Chong, L. K., Hammer-Wilson, M., Miller, K., and Siemens, A. (1979) Genetic microsurgery by laser: Establishment of a clonal population of rat kangaroo cells (PTK$_2$) with a directed deficiency in a chromosomal nucleolar organizer, *Chromosoma* 73:1.

Bessis, M., Gires, F., and Nomarski, G. (1962) Irradiation des organites cellulaires a l'aide d'un laser a rubis, *C. R. Acad. Sci.* **225**:1010.

Brenner, S. L., Liaw, L.-H., and Berns, M. W. (1980) Laser microirradiation of kinetochores in mitotic PTK$_2$ cells: Chromatid separation and micronucleus formation, *Cell Biophys.* **2**:139.

Brinkley, B. R., and Pepper, D. (1980) Tubulin nucleation and assembly in mitotic cells: Evidence for nucleic acids in kinetochores and centrosomes, *Cell Motil.* **1**:1.

Cremer, C., Cremer, T., Zorn, C., and Zimmer, J. (1978) The influence of the distribution of photolesions on the induction of chromosome shattering in Chinese hamster cells by UV-microirradiation and caffeine, *Clin. Genet.* **14**:286.

Davidson, E. H. (1968) *Gene Activity in Early Development,* Academic Press, New York.

Edwards, J. S., Chen, S.-W., and Berns, M. W. (1981) Cercal sensory development following laser microlesions of embryonic apical cells in *Acheta domesticus, J. Neurosci.* **1**:250–258.

Forer, A. (1966) Local reduction of spindle fiber birefringence in living *Nephrotoma suturalis* (Loew) spermatocytes induced by ultraviolet microbeam irradiation, *J. Cell Biol.* **25**:95.

Gould, R. R., and Borisy, G. G. (1977) The pericentriolar material in Chinese hamster ovary cells nucleates microtubule formation, *J. Cell Biol.* **73**:601.

Heidemann, S. R., Sander, G., and Kirschner, M. W. (1977) Evidence for a functional role of RNA in centrioles, *Cell* **10**:337.

Howard, R. J., and Aist, J. R. (1977) Effects of MBC on hyphal tip organization, growth, and mitosis of *Fusarium acriminatrium*, and their antagonism by D_2O, *Protoplasma* **92**:195.

Kitzes, M., Twiggs, G., and Berns, M. W. (1977) Alteration of membrane electrical activity in rat myocardial cells following selective laser microbeam irradiation. *J. Cell Physiol.* **93**:99.

Lohs-Schardin, M., Sander, K., Cremer, C., Cremer, T., and Zorn, C. (1979) Localized ultraviolet laser microbeam irradiation of early *Drosophila* embryos: Fate maps based on location and frequency of adult defects, *Dev. Biol.* **68**:533.

McBride, G. M., LaBounty, J., Adams, J., and Berns, M. W. (1974) The totipotency and relationship of seta-bearing cells to thallus development in the green alga *Coleochaete scutata*. A laser microbeam study, *Dev. Biol.* **37**:90.

McNeill, P. A., and Berns, M. W. (1981) Chromosome behavior following laser microirradiation of a single kinetochore in mitotic PTK_2 cells, *J. Cell Biol.* **88**:543–553.

Moreno, G., Lutz, M., and Bessis, M. (1969) Partial cell irradiation by ultraviolet and visible light. Conventional and laser sources, *Int. Rev. Exp. Pathol.* **7**:99.

Ohnuki, Y., Olson, R. S., Rounds, D. E., and Berns, M. W. (1972) Laser microbeam irradiation of the juxtanucleolar region of prophase nucleolar chromosomes, *Exp. Cell Res.* **71**:132.

Peterson, S. P., and Berns, M. W. (1978a) Effect of psoralen and near UV on vertebrate cells in culture: Comparison of laser with standard lamp, *Photochem. Photobiol.* **27**:367.

Peterson, S. P., and Berns, M. W. (1978b) Chromatin influence on the function and formation of the nuclear envelope shown by laser-induced psoralen photoreaction, *J. Cell Sci.* **32**:197.

Peterson, S. P., and Berns, M. W. (1978c) Evidence for centriolar region RNA functioning in spindle formation in dividing PTK_2 cells, *J. Cell Sci.* **34**:289.

Rattner, J. B., and Berns, M. W. (1976a) Centriole behavior in early mitosis of rat kangaroo cells (PTK_2), *Chromosoma* **54**:387.

Rattner, J. B., and Berns, M. W. (1976b) Distribution of microtubules during centriole separation in rat kangaroo *(Potorous)* cells, *Cytobios* **15**:37.

Rattner, J., Lifsics, J., Meredith, S., and Berns, M. W. (1976) Argon laser microirradiation of mitochondria in rat myocardial cells. VI. Correlation of contractility and ultrastructure, *J. Mol. Cell Cardiol.* **8**:239.

Salet, C., Moreno, G., and Vinzens, F. A. (1979) A study of beating frequency of a single myocardial cell. III. Laser microirradiation of mitochondria in the presence of KCN or ATP, *Exp. Cell Res.* **120**:25.

Smith-Sonneborn, J., and Plaut, W. (1967) Evidence for the presence of DNA in the pellicle of *Paramecium*, *J. Cell Sci.* **2**:225.

Strahs, K. R., and Berns, M. W. (1979) Laser microirradiation of stress fibers and intermediate filaments in non-muscle cells from cultured rat heart, *Exp. Cell Res.* **119**:31.

Strahs, K. R., Burt, J. M., and Berns, M. W. (1978) Contractility changes in cultured cardiac cells following laser microirradiation of myofibrils and the cell surface, *Exp. Cell Res.* **113**:75.

Tartof, K. D. (1974) Unequal mitotic sister chromatid exchange as the mechanism of ribosomal RNA gene magnification, *Proc. Nat. Acad. Sci. U.S.A.* **71**:1272.

Wilson, C. L., and Aist, J. R. (1967) Mobility of fungal nuclei, *Phytopathology* **57**:769.

Witt, P. N. (1969) Behavioral consequences of laser lesions in the central nervous system of *Araneus diadematus* Cl., *Am. Zool.* **9**:121.

Zirkle, R. E. (1970) Ultraviolet-microbeam irradiation of newt-cell cytoplasm: Spindle destruction, false anaphase, and delay of true anaphase, *Rad. Res.* **41**:516.

Zorn, C., Cremer, C., Cremer, T., and Zimmer, J. (1979) Unscheduled DNA synthesis after partial UV irradiation of the cell nucleus. Distribution in interphase and metaphase, *Exp. Cell Res.* **124**:111.

3

Combination of Two Physical Parameters for the Identification and Separation of Lymphocyte Subsets

FRANCIS DUMONT

I. INTRODUCTION

The elucidation of the underlying mechanisms of immunological reactions will depend for a large part on the identification and isolation of the various lymphoid cell subsets that participate in these reactions. One approach to this problem, which has recently received extensive development, involves the detection of differentiation antigens expressed at the surface of lymphocytes. Exquisite resolution of the murine lymphocyte population into functionally distinct component subpopulations has thus been achieved, particularly in the case of the thymus-dependent cell pool (Cantor and Boyse, 1977; McKenzie and Potter, 1979).

Another approach for the separation and characterization of lymphocyte subsets takes advantage of differences in the overall physical properties of the cells. There, reproducibility of experiments is mostly determined by the controllable physicochemical conditions under which measurements are done. The major physical parameters that can be used for the recognition of lymphocyte subpopulations include surface charge, buoyant density, and size. However, as emphasized by the elegant studies of Shortman (Shortman *et al.,* 1975; Shortman, 1977a), a single separation method based on any one of the above-men-

FRANCIS DUMONT • Unité de Cancérologie Expérimentale et de Radiobiologie, INSERM U95, 54500 Vandoeuvre-les-Nancy, France. Present address: Department of Immunology, Merck, Sharp and Dohme Research Laboratories, Rahway, New Jersey 07065.

tioned physical parameters is most often insufficient to characterize all the lymphocyte subsets that are present in a lymphoid organ. This is due both to the fact that for a given physical property there may be overlap among functionally different lymphocytes and to the limited resolving power of any physical separation procedure. Hence, the combination of different but complementary separation techniques appears required for a precise discrimination of lymphocyte subsets by physical criteria. A procedure of this type has been introduced by Moon *et al.* (1972) for the analysis of human bone marrow cells and has been subsequently applied by Droege *et al.* (1974) to the delineation of mouse and chicken thymus cell subsets. Basically, this procedure consists of the splitting of the cell suspension under study into distinct fractions by a preparative physical method, e.g., electrophoresis, isopycnic centrifugation, or velocity sedimentation, followed by the size distribution analysis of each cellular fraction using an electronic cell detector. The data thus obtained enable the construction of a contour map, called a "fingerprint," which visualizes various categories of cells as defined by two physical parameters. It is the purpose of this chapter to describe this "fingerprinting" technique and to review some applications demonstrating its usefulness for the characterization of mouse lymphocyte subsets.

II. METHODS

A. Preparation of Cell Suspensions

Lymphoid organs are cut into small pieces and gently disrupted in cold RPMI 1640 medium (Gibco) supplemented with 5% heat-inactivated fetal calf serum (RS). The supernatant containing dissociated lymphoid cells is filtered through gauze to remove clumps and larger tissue fragments. Cell suspension is then washed by centrifugation (400 g, 8 min) in RS at 4°C.

In the case of mouse spleen cells, it is necessary to eliminate contaminating erythrocytes. This is achieved by a modified NH_4Cl lysis technique. The cell pellet containing approximately 200×10^6 spleen cells is suspended in 1 ml of 0.17 M Tris-NH_4Cl solution (pH 7.2) and incubated for 1 min at 4°C. The cells are then immediately washed by centrifugation in 35 ml of cold RS medium. This procedure allows one to get rid of more than 80% of the erythrocytes without producing any detectable alteration in size (Fig. 1) or EPM of the nucleated cells.

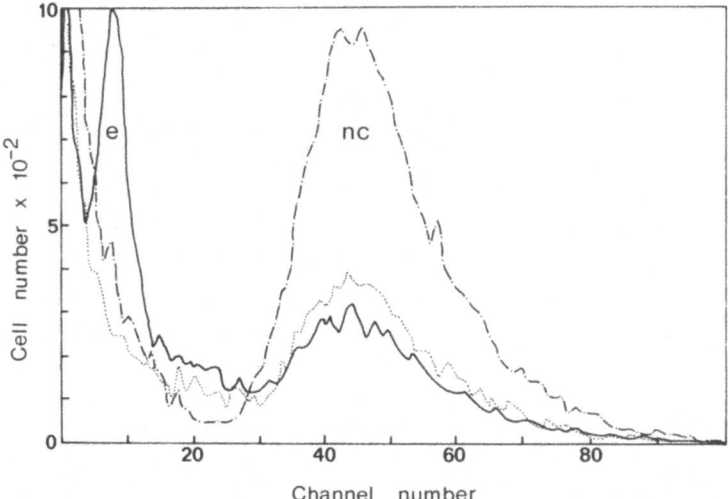

FIGURE 1. Size distribution of CBA/J mouse spleen cells before (—) and after (. . .) a 1-min incubation in Tris-NH$_4$Cl at 4°C followed by washing in RS medium. Note the disappearance of the erythrocyte peak (e) after this treatment. The same cell suspension transferred into the low-ionic-strength electrophoresis buffer, filtered through cotton wool, and kept for 45 min at 4°C in this buffer does not show any alteration in nucleated cell (nc) size distribution (–·–·–).

B. Fractionation of Cell Suspensions by Free-Flow Preparative Electrophoresis

1. Instrumentation

The free-flow preparative electrophoresis apparatus developed by K. Hannig is used (Model Vap 5, Bender and Hobein, München, Germany). This machine has been described in detail elsewhere (Hannig, 1971). We shall only recall its basic working principles (Fig. 2).

The main part of the device is a vertical separation chamber made up of two parallel glass plates separated by a gap of 0.7 mm. This chamber is 50 cm high and 12 cm wide and has 90 outlet tubings at its lower end. The electrophoresis buffer is continuously introduced to the upper rim of the separation chamber. A vertical laminar buffer flow is thus effected, the velocity of which is maintained constant by a peristaltic pump. The electric field is applied at a right angle to the electrophoresis buffer curtain by means of platinum electrodes located on both sides of the separation chamber. These are contained in electrode compartments divided from the separation chamber by ion-exchange

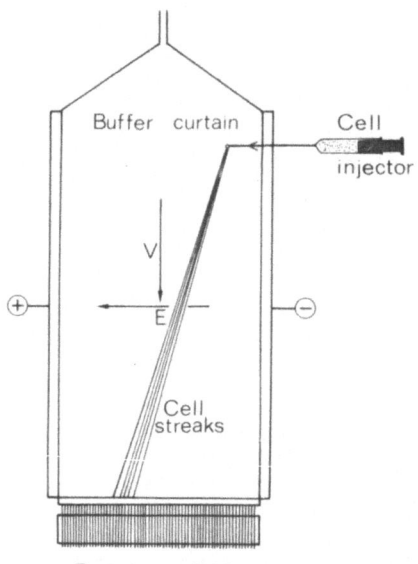

FIGURE 2. Schematic representation of the working principle of the free-flow electrophoresis (Hannig's type) apparatus. The deflection of the cells, which reflects their surface-charge density, is determined by the magnitude of the electric field (E) and by the velocity of the buffer curtain (V).

membranes. Electrolysis products generated around the electrodes are removed by a rapid circulation of electrode buffer (8 liters) through the electrode compartments. The separation chamber is cooled by an electronically controlled Freon injection system connected to the rear glass plate.

The cell suspension to be fractionated is placed in a syringe cooled by Peltier elements and continuously rotated to prevent sedimentation. It is introduced in the separation chamber through a catheter maintained in ice. The injection site is located close to the cathode at the top of the chamber.

In the absence of an electric field, the cells migrate vertically, being carried by the buffer laminar flow. When the current is raised, they are also displaced sideways toward the anode owing to their net electronegative surface charge, and this results in a slanted trajectory. Cell fractions are collected in refrigerated tubes at the bottom of the separation chamber.

The deflection of cells in the separation chamber depends on the magnitude of the electric field and on the streaming velocity of the buffer. When these factors are kept constant, the anodic migration of a cell is theoretically proportional to its surface-charge density. However, care must be exercised to minimize the influence of artifacts inherent to the principle of free-flow electrophoresis and which may alter the cell distribution pattern (Hannig *et al.,* 1975). The most important of these undesirable effects is a cathodic skewness arising as a result of the electroosmosis phenomenon. A reverse artifact, i.e., an anodic skewness, is due to the parabolic shape of the velocity profile of the buffer curtain. In theory, a situation in which these two opposite effects compensate each other is reached if the zeta potential of the walls of the separation chamber has a value identical to that of the particles to be electrophoresed (Kolin, 1960; Strikler and Sacks, 1973). In practice, for electrophoresis of cells, this can be achieved by coating the separation chamber with albumin, which possesses a zeta potential sufficiently close to that of most cellular species (Hannig *et al.*, 1975; Zeiller *et al.*, 1975). To obtain such a coating in a reproducible manner, the chamber walls are first thoroughly cleaned. Then a 5% bovine serum albumin solution (BSA, Fraction V, Miles Laboratories) is introduced and left overnight in the separation chamber. The chamber is rinsed before running the experiment with 3 liters of distilled water followed by 1 liter of electrophoresis buffer.

2. Buffers

The electrophoresis buffer is a low-ionic-strength solution (Hannig, 1971) composed of 4 mM potassium acetate, 0.015 M triethanolamine, and 0.24 M glycine, and it is rendered isoosmotic to mouse serum with 0.03 M sucrose and 0.011 M glucose. Its pH is adjusted to 7.4 by addition of 5–7 mM acetic acid.

The electrode buffer consists of 0.075 M triethanolamine and 0.04 M potassium acetate in water.

3. Experimental Procedure

For electrophoresis, cells washed in RS must be resuspended in the low-ionic-strength buffer. A step-by-step procedure is used to minimize cell agglutination that may occur during this transfer. Cells suspended in 15 ml RS medium are mixed with an equal volume of cold electrophoresis buffer. After centrifugation, the cell pellet is resuspended in 15 ml electrophoresis buffer and centrifuged again. The cells are then carefully dispersed in 3–5 ml electrophoresis buffer by several passages through a 25-gauge needle. Remaining cell clumps are removed by a rapid filtration of the suspension through 1-cm-high cotton wool plugs packed in Pasteur pipettes (von Boehmer and Shortman, 1973). Cell size distribution remains unchanged after these different steps, provided that they are all performed at +4°C (Fig. 1). Cell suspension is then brought to a concentration not exceeding 20×10^6 cells/ml and introduced into a precooled injection syringe. Electrophoretic fractionation is initiated immediately thereafter.

The operating conditions of the Hannig's machine are as follows: Electric field strength: 87–90 V/cm; intensity: 215 mA; temperature: 6°C; buffer flow rate: 450 ml/hr; sample processing rate: 3.75 ml/hr. Depending on the volume of cell suspension to be separated, sample passage duration will extend from 15 to 60 min. When large numbers of fractionated cells are needed, several inputs and harvests are made from the original cell suspension. Corresponding fractions from the successive runs are pooled at the end of the separation.

Fractions are collected in conical plastic tubes containing 2 ml RS medium at 4°C. After completion of a passage, the tubes are centrifuged and the cells are resuspended in 1 ml RS. The fractions are then kept in the cold until they are electronically counted and sized.

The preparative electrophoresis method not only enables separation of various cell fractions, but also, as mentioned above, provides information on the electrokinetic properties of these cell fractions. This information is expressed on a relative basis rather than as absolute EPM value. Thus, in all experiments referred to in this chapter, electrophoretic fractions are numbered comparatively to the peak of normal adult mouse thymocytes. Moreover, when different samples are to be compared, they are fractionated consecutively with minimal lag period between each run. Contaminating erythrocytes, which peak at electrophoretic fraction +10, also serve as an internal standard to monitor the stability of the separation.

C. Separation of Cells by Isopycnic Centrifugation in a Percoll Gradient

Percoll (Pharmacia Fine Chemicals, Uppsala, Sweden) is a newly developed medium composed of colloidal silica coated with polyvinylpyrrolidone. The use and advantages of such a medium for density gradient separation of biological particles have been reviewed by Pertoft and Laurent (1977) in a previous volume of this series. We shall describe here the preparation of a Percoll gradient suitable for isopycnic centrifugation of mouse thymocytes.

1. Formation of the Continuous Density Gradient

A solution isoosmotic with mouse serum is made by mixing 9 parts of Percoll with 1 part of $10\times$ concentrated phosphate buffered saline (PBS, pH 7.2). This solution is then diluted to 65% (v/v) with isotonic PBS to bring it to a starting density of 1.090 g/ml.

The Percoll solution is distributed in 10-ml cellulose nitrate tubes which are centrifuged at $20,000 g_{av}$ for 20 min at 4°C in a Beckmann fixed-angle (20°) rotor 50. Because of the size heterogeneity of the silica particles in Percoll, a continuous density gradient is thus generated (Pertoft, 1966). The advantage of the procedure is that up to 10 density gradients with identical shapes can be formed simultaneously.

2. Centrifugation of Cells

Layered on top of the preformed Percoll gradient are 1 to 3 × 10⁷ cells in 1 ml RS medium. Isopycnic banding of the cells is obtained by centrifugation at 2000 g for 15 min in a MSE Mistral refrigerated (4°C) centrifuge.

3. Collection of Cell Fractions

Fractions (0.3 ml) of the gradient are harvested at 4°C by means of a Buchler Auto-densiflow 11 apparatus connected through a 2-mm inner-diameter Tygon tubing to an LKB 2112 Redirac fraction collector. Fractions are then diluted with 2 ml culture medium for subsequent cell enumeration and size distribution analysis.

4. Measurement of Densities

Since self-generated Percoll gradients are not linear but S-shaped, it is important to determine the density of the various fractions collected. This is done on a parallel control gradient fractionated in the same way as above but to which cells have not been added. The refractive index of the fractions is measured in an Abbe refractometer (OPL, France). This value is directly proportional to the density of Percoll solution. Density marker beads (Pharmacia) centrifuged in an additional control gradient permit conversion of the refractive indexes to densities at 4°C.

D. Cell Enumeration and Size Distribution Analysis

1. Instrumentation

A ZBIc Coulter electronic cell detector (Coulter Electronics) equipped with an aperture of 100 μm in diameter is used. The working principle of this apparatus is described in detail in this volume (Grover et al.). A pulse height analyzer model C 1000 (Coulter Electronics) is connected to the ZBIc counter. It accumulates voltage pulses produced by passage of particles through the aperture and stores them as a function of their amplitudes, which are closely proportional to particle volumes, into a set of 100 serially arranged memory locations (channels). This analyzer includes a device (edit circuit) rejecting all pulses with imperfect shape due to particle deviations from aperture axial flow or to coincident passage. Spectra of count vs. channel number, i.e., size distribution histograms, are generated by the C 1000 analyzer and are visualized on an oscilloscope display. These can be recorded on a paper sheet by means of an XY plotter (HR 2000, Houston Instruments).

The ZBIc counter coupled to the C 1000 analyzer can measure relative cell volumes with a high degree of accuracy (better than 1%). Conversion to absolute cell volumes is done after calibration of the instrument with standard particles 3.66 and 20.8 μm in diameter (Coulter Electronics) and with thymocytes from adult mice. This enables one to determine the volume of the cells in the medium in which they are suspended. It must be stressed, however, that this value probably differs by a significant factor from the true physiological volume of the cells.

A Bio/Physics Cytograph can also be utilized as an alternate instrument for estimations of cell volume distributions intended at fingerprint analysis (Boersma et al., 1979). In this case, small-angle forward light-scatter signals are measured and integrated. These signals are determined by the areas of individual cell silhouettes.

2. Experimental Procedure

For analysis of murine lymphoid cells, the ZBIc counter is usually operated with an aperture current of 1.414 mA and an amplification factor of 2. The base channel threshold of the C 1000 analyzer is set at channels 10–15 and the window width at 70–100. Under these conditions, the size peak of normal lymphocytes appears between channels 20 and 40. Cell debris and erythrocytes are distributed below channels 5–15 (Fig. 3).

In order to measure the relative number of cells contained in each fraction resulting from physical separation, the C 1000 analyzer is operated in the preset time mode. In this position, an internal clock stops pulse accumulation after a given time which determines the volume of sample analyzed. Counting time length of 40 sec corresponding to a sample volume of 100 μl is most often used. The sample is appropriately diluted so that after dispersion in 20 ml of 0.15 M NaCl solution, counting rate in the C 1000 analyzer does not exceed 500 cells/ sec. In this condition, counting error caused by coincident passage of particles is less than 3%. Size distribution is monitored during analysis, and count range scale (K on the y axis) is switched so as to permit optimal visualization of the pulses between a lower channel situated at the limit between debris and nucleated cells which can be estimated visually and an upper channel at 98 (Fig. 3). This method of measurement provides relative but not absolute cell counts; because of the selectivity of the particle-sizing circuitry in the C 1000 analyzer, approximately 30% of the pulses generated by the ZBIc counter are lost. However, taking into account the initial dilution of the sample, the counts thus obtained correlate well with those determined in a standard hemocy-

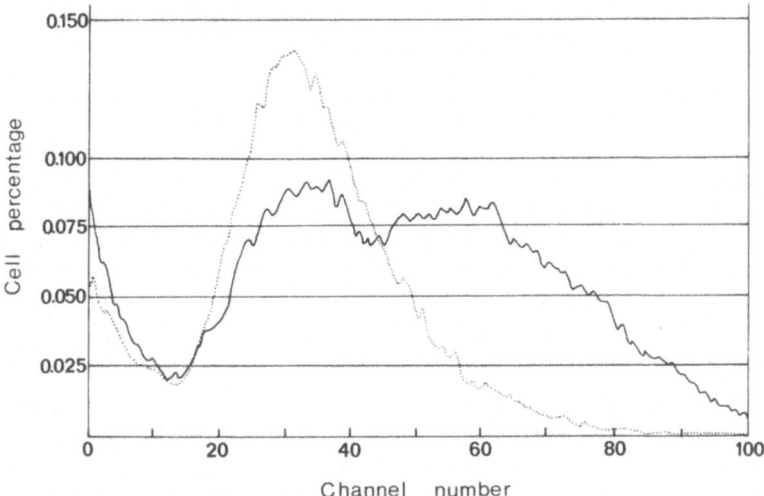

FIGURE 3. Size distribution profiles of two different electrophoretic fractions of NZB mouse spleen cells. The levels corresponding to increasing percentages of all cells recovered in this particular experiment are indicated. The intersections of these levels with the size spectra make it possible to determine the channel numbers that will be used for the construction of isopercentage curves.

tometer ($R = 0.87$). Count values measured by this method are reproducible within $\pm 1\%$ (S.E.M. of 40 determinations).

After completion of the analysis of a sample, the spectrum of count vs. C 1000 channel number of this sample is recorded on transparent paper. The same operation is repeated for each sample. One thus gets a series of curves which represent the size distribution characteristics of the various cellular fractions yielded by the separation procedure.

E. Construction of the Fingerprints

Much of the information contained in the size distribution profiles can be depicted as a contour map (fingerprint) in which cell EPM (or density) and electronic size are the independent variables and relative cell frequency is the dependent variable (Moon *et al.*, 1972). We use a simple semigraphical method for the construction of such contour maps.

1. The relative cell counts (n) of all fractions are added up. This gives us a value (N) proportional to the total number of cells recovered after physical separation.

2. Knowing N, it is easy to convert the full count range value (K) of each size distribution spectrum into a percentage (P) of all cells recovered.

$$P = \frac{K}{N} \times 100$$

Figure 3 shows an example of such a conversion.

3. Levels corresponding to a given percentage (p) of all cells recovered can thus be determined on the count axis of the size distribution spectrum. We arbitrarily consider the following percentages: 0.025, 0.05, 0.075, 0.10, 0.15, 0.20, 0.25, 0.30, 0.35, 0.40, 0.45, 0.50, 0.60, 0.70 . . . up to 1.20.

4. Each of these p levels will or will not intersect the size distribution curve. When it does, at least two points are defined by the intersections. The projections of these points onto the x axis of the size distribution spectrum give the channel numbers for the corresponding p level (Figs. 3 and 4).

Steps (3) and (4) are repeated on the size distribution profile of each cellular fraction. For a given p value, one thus gets a series of isopercentage points, each being characterized by a fraction number and a C 1000 channel number (electronic size).

5. The isopercentage points are plotted and joined together to form lines of constant relative cell frequency. This results in a contour map, called a fin-

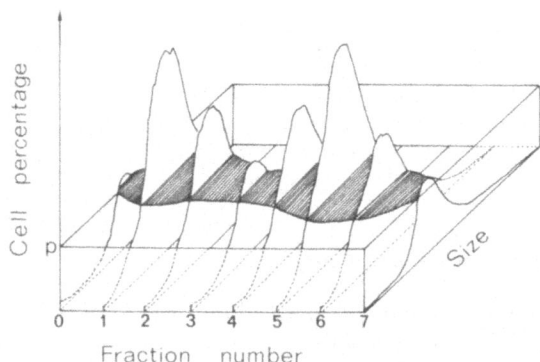

FIGURE 4. Schematic representation of the size distribution profiles of several adjacent cell fractions separated by electrophoresis (or isopycnic centrifugation). This shows how the plane corresponding to any given percentage (p) of all cells recovered will intersect the size spectra. A contour line can be drawn which delimits the hatched area on the figure. Contours corresponding to other cell percentages will be obtained similarly. The final fingerprint is constructed by projecting all the contours onto the horizontal plane.

gerprint, in which the outer line corresponds to the lowest cell percentage taken into consideration (0.025%) and the inner lines correspond to cell percentages of increasing order.

The method here described for the construction of fingerprints can be carried out in a laboratory in which computer facilities are not available, but it is relatively time-consuming. Obviously, use of a computer directly interfaced to the electronic cell detector would permit faster analysis of the size distribution data and fingerprint construction. This has been done by Dr. R. G. Miller and his collaborators at the Ontario Cancer Institute in Toronto. Their computer programs enable not only rapid fingerprint generation but also comparison of individual fingerprints to determine regions of difference between them (Wiseman et al., 1976).

F. Interpretation of the Fingerprints

A fingerprint provides information on the degree of heterogeneity, as defined by two physical parameters, of the cell suspension under study. This information is derived from the analysis of a large number of cells (approximately 4×10^5) and is thus statistically highly valuable. When present in excess, a physical cell type or subpopulation will appear on the fingerprint as a spot surrounded by concentric contour lines (cluster). The modal EPM (or buoyant density) and modal electronic size of this subpopulation can be read directly from the fingerprint. The existence of minor physical subpopulations can also be inferred from the overall shape of the fingerprint. The modal physical properties characterizing such subpopulations can be determined, although with lesser accuracy than in the case of a separate cluster.

Assuming that any given cell subpopulation has gaussian distribution with respect to both electrokinetic properties (or buoyant density) and size, the semigraphical resolution method developed by Bhattacharya (1967) can be used for quantification of the various physical cell categories recognized on a fingerprint. Then the relative number of cells with the modal EPM (or buoyant density) and the modal volume of a cluster can be taken as a measure of the proportion of the subpopulation making up this cluster. This method provides satisfactory estimations of the relative proportions of distinct clusters but becomes approximate when the clusters extensively overlap and in the case of minor subsets. Computer-assisted mathematical decomposition permits more precise quantification of the physical subsets revealed by a fingerprint (Wiseman et al., 1976; Droege, 1976). Alternatively, the areas within the contours delimiting a cluster can be measured by means of a Quantimet image analyzer (Leitz) to calculate the frequency of the cells belonging to this cluster.

III. APPLICATIONS

A. Fingerprint Analysis of Murine Thymocytes

It has long been recognized that the lymphocyte population contained in the thymus is heterogeneous. This heterogeneity reflects the occurrence of as yet poorly understood differentiation processes leading from immature precursor cells to fully immunocompetent T lymphocytes (Shortman, 1977b). A prerequisite for a better understanding of such processes is certainly to identify and separate from each other cell subpopulations that might represent stages in the intrathymic lymphocyte developmental pathway(s). As will be shown here, the combination of electrophoretic or density gradient fractionation procedures with electronic cell sizing allows us to distinguish easily such subpopulations.

1. Identification of Physical Subpopulations of Lymphocytes in the Normal Mouse Thymus

Figure 5 presents the electrophoretic (a) and size (b) distribution profiles of thymus cells from 3-month-old CBA mice. Although both profiles showed a single peak, their assymetric shape suggested the existence of physical heterogeneity within the thymocyte population. Thus, in the case of the electrophoretic distribution, a slight but definite anodic tailing was apparent. The size spectrum also exhibited a distinct skewness toward larger volume. However, the fingerprint constructed after conjunction of these two parameters (Fig. 5c) clearly disclosed a greater degree of heterogeneity than either of the two physical properties considered alone. At least four regions corresponding to different physical cell types can be distinguished on this pattern. The predominant thymocyte subset (65%), which appeared as a spot on the fingerprint, possessed a modal volume of 103 μm^3. Its peak electrophoretic fraction was arbitrarily taken as relative EPM origin (electrophoretic fraction 0). On the cathodic side of the fingerprint (electrophoretic fractions -2 to -4), the cells displayed a slightly smaller modal volume (95 μm^3) than the major cluster and might thus be regarded as a distinct physical subset. These cells represented approximately 15% of the total thymocyte population. In the anodic part of the fingerprint, a subset of larger cells (volume: 150–250 μm^3), the frequency of which did not exceed 5%, was visible as a tongue extending upward around electrophoretic fractions $+2$ to $+3$. A fourth subset of cells, with a volume of 125–130 μm^3 and accounting for about 15% of the whole population, could be seen in the faster EPM region (electrophoretic fractions $+4$ to $+9$). For the sake of clarity, these physical subpopulations of thymocytes were designated as Th1, Th2, Th3, and Th4 cells toward increasing EPM.

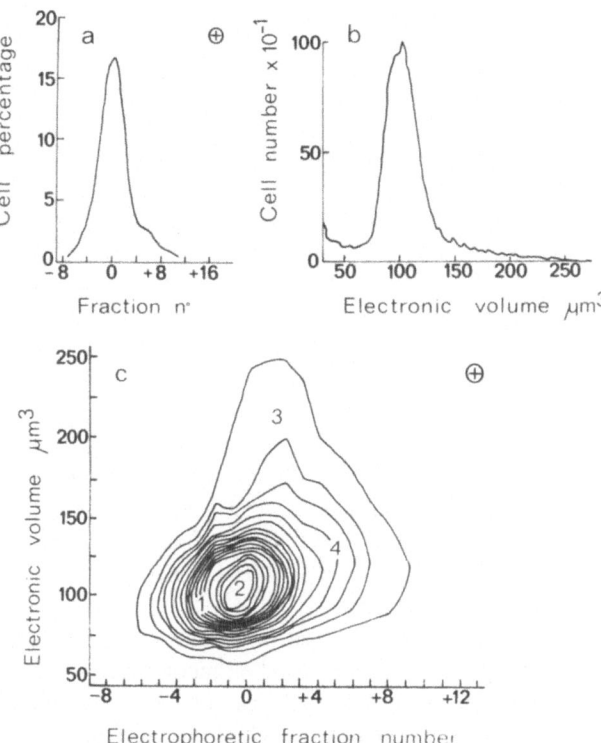

FIGURE 5. (a) Electropherogram and (b) size distribution profile of thymus cells from adult CBA/J mice. (c) Fingerprint constructed after combination of these two parameters. The four physical cell types described in the text are located by numbers on this fingerprint.

2. Distinctive Sensitivities of the Physical Subpopulations of Thymocytes to Immunosuppressive Drugs

The distinction of four cell subpopulations in the adult mouse thymus was further justified by fingerprint analysis of thymus cells from mice treated with thymolytic drugs.

Administration of glucocorticosteroid hormones in pharmacological doses has been widely used as a method to select for a subpopulation of thymocytes—termed steroid-resistant cells—the immunological properties of which have been well characterized (Blomgren and Andersson, 1969). As shown in Fig. 6, such steroid-resistant thymocytes, obtained 2 days after i.p. injection of CBA mice with hydrocortisone acetate, appeared physically identical to the Th4 subset already detectable in the normal thymus. However, since after steroid treat-

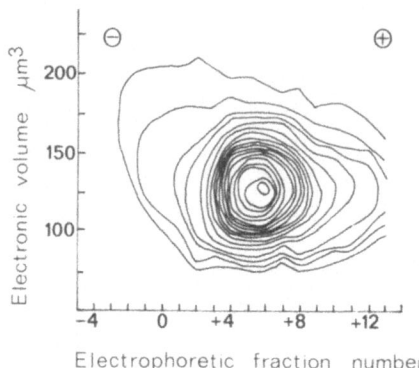

FIGURE 6. Size-vs.-EPM fingerprint of CBA/J mouse thymus cells two days after i.p. injection of hydrocortisone acetate (125 mg/kg of body weight).

ment only 5% of the original thymus cell number was recovered, it is clear that not all but only 35% of Th4 cells present in the normal thymus are actually steroid-resistant (Dumont and Robert, 1976). Part of the Th4 subset is thus steroid-sensitive as well as the majority of Th1, Th2, and Th3 cells which were no longer visible on the fingerprint.

Cyclophosphamide, a well-known antimitotic and immunosuppressive drug, was also found to exert a differential effect on thymocyte subpopulations. Two days after injection of a sublethal dose of this drug, thymus cellularity was reduced to 15% of the control value (Dumont and Barrois, 1975). Figure 7 demonstrates that at this time, most Th2 and Th3 cells had disappeared while Th1 and Th4 cells remained apparent as two well-separated clusters on the fingerprint.

Therefore, the thymocyte subpopulations identifiable with physical

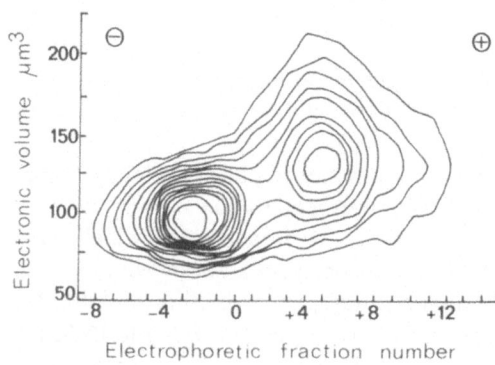

FIGURE 7. Size-vs.-EPM fingerprint of CBA/J mouse thymus cells two days after i.p. injection of cyclophosphamide (300 mg/kg of body weight). Two cell subpopulations are visible (Th1 and Th4).

parameters can also be distinguished by their sensitivity to immunosuppressive drugs. This suggests that they differ in their metabolic state or position in the cell cycle and are thus of biological significance.

3. Peanut-Agglutinin-Binding Properties of the Physical Subpopulation of Thymocytes

Another set of evidence which legitimates the distinction of several thymocyte subpopulations was provided by experiments aimed at relating physical cell characteristics with the expression of surface receptors for peanut-agglutinin (PNA), a lectin with specificity for D-galactosyl terminal residues. Such PNA receptors have recently been proposed as markers for thymocyte subsets. As demonstrated by Reisner et al. (1976), thymocytes could be separated by PNA-mediated aggregation followed by sedimentation for 45 min in 20% fetal calf serum (FCS) into fractions with different affinities for the lectin. After removal of cell-bound PNA by incubation in D-galactose-containing medium, the cells collected in these fractions were submitted to fingerprint analysis (Dumont and Nardelli, 1979). Thymocytes not agglutinable by PNA which remained on top of FCS were found to belong mostly to the Th4 subset. In contrast, those which due to strong agglutination by PNA settled to the bottom of FCS included a majority of Th1 and Th2 cells and some Th3 cells. Cells harvested in the intermediate region of FCS contained predominantly Th2 and Th3 cells with a significant contamination by Th4 cells. Such data are consonant with the notion that PNA$^-$ and PNA$^+$ thymocytes correspond respectively to steroid-resistant and steroid-sensitive subsets (London et al., 1978). They clearly indicated that the subpopulations of thymocytes which we discriminated on the basis of physical criteria also differ in the expression of surface receptors for PNA. This was further substantiated by the evaluation, using fluorescence microscopy, of the fixation of fluorescein (Fitc)-conjugated PNA to the membrane of electrophoretically fractionated thymocytes that revealed an inverse relationship between EPM and the amount of accessible PNA binding sites (Dumont and Nardelli, 1979).

4. Functional Properties of the Physical Subpopulations of Thymocytes

Earlier studies by Zeiller et al. (1974) have demonstrated that electrophoretically separated high-mobility thymocytes display the graft-vs.-host reactivity and helper activity for antibody production that are typical of immunocompetent T lymphocytes, whereas the low-mobility thymocytes appear devoid of such capabilities. In keeping with these data, we found the in vitro proliferative responses to the lectins PHA and Con A, which represent

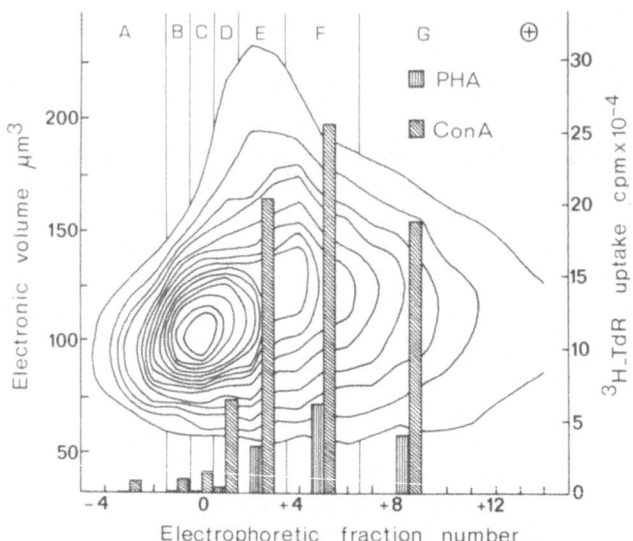

FIGURE 8. Size-vs.-EPM fingerprint of thymus cells from 3-month-old female SJL/J mice. The extent of mitogenic reactivities of the cells in the various individual or pooled electrophoretic fractions are indicated by the vertical bars. This was assessed by measurement of [³H]-thymidine ([³H]-TdR) incorporation after 2 days of cultivation in microplates.

another measure of T-lymphocyte immunocompetence (Stobo, 1972), to be principally expressed by Th4 thymocytes (Fig. 8). Cells of the low-mobility subsets were completely refractory to PHA stimulation. Nevertheless, such cells—and particularly those making up the anodic part of the Th2 cluster (pool D)—proved able to react significantly to Con A. The possible participation of Th3 cells in these mitogenic reactions was, however, difficult to evaluate. Examination of this point would require further purification of electrophoretically enriched Th3 cells by another separation technique based on cell size differences such as velocity sedimentation. Several data have suggested that the Th3 subset includes spontaneously proliferating cells which probably act as precursors for the other thymocyte subpopulations (Dumont, 1978a).

5. Density Gradient Separation of Electrophoretically Fractionated Thymocytes

Isopycnic centrifugation in a bovine serum albumin (BSA) gradient has been successfully utilized for the separation of thymocyte subsets (Shortman *et al.,* 1972a) and their mapping as size vs. buoyant density fingerprint (Droege *et al.,* 1974; Boersma *et al.,* 1979). Application of this method to the analysis of the cellular composition of the thymus from newborn and adult normal or drug-treated mice has permitted Droege and Zucker (1975) to define four physical classes of thymocytes seemingly superposable to those recognized by combination of size and EPM.

We used centrifugation in a continuous Percoll gradient (Pertoft and Laurent, 1977) as an alternate procedure for the fractionation of thymocytes according to their buoyant density. Size distribution analysis of the various fractions collected was performed, which enabled construction of size vs. buoyant density fingerprints. A typical fingerprint thus obtained with SJL/J mouse thymocytes is illustrated in Fig. 9. Two cell clusters were conspicuous on this pattern. The predominant cluster, with a mean density around 1.089 g/ml, possessed the size characteristics of the major low-EPM subset (Th2 cells). The other one, accounting for 18% of total cells, peaked at a density of 1.073 g/ml and appeared similar in modal size to the high-EPM Th4 cells.

The density distribution of electrophoretically separated thymocytes was also investigated by this method. Adjacent electrophoretic fractions were pooled so as to correspond to the four physical cell types identifiable on a size-vs.-EPM fingerprint. These cell pools were then fractionated in parallel by cen-

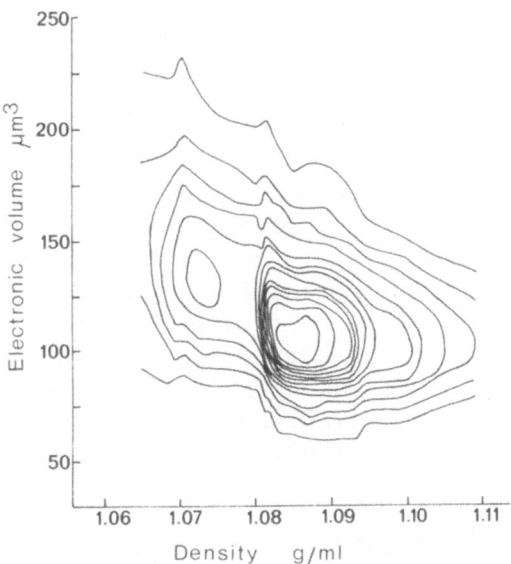

FIGURE 9. Size-vs.-buoyant-density fingerprint of thymus cells from a 3-month-old SJL/J female mouse. Cells were fractionated by isopycnic centrifugation in a continuous Percoll gradient.

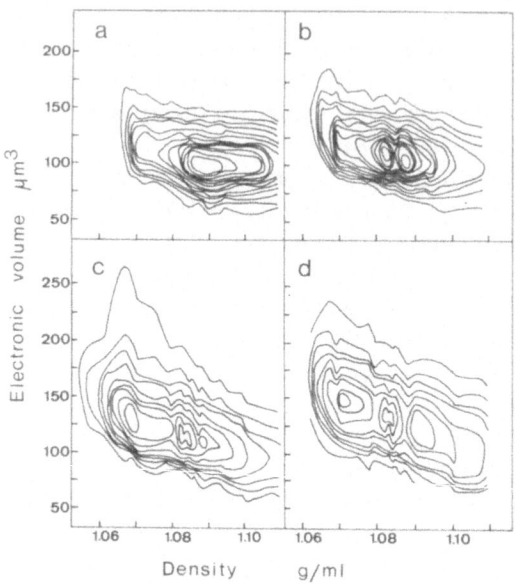

FIGURE 10. Size-vs.-buoyant-density fingerprints of electrophoretically separated thymus cells from 3-month-old SJL mice. Pool of electrophoretic fractions (a) −4 to −2; (b) −1 to 0; (c) +1 to +3; (d) +4 to +12.

trifugation to equilibrium in identical Percoll density gradients. Figure 10 shows the fingerprints resulting from such an experiment. It is clear that each of the various cell pools, although relatively homogeneous with respect to EPM, was indeed heterogeneous with regard to buoyant density and size. This was particularly striking in the case of the cells from the high-EPM region (pools C and D), the fingerprints of which regularly presented 2–3 distinct clusters. Such findings are quite compatible with our previous observations made using a reverse approach and indicating that thymocyte subpopulations with a given buoyant density are electrophoretically heterogeneous (Sabolovic and Dumont, 1973; Dumont and Sabolovic, 1973). Although the thymocyte subpopulations isolated on the basis of buoyant density differences certainly overlap with those separated by electrophoresis, as suggested by Droege and Zucker (1975) and by the above-mentioned data, the homology is not absolute. In fact, each of the four physical subpopulations of thymocytes distinguished by size-vs.-EPM fingerprint analysis, and especially the Th4 subset, appears to be composed of several density subsets. This strongly suggests that many more than four physical categories of cells constitute the thymocyte population of normal adult mice. This conclusion was also recently reached by Josefowicz *et al.* (1977), who combined velocity sedimentation separation and high-resolution electrophoretic light-scattering analysis to identify thymocyte subpopulations. Further studies should aim at determining the functional significance of such thymocyte categories defined by subtle differences in their physical properties.

6. Physical Subpopulations of Lymphocytes in the Pathological Mouse Thymus

The fingerprint procedure also proved useful for characterizing and monitoring the alterations in the cellular constitution of the thymus that occur under certain pathological conditions. Thus, the development of thymus-borne leukemia in AKR mice could be identified as the unrefrained hyperplasia of a cell compartment physically similar to the normal Th3 subset (Dumont, 1978b).

Modifications of another kind were disclosed during aging in SJL/J mice, a strain known to suffer from reticulum cell sarcomas arising spontaneously with a high incidence (Murphy, 1969) and associated with a number of immunological disturbances (Cinader *et al.,* 1978; Crowle *et al,* 1978). Although the thymocyte fingerprint of young adult SJL/J mice (Fig. 5) was comparable to that of normal mouse strains (CBA/J, C3H), a marked expansion of the Th4 subset could be noted from the age of 6 months onward. At the age of 16 months, these cells accounted for up to 75% of the total thymocyte population. By the same time, the typical low-EPM cluster regressed and a new physical type of lymphocytes characterized by a lower EPM (peak at electrophoretic fraction -4) and a modal volume larger than Th2 cells was found to emerge. In the oldest mice examined, these cells appeared as a well-distinct cluster on the fingerprint and reached a proportion as high as 20%. Evidence was obtained that these cells are devoid of T-lymphocyte markers (but possess B-lymphocyte attributes such as the presence of surface immunoglobulins) and of receptors for complement (Dumont, 1978b, 1980, unpublished results).

A very similar situation was found in the thymus of NZB $\times$ SJL/J F_1 hybrid mice. There, the thymic abnormalities were still more pronounced and occurred earlier in life than in SJL/J mice. They were encountered in female but not in male individuals, which indicates that their appearance is influenced by sex-linked factors (Dumont and Robert, 1980). A typical size-vs.-EPM fingerprint of thymus cells from a 12-month-old female NZB $\times$ SJL/J mouse is depicted in Fig. 11. While the low-EPM cluster was no longer detectable on this pattern, two other electrophoretic cell classes were evident. The predominant one, which represented 70% of all thymus cells, displayed the physical properties of Th4 cells. The other one corresponded to a B-lymphocyte popu-

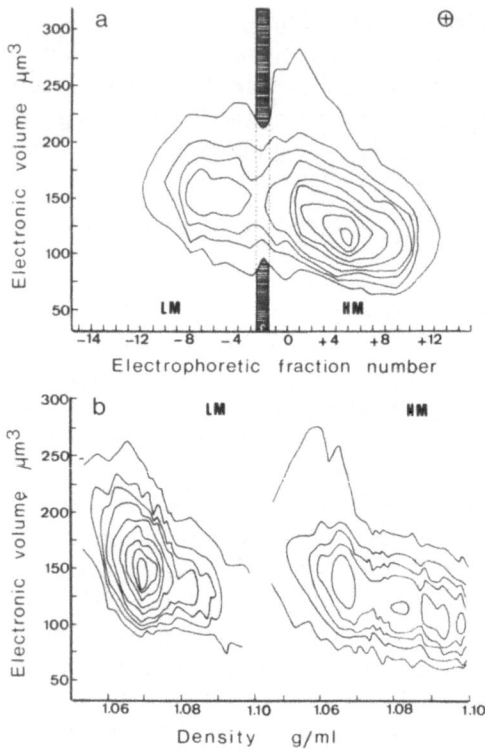

FIGURE 11. (a) Size-vs.-EPM fingerprint of thymus cells from a 12-month-old NZB × SJL F_1 female mouse. Cells from the LM and HM regions were pooled separately and fractionated in parallel Percoll density gradients. (b) Size-vs.-buoyant-density fingerprints of these two electrophoretic cell classes.

lation (as characterized by a set of surface and functional markers; Dumont *et al.,* 1981b) endowed with a lower EPM than normal thymocytes and a modal volume around 150 μm^3. The aspect of these two clusters suggested some further physical heterogeneity. This point was made clear after isopycnic centrifugation of both groups of cells in Percoll gradients. Then two density subsets could be recognized within the Th4 subpopulation. In the case of the B-lymphocyte population, the size-vs.-buoyant-density fingerprint showed a single cluster with a tail in the denser region. These data provide another example of how association with an additional physical parameter (buoyant density) can increase the resolving power of the fingerprint technique.

B. Fingerprint Analysis of Murine Splenocytes

1. Identification of Physical Subpopulations of Splenocytes in Normal CBA/J Mice

Electrophoretic fractionation of erythrocyte-depleted spleen cell suspensions from adult CBA/J mice regularly resulted in a bimodal distribution profile. The two electrophoretic cell classes thus separated appeared as well-distinct clusters on the fingerprint (Fig. 12a). The first cell cluster, accounting for 53% of the total splenocyte population, peaked by four electrophoretic fractions slower than thymocytes and was arbitrarily designated as the low-mobility (LM) subpopulation. The other one, representing 44% of total cells, centered around electrophoretic fraction +6 and was called the high-mobility (HM) subpopulation. The inclined oval shape of these two cell clusters indicated that both were heterogeneous as far as EPM and size are concerned. Thus, a cathodic, slow LM cell subset with a modal volume around 115–120 μm^3 might be roughly distinguished from a fast LM cell subset located in the anodic part of the LM cluster and exhibiting a larger modal volume (130–135 μm^3). Similarly, a slow and a fast region might be recognized within the HM cluster. However, in this case the size-vs.-EPM relationship was the reverse of that noticed for LM cells, namely, slow HM cells appeared of modal volume larger (135–140 μm^3) than fast HM cells (120 μm^3). Obviously, such subsets defined as areas of the fingerprint may not represent discrete entities. Since there was a gradual shift in the modal volume of the individual fractions constituting each cluster, a greater number of overlapping, physically different cell subpopulations might exist. In addition to the major cell clusters, a bulge was clearly visible in the middle part of the fingerprint, reflecting the presence of a minor subpopulation (3%) of large cells (175–225 μm^3) with intermediate EPM. Such observations are quite in keeping with earlier data obtained after combination of buoyant density separation and electrophoretic analysis of spleen cells (Dumont, 1974).

The assessment of various antigenic and functional markers has demonstrated that the LM and HM subpopulations of mouse splenocytes correspond respectively to B and T lymphocytes (Nordling et al., 1972; Wioland et al., 1972; Andersson et al., 1973; Zeiller et al., 1974; Platsoucas and Catsimpoolas, 1979). Some evidence has also been presented which suggests that the physical subsets here identified by the fingerprinting technique represent subpopulations of B and T lymphocytes. Thus, slow LM cells have been characterized as memory B cells, whereas fast LM cells have been shown to comprise less mature or activated forms of the B-cell lineage (Schlegel et al., 1975; Zeiller et al., 1976). Recently, we have demonstrated that slow LM cells

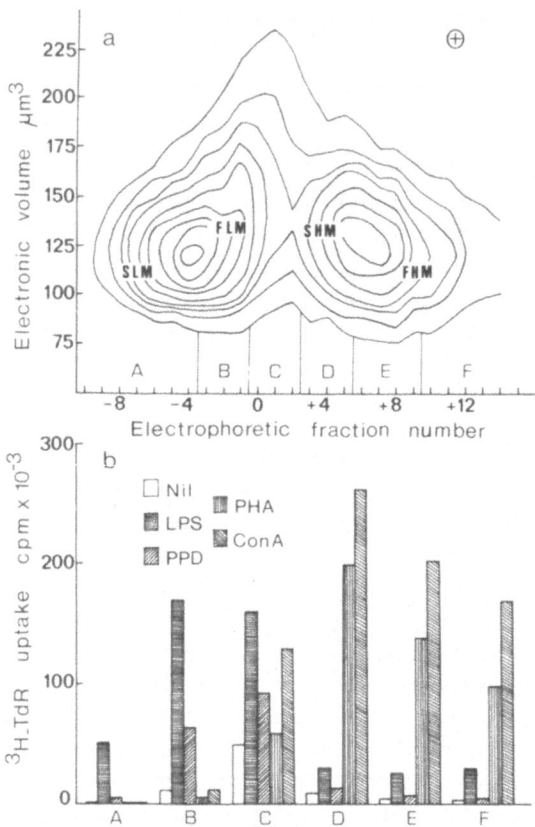

FIGURE 12. (a) Size-vs.-EPM fingerprint of spleen cells from 3-month-old female CBA/J mice. The LM and HM subpopulations are visible as well-distinct clusters. Within each of these clusters, slow (S) and fast (F) EMP regions may be distinguished. (b) Mitogenic reactivities of the cells in the various electrophoretic fractions pooled as indicated in (a). Cells were cultivated for 48 hr in microplates without mitogen or in the presence of optimal concentrations of B- (LPS, PPD) or T- (PHA, Con A) cell mitogens. Cultures were pulsed with [^{3}H]-TdR for the last 24 hr of incubation.

express a high density of sIgD but a low density of sIgM in contrast to fast LM cells, which display a low density of sIgD and larger amounts of sIgM (Dumont *et al.,* 1979). On the other hand, slow HM cells have been shown to include newly formed T cells (Häyry *et al.,* 1975) with a higher density of surface Thy-1 antigen than fast HM cells (Zeiller *et al.,* 1975b; Dumont *et al.,* 1981a).

As demonstrated in Fig. 12b, the various physical categories of CBA/J mouse splenocytes also differ in their rate of spontaneous or mitogen-induced proliferation *in vitro* as evaluated by measurement of [³H]thymidine (TdR) incorporation. Fractions pooled from the intermediate EPM range (pool C) were found to exhibit a high degree of [³H]-TdR uptake in the absence of mitogen, suggesting that this region is enriched for spontaneously dividing elements. The repartition of the reactivities to the B-lymphocyte mitogens LPS and PPD and to the T-lymphocyte mitogens PHA and Con A was consistent with the predominant B-lymphocyte nature of LM cells and predominant T-lymphocyte nature of HM cells. The weak LPS response which was repeatedly observed in the HM fractions probably denotes the existence of a minor B-cell subset contaminating the T-cell region as previously reported by von Boehmer *et al.* (1974) and Zeiller *et al.* (1976). Most interesting was the finding that slow LM cells reacted to a lesser extent to LPS and PPD than did fast LM cells. Moreover, although PHA- and Con-A-reactive T-cell subsets (Stobo, 1972) could not be separated by electrophoresis (Shortman *et al.*, 1975), fast HM cells (pools E and F) evidently tended to respond to both mitogens less vigorously than slow HM cells (pool D).

2. Fractionation on Nylon-Wool Columns of the Physical Subpopulations of Splenocytes

Differential adhesiveness of cells to nylon fiber was used as an adjunct parameter for increasing the resolution of the size-vs.-EPM analysis of the splenocyte population. Spleen cell suspensions from CBA/J mice were incubated for 45 min at 37°C in columns of nylon wool (Fenwal Leukopak, Travenol Laboratories) prepared as described by Julius *et al.* (1973). Three cellular fractions were eluted from these columns and were subsequently submitted to electrophoretic separation and electronic sizing.

A first fraction was collected upon slowly washing (1 drop/sec) the column with RS medium at 37°C. Such a nonadherent population has been shown to be enriched for T lymphocytes (Julius *et al.*, 1973; Trizio and Cudkowicz, 1974). Accordingly, its fingerprint (Fig. 13a) was found to present a prevalent HM cell cluster (90%). The overall shape of this cluster would again justify the distinction of a slow and a fast HM region possessing different modal volumes. The minor LM cluster also visible on this fingerprint was characteristic of small, slow-moving B cells.

The second fraction, made up of weakly adherent cells, was recovered by further washing the column with a rapid flow of warm RS medium. As illustrated in Fig. 13b, both LM and HM cells were detectable on the fingerprint and represented 35 and 55%, respectively, of this fraction. Interestingly, a third cluster (10%) of large cells was conspicuous in the intermediate EPM zone. These cells were identified on cytocentrifuge smears as being mainly of granulocytic nature.

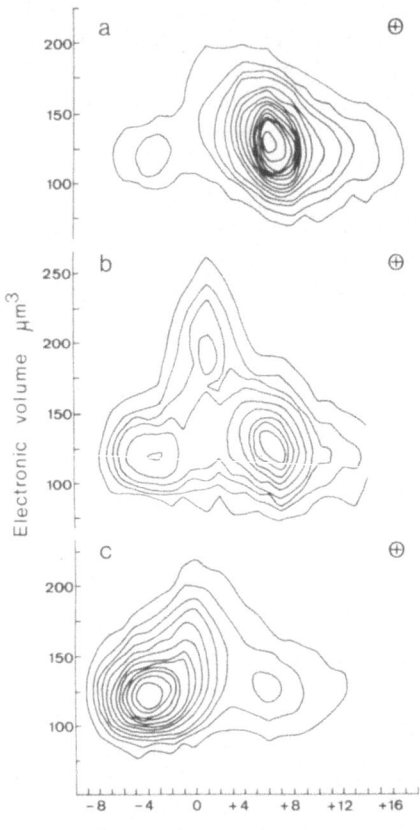

FIGURE 13. Size-vs.-EPM fingerprints of CBA/J spleen cells separated by differential adhesiveness to nylon wool. (a) Nonadherent cells; (b) weakly adherent cells; (c) strongly adherent cells.

The last fraction of firmly adherent cells was eluted by repeated compression and expansion of the nylon wool and flushing with cold (4°C) RS medium. In agreement with its predominant B-lymphocyte content (Handwerger and Schwartz, 1974), up to 80% of the cells in this fraction segregated in an apparent LM cluster (Fig. 13c). This cluster clearly exhibited in its anodic size a trail oriented upward, suggesting the existence of fast LM cells with larger size than slow LM cells. The yield of these various fractions was consistent with the initial frequency of LM and HM cells in the total splenocyte population, which indicates that no selective loss of cells occurred during the nylon-wool separation procedure.

These data support the notion that LM and HM splenocytes represent mostly B and T lymphocytes. The identity of the minor LM and HM subsets contaminating the nonadherent and adherent fractions, respectively, was not directly examined. However, such contaminations appear consistent with the presence of a small proportion (5–10%) of B lymphocytes in the nonadherent population (Julius *et al.,* 1973) and of up to 15% of T lymphocytes in the adherent population (Handwerger and Schwartz, 1974).

3. Differential Sensitivities of the Physical Subpopulations of Splenocytes to Immunosuppressive Drugs

As clearly exemplified in the case of thymocytes, treatment of mice with drugs selectively destroying certain lymphocyte subpopulations may serve as a dissecting tool for the characterization of these and the surviving subpopulations. In this respect, we have particularly studied the effects of hydrocortisone and of cyclophosphamide.

The administration of hydrocortisone (250 mg/kg of body weight), which reduced spleen cellularity to approximately 25% of the normal value, was found to diminish the absolute numbers of both LM and HM cells (Dumont and Barrois, 1977; Dumont and Bischoff, 1978). However, LM cells were more affected than HM cells, and this led to a proportional increase (up to 75%) of the latter population. Interestingly, the whole fingerprint of splenocytes from hydrocortisone-treated animals was displaced downward as compared to the normal pattern (Fig. 14a). This suggests that the drug acted similarly on both lymphocyte compartments by removing the larger elements. Moreover, since the HM cluster of hydrocortisone-resistant splenocytes shifted by one to two fractions toward more anodic EPM, it seems likely that the T cells spared by the drug belong to the fast LM subset. Some granulocytes also survived hydrocortisone treatment and made up a well-individualized cluster with large modal volume and intermediate EPM.

Cyclophosphamide, injected in a sublethal dose (300 mg/kg of body weight), exerted a still more drastic spleen depletion than did hydrocortisone. Two days after the treatment, only 5–10% of the initial cell number remained in the spleen. The remaining cells have been shown to be essentially T lymphocytes (Poulter and Turk, 1972; Dumont, 1974). Accordingly, the fingerprint of cyclophosphamide-resistant splenocytes exhibited a major cluster in the HM region (Fig. 14b). However, a small bulge in the fast LM range— representing 2–3% of the recovered cells—and a cluster of large granulocytic cells were also visible.

Thus, despite quantitative differences, both drugs tested preferentially eliminated cells of the LM cluster. A substantial part of the HM compartment was nevertheless also removed by these drugs. Particularly in the case of hydrocortisone, there was a tendency for the selective disappearance of those larger

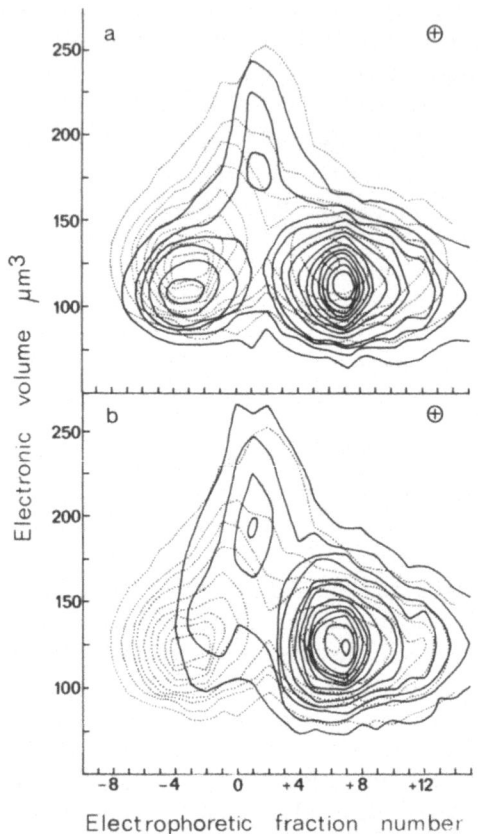

FIGURE 14. Size-vs.-EPM fingerprints of spleen cells from CBA/J mice treated 2 days previously with either (a) hydrocortisone acetate (250 mg/kg of body weight) or (b) cyclophosphamide (300 mg/kg of body weight). Each of these fingerprints is superimposed to that of control CBA/J spleen cells (· · ·).

cells constituting the slow HM region. This point would fit with the idea that the slow HM area contains short-lived T lymphocytes (Häyry *et al.,* 1975; Dumont and Bischoff, 1977).

4. Strain Dependence of the Size-vs.-EPM Fingerprint Pattern of Splenocytes

All the data on the physical subpopulations of splenocytes mentioned above concerned CBA/J mice. A number of other mouse strains were also investigated in this respect. As shown by Figs. 15 and 16, the size-vs.-EPM fingerprint pattern turned out to be quite variable from one strain to another. Each of the experiments was repeated on several instances (4–8 times), and the interstrain differences were always found to be greater than the intrastrain individual variations (except with mice of the B10 family, Fig. 16). The differences bore not only on the relative frequencies of LM and HM cells but also on the electrophoretic separability and modal volume of these two clusters. This strongly suggests that genetic factors determine the physical properties of mouse splenocytes. Whether such factors control the proportions of physically similar B and T lymphocyte subpopulations, irrespective of strain, to those defined in CBA/J mice, or whether they influence the physical characteristics of these subpopulations, remains as yet unknown. At any rate, since the splenocyte fingerprints of mice with the same H-2 haplotype but different background genomes such as DBA/2, Balb/c, NZB (Fig. 15), and B10.D2 (Fig. 16) were strikingly different, whereas those of the H-2 congenic strains on C57B/10 background were very similar (Fig. 16), it appears likely that the gene(s) involved in this control belong(s) to non-H-2 areas. This would stand

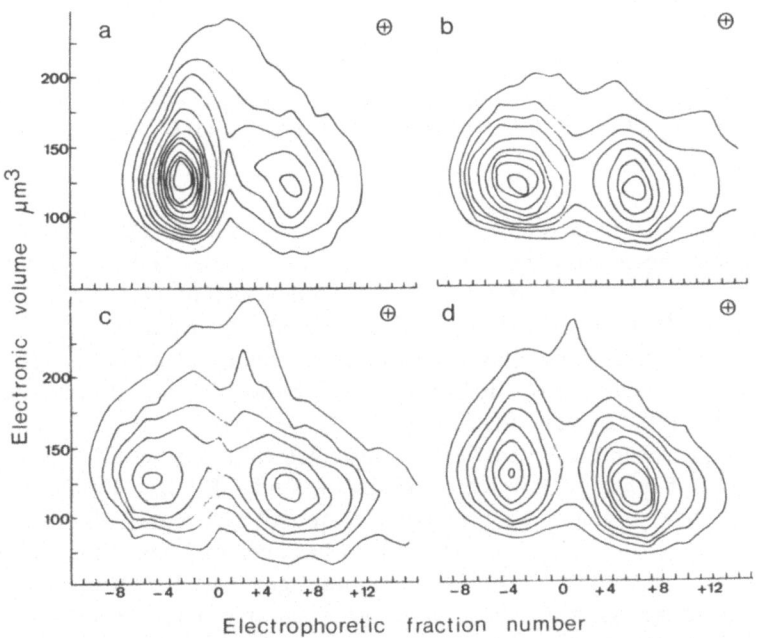

FIGURE 15. Size-vs.-EPM fingerprints of spleen cells from 3-month-old female mice of the following strains: (a) DBA/2 (H-2^d); (b) Balb/c (H-2^d); (c) NZB (H-2^d); and (d) SJL/J (H-2^s).

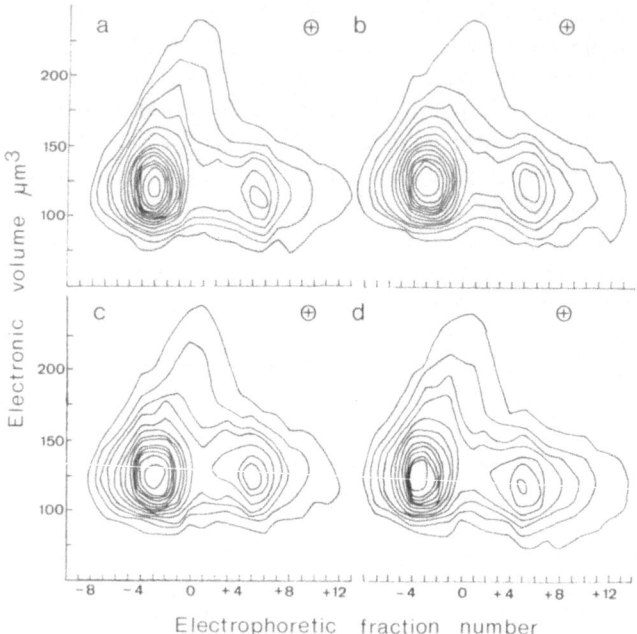

FIGURE 16. Size-vs.-EPM fingerprints of spleen cells from 3-month-old female H-2 congenic mice. (a) B10.Br (H-2^k); (b) B10.D2 (H-2^d); (c) B10.M (H-2^f); (d) B10.S (H-2^s).

in contrast with data reported by Donner and Wioland (1976), which suggested that the electrokinetic properties of mouse spleen cells are H-2-dependent. However, this point has been recently questioned by Platsoucas and Catsimpoolas (1979), who studied the electrophoretic distribution of splenocytes from several mouse strains and concluded that the influence of the H-2 complex on the surface charge of lymphocytes is probably limited in comparison with that exerted by non-H-2 gene(s).

Another example of genetically determined variations in the size-vs.-EPM fingerprint of splenocytes was provided by the study of CBA/N mice and their hybrids. There, a genetic defect leading to quantitative and qualitative abnormalities of the B-lymphocyte compartment has been well characterized (Scher *et al.,* 1975; Ahmed *et al.,* 1977). The mutant gene in CBA/N mice has been shown to be recessive and located on the X chromosome (Amsbaugh *et al.,* 1972). Then, when a female CBA/N (xx) is mated with a male (XY) from a normal strain, e.g., DBA/2, the female offspring of this cross (xX) will be phenotypically normal, whereas the male offspring (xY) will bear the same abnormality as CBA/N mice. The use of such F_1 hybrid mice allowed for an adequate system to investigate the influence of the CBA/N defect on the fingerprint pattern since female F_1 littermates could serve as age-matched controls for immunologically aberrant males. In Fig. 17, typical splenocyte fingerprints of 3-month-old female and male CBA/N × DBA/2 F_1 hybrid mice are superimposed. The LM and HM clusters were conspicuous in both patterns. However, while the LM population prevailed in female mice (57 ± 3%), its frequency did not exceed 40% in the males. Consideration of spleen cellularities indicated that this anomaly reflected a true numerical deficiency of LM cells and not a relative increase of HM cells. Moreover, although the HM clusters of the hybrids of both sexes were physically much alike, slight differences could be repeatedly noticed regarding LM cells. Thus, in male hybrid mice, the LM cluster migrated faster by one to two electrophoretic fractions and possessed a larger modal volume than in their female littermates. Such differences might reflect a selective absence or depletion of cells belonging to the slow LM splenocyte subset in the defective CBA/N × DBA/2 F_1 males. Application of the fingerprint procedure to the analysis of Peyer's patch lymphocytes in CBA/N mice and their hybrids similarly led us to conclude that the X-linked mutation of these mice results in the lack of a subpopulation of small B lymphocytes with low anodic EPM (Dumont *et al.*, 1979). It must be stressed, however, that these data do not necessarily imply that the fast LM cells present in CBA/N mice and the defective male hybrids are functionally identical to those encountered in normal adult mice.

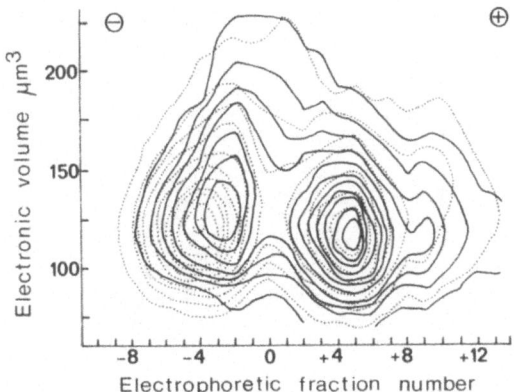

FIGURE 17. Size-vs.-EPM fingerprint of spleen cells from 3-month-old male CBA/N × DBA/ 2 F_1 mice (———) which bear an X-linked B-cell deficiency as compared to that of spleen cells from their normal female littermates ($\cdot \cdot \cdot$).

C. Fingerprint Analysis of Murine Lymph Node Cells

1. Normal Situation

Figure 18 shows a typical size-vs.-EPM fingerprint as obtained with lymph node cells from adult CBA/J mice. Two electrophoretic subpopulations are clearly visible. As in the spleen, these subpopulations have been demonstrated to correspond to B (LM cluster) and T (HM cluster) lymphocytes (Nordling *et al.*, 1972; Andersson *et al.*, 1973). In contrast to this bimodal pattern also observable in C3H mice (Dumont, 1977), Droege (1976) has reported a trimodal fingerprint distribution of lymph node cells in C57 B1/6 mice showing, besides a minor LM cluster, two HM subpopulations with similar size but distinct electrokinetic behaviors. It is not clear whether this discrepancy is related to technical factors or reflects strain-associated differences in the physical properties of lymph node cells as those mentioned above for splenocytes.

In C3H mice, modifications of the lymph node cell fingerprint were found to take place during postnatal development, which suggested that, at least in this strain, the physical properties of lymphocytes might be indicators of their state of maturation (Dumont, 1977).

2. Pathological Alterations

a. Changes Accompanying Reticulum Cell Sarcoma Development in SJL/ J Mice. SJL/J mice develop relatively early in life a high incidence of reticulum cell sarcoma (RCS), which arises primarily in abdominal lymph nodes and

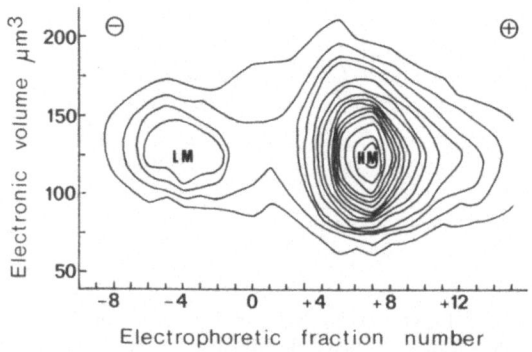

FIGURE 18. Size-vs.-EPM fingerprint of axillary and brachial lymph node cells from 3-month-old female CBA/J mice.

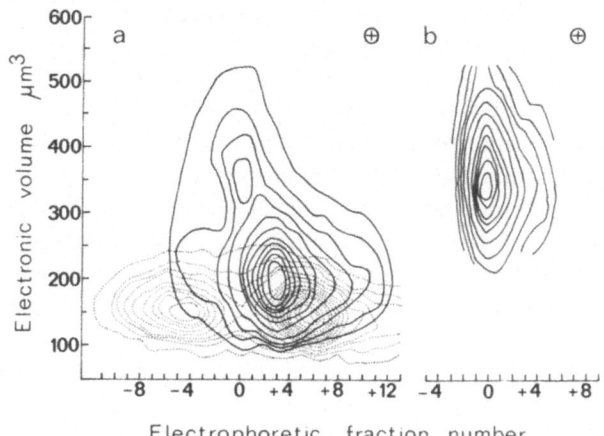

FIGURE 19. (a) Size-vs.-EPM fingerprint of a hypertrophied (————) as compared to a normal-sized (· · ·) mesenteric lymph node from 10-month-old SJL mice. (b) Size-vs.-EPM fingerprint of a SJL RCS-derived cell line (LC2) growing *in vitro*.

Peyer's patches (Murphy, 1969) and which is believed to originate from the B-cell lineage.

We have applied the fingerprint technique to the analysis of the cellular composition of mesenteric lymph nodes from 10 to 12-month-old SJL/J mice at different stages of RCS involvement (Robert and Dumont, 1979). The fingerprint of normal-sized lymph nodes regularly presented two well-individualized clusters separated by 10–11 electrophoretic fractions and with comparable modal volumes (125–130 μm^3). These LM and HM clusters represented, on the average, 35 and 65%, respectively, of the total lymph node cell population. In lymph nodes of increasing size, the fingerprint pattern was at first little modified but then became markedly altered. As exemplified by Fig. 19a, in the ultimate stages of RCS-induced lymph node enlargement, a predominant cluster (80%) was visible in the HM region. These cells, however, were different from the normal HM subpopulation by both their slower EPM and their larger modal size (150–175 μm^3). The evaluation of several markers demonstrated these cells to be Thy-1$^+$, Lyt-1$^+$, and α-naphthyl esterase$^+$ and thus of probable T-lymphocyte nature. A second subset was most often apparent as a minor cluster with a volume still larger (300 μm^3) than the former subpopulation. These cells were found to lack T-lymphocyte attributes but to exhibit the Ia antigen. Interestingly, they appeared physically very similar to cells from an RCS-derived line (LC2) growing *in vitro* and probably representing malignant elements (Fig. 19b).

These data would suggest that lymph node hypertrophy in primary SJL/J lymphomas mainly results from a massive *in situ* proliferation of Lyt-1$^+$ T cells, possibly reflecting an autoimmune reaction against Ia$^+$ malignant cells such as that reported to occur *in vitro* by Ponzio *et al.* (1977).

    b. Characterization of Lymphoproliferative Disease of MRL/l Mice. Mice of the MRL/l strain bear an autosomal gene (lpr) which determines the appearance by 3–4 months of a pronounced and generalized lymphadenopathy (Murphy and Roths, 1978). Most lymph nodes then become enlarged by up to 100 times normal size. The major cell type present in such hyperplastic nodes has been identified as a Thy-1$^+$ lymphocyte. It appeared of interest to further characterize this cell type on the basis of physical criteria

    The size-vs.-EPM fingerprint of the cells dissociated from a hypertrophied MRL/l mesenteric node is depicted in Fig. 20. This pattern strikingly contrasts with that observed for nonenlarged lymph nodes from 3-week-old MRL/l mice, which presented a normal bimodal distribution. Thus, the cells from hypertrophied lymph nodes were relatively homogeneous with respect to electrokinetic properties and migrated as a single peak in the HM region. However, this peak was slower than the control HM (T-cell) subpopulation by three to four electrophoretic fractions. Moreover, although the modal volume of these cells was close to that of normal T lymphocytes, a skewness toward larger volumes was clearly apparent on the fingerprint.

    These results, which were repeatedly obtained in a series of 15 experiments, support the notion that the cells expanding in enlarged MRL/l lymph

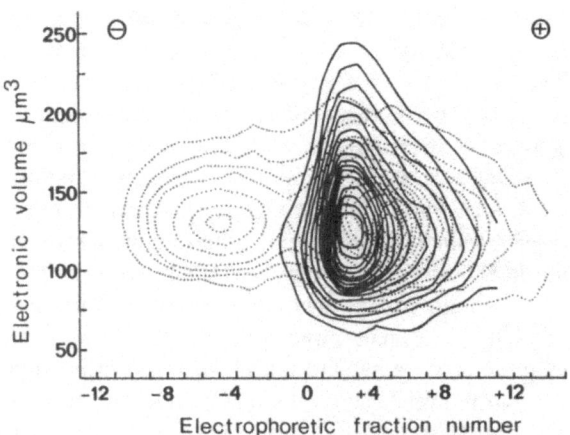

FIGURE 20. Size-vs.-EPM fingerprints of lymph node cells from MRL/lpr mice. Comparison of a hypertrophied mesenteric node from a 3-month-old mouse (————) with a normal-sized one from a 3-week-old mouse ($\cdot \cdot \cdot$).

nodes are T lymphocytes. However, they also indicate that these cells are physically distinct from the normal T-lymphocyte subpopulation. Indeed, although these cells stained heavily for α-naphthyl esterase activity, another T-lymphocyte marker, they were found poorly responsive to the T-lymphocyte mitogens PHA and Con A, which suggests that they also differ functionally from normal T lymphocytes (Dumont, unpublished observations).

IV. CONCLUSION

The combination of two physical methods measuring independent parameters provides a powerful tool for the discrimination of several categories of murine lymphoid cells differing in some of their biological properties. This procedure proved useful for analyzing the cellular composition of the thymus, the spleen, and other lymphoid organs under various experimental conditions. Although the physical subsets of lymphocytes thus delineated were actually fractionated only according to a single parameter, i.e., surface charge or buoyant density, their mapping as two-dimensional fingerprints should help investigators to elaborate on a strategy leading to a higher degree of purification. Thus, once identified on a fingerprint, a cell type of interest might be isolated by substituting the electronic sizing analytical technique for a preparative technique relying on size differences such as velocity sedimentation or centrifugal elutriation. The resolution of this procedure could also be improved by additional separation methods, e.g., nylon-wool adherence or lectin-induced differential agglutination. It might be predicted that association of the fingerprint technique with more refined means of cell characterization, such as offered by fluorescence-activated cell sorting, will contribute to the unraveling of the complexity of lymphocyte differentiation pathways.

ACKNOWLEDGMENTS. This work was supported by a grant from the FEGE-FLUC and by INSERM (ATP 71-78-103 and CRL 79-5-020-2). We thank Misses Josiane Bara and Martine Labrune for their help in the preparation of this manuscript.

REFERENCES

Ahmed, A., Scher, I., Sharrow, S. O., Smith, A. H., Paul, W. E., Sachs, D. H., and Sell, K. W. (1977) B-lymphocyte heterogeneity: Development and characterization of an alloantiserum which distinguishes B-lymphocyte differentiation alloantigens, *J. Exp. Med.* 145:101.

Amsbaugh, D. F., Hansen, C. T., Prescott, B., Stashak, P. W., Barthold, D. R., and Baker, P. J. (1972) Genetic control of the antibody response to Type III pneumococcal polysaccharide in mice. I. Evidence that an X-linked gene plays a decisive role in determining responsiveness, *J. Exp. Med.* **136**:931.

Andersson, L. C., Nordling, S., and Häyry, P. (1973) Fractionation of mouse T and B lymphocytes by preparative cell electrophoresis. Efficiency of the method, *Cell. Immunol.* **8**:235.

Bhattachraya, C. G. (1967) A simple method of resolution of a distribution into gaussian components, *Biometrics* **23**:115.

Blomgren, H., and Andersson, B. (1969) Evidence for a small pool of immunocompetent cells in the mouse thymus, *Exp. Cell Res.* **34**:185.

Boehmer, H. von, and Shortman, K. (1973) The separation of different cell classes from lymphoid organs. IX. A simple and rapid method for removal of damaged cells from lymphoid cell suspensions, *J. Immunol. Methods* **2**:293.

Boehmer, H. von, Shortman, K., and Nossal, G. J. V. (1974) The separation of different cell classes from lymphoid organs. X. Preparative electrophoretic separation of lymphocyte subpopulations from mouse spleen and thoracic duct lymph, *J. Cell. Physiol.* **83**:231.

Boersma, W., Betel, I., and Van der Westen, G. (1979) Thymic regeneration after dexamethasone treatment as a model for subpopulation development, *Eur. J. Immunol.* **9**:45.

Cantor, H., and Boyse, E. A. (1977) Lymphocytes as models for the study of mammalian cellular differentiation, *Immunol. Rev.* **33**:105.

Cinader, B., Paraskevas, F., and Koh, S. (1978) An age-dependent decline in suppressor capacity of SJL/J mice, *Cell. Immunol.* **40**:445.

Crowle, A. J., Jacobson, S., and May, M. (1978) Resistance of SJL mice to immunosuppression by antibodies, *Immunol. Commun.* **7**:417.

Donner, M., and Wioland, M. (1976) Possible involvement of H-2 region in distribution control of T and B lymphocytes of spleen and in surface characteristics of T lymphocytes, *Folia Biol.* **22**:51.

Droege, W. (1976) Early T cells and late T cells; suggestive evidence for two T cell lineages with separate developmental pathways, *Eur. J. Immunol.* **6**:763.

Droege, W., and Zucker, R. (1975) Lymphocyte subpopulations in the thymus, *Transplant. Rev.* **25**:3.

Droege, W., Zucker, R., and Jauker, U. (1974) Cellular composition of the mouse thymus: Developmental changes and the effect of hydrocortisone, *Cell. Immunol.* **12**:173.

Dumont, F. (1974) Destruction and regeneration of lymphocyte populations in the mouse spleen after cyclophosphamide treatment, *Int. Arch. Allergy Appl. Immunol.* **47**:110.

Dumont, F. (1975) Electrophoretic mobility and surface immunoglobulin of albumin gradient fractionated mouse spleen cells, *Immunology* **28**:731.

Dumont, F. (1977) Physical properties of mouse peripheral lymph node cells: Changes during development, *Ann. Immunol.* **128C**:1053.

Dumont, F. (1978a) Physical subpopulations of mouse thymocytes: Changes during regeneration subsequent to cortisone treatment, *Immunology* **34**:841.

Dumont, F. (1978b) Electrophoretic separation and characterization of lymphocyte subpopulations in the normal and pathological mouse thymus, in: *Electrophoresis '78* (N. Catsimpoolas, ed.), Elsevier/North-Holland, Amsterdam, pp. 357–372.

Dumont, F. (1980) Lymphocyte subpopulations in the thymus of SJL/J mice: Age-related alterations and the effect of spontaneous reticulum cell sarcoma development, *J. Clin. Lab. Immunol.* **3**:51.

Dumont, F., and Barrois, R. (1975) Effect of treatment with cyclophosphamide on the electrophoretic mobility and mitogen responsiveness of mouse thymus cells, *Biomedicine* **23**:391.

Dumont, F., and Barrois, R. (1977) Electrokinetic properties and mitogen responsiveness of mouse splenic B and T lymphocytes following hydrocortisone treatment, *Int. Arch. Allergy Appl. Immunol.* **53**:293.

Dumont, F., and Bischoff, P. (1977) Developmental changes in the electrophoretic mobility of B and T cell populations in the mouse spleen, *Ann. Immunol.* **128C**:771.

Dumont, F., and Bischoff, P. (1978) Differential effect of hydrocortisone on lymphocyte populations in the mouse spleen, *Biomedicine* **29**:28.

Dumont, F., Habbersett, R., and Ahmed, A. (1981a) Characterization of the surface phenotype of electrophoretically fractionated mouse lymphocytes by flow microfluorometry analysis, in: *Electrophoresis 81* (P. Arnaud, ed.), Walter de Gruyter, New York, pp. 831–840.

Dumont, F., and Nardelli, J. (1979) Peanut agglutinin (PNA)-binding properties of murine thymocyte subpopulations, *Immunology* **37**:217.

Dumont, F., and Robert, F. (1976) Dose-related effect of hydrocortisone treatment on the electrokinetic properties and mitogen responsiveness of mouse thymocytes, *Int. Arch. Allergy Appl. Immunol.* **51**:482.

Dumont, F., and Robert, F. (1980) Age- and sex-dependent thymic abnormalities in NZB × SJL Fl hybrid mice, *Clin. Exp. Immunol.* **41**:63.

Dumont, F., Robert, F., and Gerard, H. (1981b) Abnormalities of the thymus in aged (NZB × SJL)Fl mice: separation and characterization of intrathymic T cells, B cells, and plasma cells, *J. Immunol.* **126**:2450.

Dumont, F., and Sabolovic, D. (1973) Heterogeneity of the hydrocortisone-resistant cell subpopulation in the mouse thymus, *Biomedicine* **19**:257.

Dumont, F., Ahmed, A., and Habbersett, R. (1979) Electrokinetic properties of murine B lymphocyte subpopulations characterized by their surface-immunoglobulin phenotype, in: *Cell Electrophoresis: Clinical Application and Methodology* (A. W. Preece and D. Sabolovic, eds.), Elsevier/North-Holland, Amsterdam, pp. 157–168.

Dumont, F., Bischoff, P., and Ahmed, A. (1979) Biophysical characterization of Peyer's patch lymphocytes in the B-cell deficient CBA/N mice and in their Fl hybrids, *Cell Biophys.* **1**:293.

Handwerger, B. S., and Schwartz, R. H. (1974) Separation of murine lymphoid cells using nylon wool columns. Recovery of the B-cell enriched population, *Transplantation* **18**:544.

Hannig, K. (1971) Free-flow electrophoresis, in: *Methods in Microbiology,* Volume 5B (J. R. Norris and D. W. Ribbons, eds.), Academic Press, London, pp. 513–548.

Hannig, K., Wirth, H., Meyer, B. H., and Zeiller, K. (1975) Free-flow electrophoresis. I. Theoretical and experimental investigations of the influence of mechanical and electrokinetic variables on the efficiency of the method, *Hoppe-Seyler's Z. Physiol. Chem.* **356**:1225.

Häyry, P., Andersson, L. C., Gahmberg, C., Roberts, P., Ranki, A., and Nordling, S. (1975) Fractionation of immunocompetent cells by free-flow cell electrophoresis, in: *Lymphocytes and Their Cell Membranes* (M. Schlesinger, ed.), Academic Press, New York, pp. 71–90.

Josefowicz, J. Y., Ware, B. R., Griffith, A. L., and Catsimpoolas, N. (1977) Physical heterogeneity of mouse thymus lymphocytes, *Life Sci.* **21**:1483.

Julius, M. H., Simpson, E., and Herzenberg, L. A. (1973) A rapid method for the isolation of functional thymus-derived murine lymphocytes, *Eur. J. Immunol.* **3**:645.

Kolin, A. (1960) Continuous electrophoresis fractionation stabilized by electromagnetic rotation, *Proc. Nat. Acad. Sci. U.S.A.* **46**:509.

London, J., Berrih, S., and Bach, J. F. (1978) Peanut agglutinin. I. A new tool for studying T lymphocyte subpopulations, *J. Immunol.* **121**:438.

McKenzie, I. F. C., and Potter, T. (1979) Murine lymphocyte surface antigens, *Adv. Immunol.* **27**:179.

Moon, R., Phillips, R. A., and Miller, R. G. (1972) Sedimentation and volume analysis of human bone-marrow, *Ser. Haematol.* **5**:163.

Murphy, E. D. (1969) Transplantation behavior of Hodgkin's like reticulum cell neoplasms of strain SJL/J mice and results of tumor reinoculation, *J. Nat. Cancer Inst.* **43**:797.

Murphy, E. D., and Roths, J. B. (1978) Autoimmunity and lymphoproliferation: induction by mutant gene *lpr*, and acceleration by a male associated factor in strain BXSB, in *Genetic Control of Autoimmune Disease* (N. R. Rose, P. E. Bigazzi, and N. L. Warner, eds.) pp. 207–220, Elsevier/North-Holland, Amsterdam.

Nordling, S., Andersson, L. C., and Häyry, P. (1972) Separation of T and B lymphocytes by preparative cell electrophoresis, *Eur. J. Immunol.* **2**:405.

Pertoft, H. (1966) Gradient centrifugation in colloidal silica-polysaccharide media, *Biochim. Biophys. Acta* **126**:594.

Pertoft, H., and Laurent, T. C. (1977) Isopycnic separation of cells and cell organelles by centrifugation in modified colloidal silica gradients, in: *Methods of Cell Separation,* Volume 1 (N. Catsimpoolas, ed.), Plenum Press, New York, pp. 25–65.

Platsoucas, C. D., and Catsimpoolas, N. (1979) Separation of T and B lymphocytes from various mouse strains by density gradient electrophoresis, *Cell Biophys.* **1**:161.

Ponzio, N. M., David, C. S., Shreffler, D. C., and Thorbecke, G. J. (1977) Properties of reticulum cell sarcomas in SJL/J mice. V. Nature of RCS antigen which induces proliferation of normal SJL/J T cells, *J. Exp. Med.* **146**:132.

Poulter, L. W., and Turk, J. L. (1972) Proportional increase in the theta-carrying lymphocytes in peripheral lymphoid tissue following treatment with cyclophosphamide, *Nature (New Biol.)* **238**:17.

Reisner, Y., Linker-Israeli, M., and Sharon, N. (1976) Separation of mouse thymocytes into two subpopulations by the use of peanut agglutinin, *Cell. Immunol.* **25**:129.

Robert, F., and Dumont, F. (1979) Surface properties of lymphoid cells in spontaneous reticulum cell sarcoma from SJL/J mice, *Br. J. Cancer* (abstr).

Sabolovic, D., and Dumont, F. (1973) Separation and characterization of cell subpopulations in the thymus, *Immunology* **24**:601.

Scher, I., Ahmed, A., Strong, D. M., Steinberg, A. D., and Paul, W. E. (1975) X-linked B lymphocyte defect in CBA/N mice. I. Studies of the function and composition of spleen cells, *J. Exp. Med.* **141**:788.

Schlegel, R. A., Boehmer, H. von, and Shortman, K. (1975) Antigen-initiated B-lymphocyte differentiation. V. Electrophoretic separation of different subpopulations of AFC-progenitors for unprimed IgM and memory IgG responses to the NIP determinant, *Cell. Immunol.* **16**:203.

Shortman, K. (1977a) The separation of lymphoid cells on the basis of physical parameters: Separation of B- and T-cell subsets and characterization of B-cell differentiation stages, in: *Methods of Cell Separation,* Volume 1 (N. Catsimpoolas, ed.), Plenum Press, New York, pp. 229–249.

Shortman, K. (1977b) The pathway of T-cell development within the thymus, in: *Progress in Immunology III* (T. E. Mandel, ed.), North-Holland, Amsterdam, pp. 197–205.

Shortman, K., Brunner, K. T., and Cerottini, J. C. (1972a) Separation of stages in the development of the T cells involved in cell-mediated immunity, *J. Exp. Med.* **135**:1375.

Shortman, K., Cerottini, J. C., and Brunner, K. T. (1972b) The separation of subpopulations of T and B lymphocytes, *Eur. J. Immunol.* **2**:313.

Shortman, K., Boehmer, H. von, Lipp, J., and Hopper, K. (1975) Subpopulations of T-lymphocytes. Physical separation, functional specialisation and differentiation pathways of subsets of thymocytes and thymus-dependent peripheral lymphocytes, *Transplant. Rev.* **25**:163.

Stobo, J. D. (1972) Phytohemagglutinin and concanavalin A: Probes for murine T cell activation and differentiation, *Transplant. Rev.* **11**:60.

Strikler, A., and Sacks, T. (1973) Focusing in continuous-flow electrophoresis systems by electrical control of effective wall zeta potentials, in: *Isoelectric Focusing and Isotachophoresis* (N. Catsimpoolas, ed.), New York Academy of Sciences, New York, pp. 497–514.

Trizio, D., and Cudkowicz, G. (1974) Separation of T and B lymphocytes by nylon wool columns: Evaluation of efficacy by functional assays *in vivo, J. Immunol.* **113**:1093.

Wioland, M., Sabolovic, D., and Burg, C. (1972) Electrophoretic mobilities of T and B cells, *Nature (New Biol.)* **237**:274.

Wiseman, L. L., Senn, J. S., Miller, R. G., and Price, G. B. (1976) Stem cell characterization of neutropenia: Velocity sedimentation and mass culture analysis, *Br. J. Cancer* **34**:46.

Zeiller, K., Pascher, G., Wagner, G., Liebich, H. G., Holzberg, E., and Hannig, K. (1974) Distinct subpopulations of thymus-dependent lymphocytes. Tracing of the differentiation pathway of T cells by use of preparatively electrophoretically separated mouse lymphocytes, *Immunology* **26**:995.

Zeiller, K., Löser, R., Pascher, G., and Hannig, K. (1975a) Free-flow electrophoresis. II. Analysis of the method with respect to preparative cell separation, *Hoppe-Seyler's Z. Physiol. Chem.* **356**:1225.

Zeiller, K., Schindler, R. K., and Liebich, H. G. (1975b) The T lymphocyte surface in development. A study of the electrokinetic, antigenic and ultra-structural properties of T lymphocytes in mouse thymus and lymph nodes, *Israel J. Med. Sci.* **11**:1242.

Zeiller, K., Pascher, G., and Hannig, K. (1976) B lymphocyte subpopulations in the mouse spleen. A study of the differentiation pathway using free flow electrophoretically separated subpopulations of direct PFC progenitor cells, *Immunology* **31**:863.

4

Electrical Sizing of Cells in Suspension

N. B. GROVER, S. A. BEN-SASSON, AND J. NAAMAN

I. INTRODUCTION

The properties of many materials are strongly influenced by particle size, and the size of a living cell is often a reflection of its physiological state. Thus the development by Coulter (1953, 1955) of an electric transducer for detecting the size of a particle suspended in an electrolytic medium was of major importance as it provided for the first time a rapid and convenient approach to particle size analysis on large samples.

In this review we shall discuss electrical sizing from a practical point of view and will restrict theory to those areas where it helps to provide an intuitive grasp of the concepts involved; for rigorous derivations, the reader will be referred to the original literature.

The Coulter transducer consists of a small cylindrical orifice through which the suspension is pumped, and Section II opens with a description of the electric and hydrodynamic fields within such an orifice in the absence of a particle. This is followed by a consideration of the changes that take place when a cell traverses the orifice and includes the influence of cell volume, shape, orientation, and resistivity.

The next section describes a measuring system compatible with the preceding theoretical considerations and provides specifications for the transducer and the various amplifiers and discriminators; the question of data storage and processing is treated briefly.

Section IV is a practical guide to instrument operation and contains sub-

N. B. GROVER, S. A. BEN-SASSON, AND J. NAAMAN • The Hubert H. Humphrey Centre for Experimental Medicine and Cancer Research, The Hebrew University–Hadassah Medical School, Jerusalem, Israel.

sections on sample preparation, orifice dimensions, amplifier settings, and calibration. The final two sections are concerned with applications other than direct volume determination, including the use of high electric fields and cell sorting.

II. THEORY

The transducer, depicted schematically in Fig. 1a, consists in essence of an orifice (lower arrow) through which the cell suspension is pumped and a pair of electrodes connected to a power supply. Because cells are effectively nonconducting, whenever one of them enters the orifice, displaying an equal volume of electrolyte, it causes an increase in the electrical resistance. If a constant electric current is maintained between the measuring electrodes, this resistance change will give rise to a corresponding increase in voltage (if the voltage is held constant, the current will decrease). We would expect any such change to be proportional to cell volume, and so it is, to a first approximation, provided certain conditions are met. These conditions are considered first, and

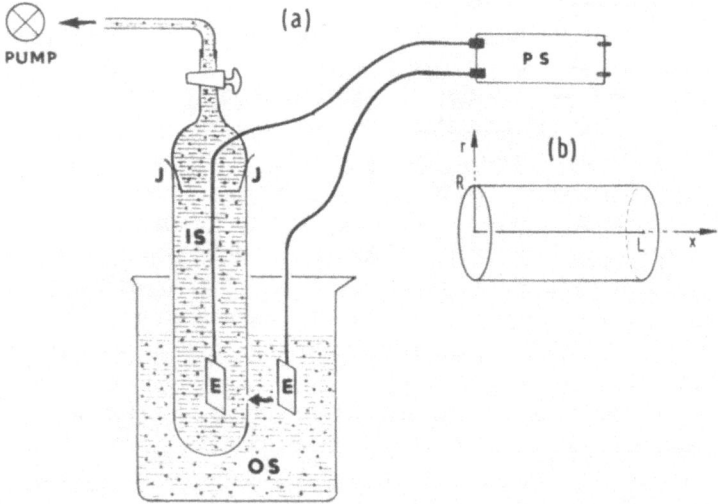

FIGURE 1. (a) Schematic diagram of electric transducer showing outer (OS) and inner (IS) cell suspensions, orifice (lower arrow), measuring electrodes (E), ground-glass joint (J), direction of fluid flow (arrows), mechanical pump, and power supply (PS). (b) Notation for orifice dimensions. R: Cross-sectional radius; r: radial coordinate; L: length of cylinder; x: axial coordinate.

then an explicit expression for the proportionality coefficient is presented in terms of measurable geometric and physical parameters.

A. Electric Field

In order to understand the relationship between resistance change and cell size, it is first necessary to consider the characterisitics of the electric field within an orifice the diameter of which is far from negligible with respect to its length (Fig. 1b), so that edge effects must be taken into account. The results are best presented graphically; Fig. 2 is a plot of the electric field E as a function of the distance x from the mouth of the orifice at various distances r from the orifice axis. The solution is of course symmetrical about the center of the orifice; it is valid only when the orifice is sufficiently long to permit the field to attain a value that is independent of position. From the figure this is seen to occur at distances greater than the radius, so that the total length of the orifice must be greater than its diameter.

Because the electric field is not uniform except in the vicinity of the center of the orifice, the changes in current (or voltage) produced by a cell during its

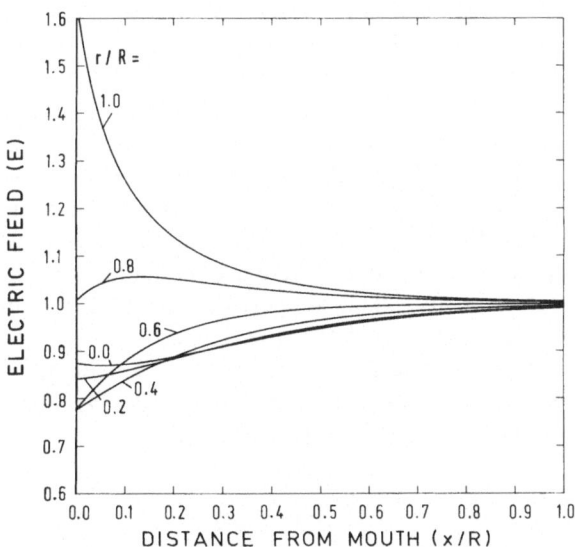

FIGURE 2. Electric field E in a cylindrical orifice as a function of distance x from mouth of orifice at various distances r from orifice axis. The field is expressed relative to its value far from the mouth ($x \gg R$); lengths are expressed in units of orifice radius R.

passage through the orifice will depend in large measure on the path it follows. This is determined almost entirely by the hydrodynamic field, the contribution due to electrophoresis being very small.

B. Hydrodynamic Field

The hydrodynamic field inside the orifice is created by the pump (Fig. 1a). Pumping is necessary in order to remove the electrolyte from within the orifice before it becomes overheated and in order to increase the flow rate of the cells and confine the duration of an experiment to reasonable values. Flow rates are normally sufficiently low for laminar flow to prevail and one would expect the velocity profile to be parabolic. But parabolic (Poiseuille) flow requires a certain distance over which to develop, and this distance usually exceeds the actual length of the orifice. It is thus necessary to consider the hydrodynamics in the region preceding complete parabolic flow. As an approximation, one can assume that the velocity is uniform across the mouth of the orifice and zero at the walls. The thickness of the layer adjacent to the walls in which the flow is retarded as a result of the finite viscosity of the fluid, the so-called boundary layer, increases downstream until (for sufficiently long tubes) it becomes equal to the radius of the orifice; from that distance on, we have complete parabolic flow.

Figure 3 is a plot of axial fluid velocity u as a function of distance r for various values of the dimensionless distance ξ from the mouth of the orifice. Note that the boundary layer retards more and more of the flow as ξ increases and since the total flux over the cross section is constant, the velocity in the core region increases as a function of x and is always greater than the average velocity $\bar{u}$. The velocity gradient within the boundary layer is substantial in the radial direction and will cause orientation of nonspherical cells and a migration of all particles toward the core region. This migration increases with flow rate, cell size, and distance from the axis. There are numerous considerations, both theoretical and experimental, which support the idea that in many cases cells may concentrate exclusively within the core region.

Having described the electric and hydrodynamic fields within the orifice in the absence of a cell, we now turn to an examination of the transduction of cell volume into voltage (or current) change.

C. Cell Volume

Most intact animal cells behave in suspension as rigid, nonconducting spheres. [The case of nonspherical particles is considered in the next subsection, the question of membrane leakiness in II.E; deformation is not discussed, and the interested reader is referred to the literature (Grover *et al.,* 1969a).]

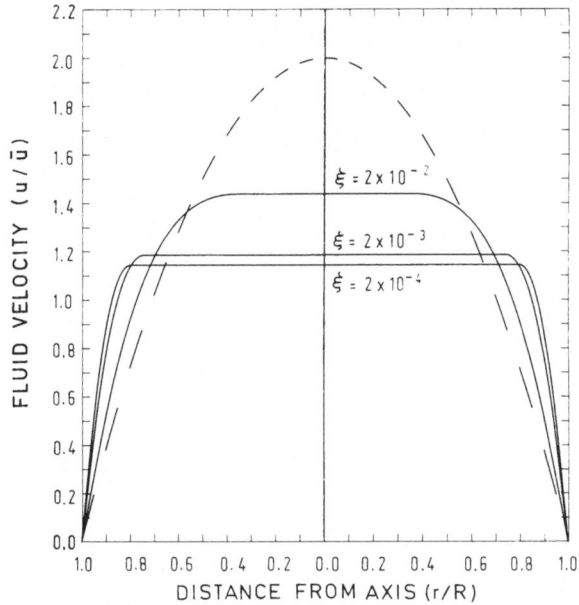

FIGURE 3. Fluid velocity u as a function of fractional distance r/R from orifice axis for various values of the dimensionless parameter $\xi \equiv \mathcal{R}^{-1}x/R$, where $\mathcal{R}$ is tube Reynolds number and x/R is distance from mouth of orifice in units of orifice radius R. Velocity is expressed relative to mean linear velocity $\bar{u}$. Dashed curve is velocity profile for complete parabolic flow.

For such cells, we have that the relative voltage (or current) change within the region of uniform electric field is simply $1.5v/(V - v)$, where v is the volume of the cell and V that of the orifice. In all practical situations, $v \ll V$, and the relative change in voltage (or current) is proportional to the relative volume of the cell, as expected. The proportionality coefficient, however, is not unity but 1.5. For nonspherical particles, this coefficient, usually termed the shape factor and represented by γ, is no longer constant but becomes a function of the shape of the cell and its orientation in the electric field.

D. Shape and Orientation

When a cell of volume v enters an orifice of volume V, it not only displaces an equal volume of electrolyte but also distorts the electric field in its vicinity. The former gives rise to a signal proportional to v/V, the latter to the proportionality factor's being greater than unity.

Figure 4a is a plot of the shape factor γ for a prolate (rod-shaped) spheroid with its major axis aligned parallel (solid curve) or perpendicular

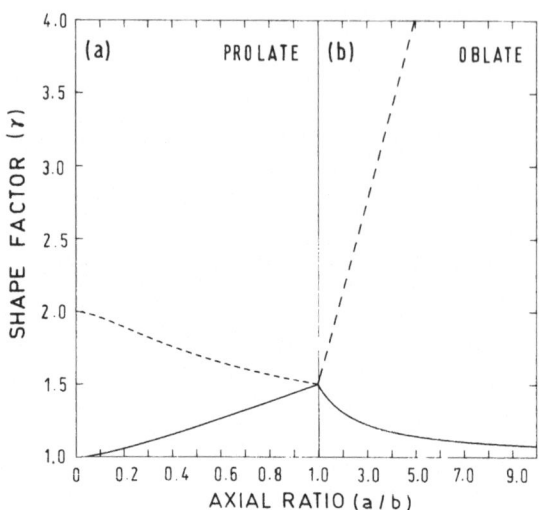

FIGURE 4. Shape factor γ for a spheroid (axes: a,a,b) aligned with major axis (solid curve) or minor axis (dashed curve) parallel to the electric field, as a function of the axial ratio a/b.

(dashed curve) to the electric field (that is, to the orifice axis); the corresponding curves for an oblate (disk-shaped) spheroid are shown in Fig. 4b. A simple trigonometric expression provides values for intermediate orientations (Grover *et al.,* 1972).

It is clear from the figures that the greater the eccentricity of a cell, the less the distortion of the electric field (the smaller the shape factor) when the cell lies parallel to it and the greater the distortion when it is perpendicular; hydrodynamic considerations generally favor the former.

E. Membrane Conductance

In the event that the cell membrane is not a perfect insulator, the relative voltage change produced by the cell during its passage through the orifice will be less: instead of $\gamma v/(V - v)$, the signal will now be $\gamma v/(cV - v)$ or, to a very good approximation, $\gamma v/cV$. Thus c is the factor by which cell size appears to decrease due to membrane conductivity; its actual value depends on ω, the ratio of cell resistivity to electrolyte resistivity, and on γ: $c = 1 + \gamma/(\omega - 1)$.

Figure 5 is a plot of $1/c$ as a function of ω for cells of various shapes and orientations and illustrates the effect of decreasing cellular resistivity. Clearly, when the conductivity of the cell reaches that of the electrolyte, the signal vanishes; when it exceeds it, the signal becomes inverted.

The effective resistivity of a cell depends on its internal resistivity, on its dimensions, and on the resistivity of its membrane (Cole, 1928); increased cell conductivity normally implies membrane damage (Grover *et al.,* 1982).

F. Electrical Size

The term "electrical size" is usually reserved for the quantity actually sensed by the electrical transducer, namely $\gamma v/c$. Relative electrical size, on the other hand, denotes the ratio between the true electrical size of a cell and that of a nonconducting particle of identical geometrical volume v and shape factor γ, and is given by $1/c$. An equivalent spherical volume is sometimes introduced in order to characterize irregularly shaped cells and refers to the volume of an insulating sphere having the same electrical size; numerically, this is just $\gamma v/1.5c$. (One should bear in mind that the values for γ quoted in the literature almost invariably apply to the hydrodynamically favored orientations.)

III. INSTRUMENTATION

The measuring system can conveniently be divided into five sections: the transducer, the amplifiers, the signal selector system, data sorting and storage devices, and auxiliary units.

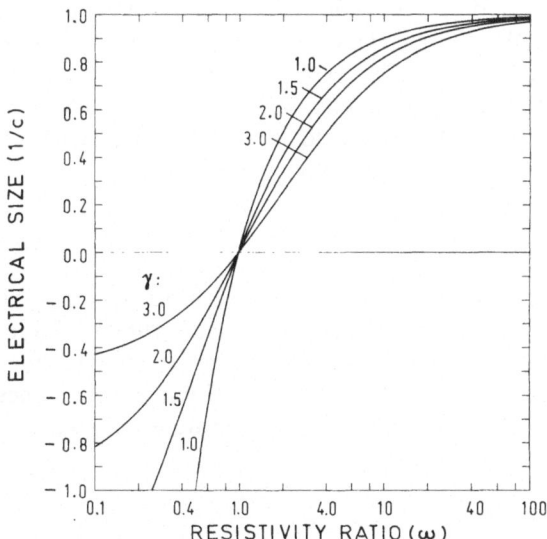

FIGURE 5. Relative electrical size $1/c$ as a function of resistivity ratio ω for various values of the shape factor γ. (Note semilogarithmic scale.)

A. The Transducer

The transducer consists of a ruby (sapphire) ring jewel with a small hole mounted on a pyrex tube sealed at one end and with a ground joint at the other (Fig. 1a). Theoretical considerations (Grover *et al.*, 1969a) require that the hole be cylindrical and longer than it is wide. The orifice dimensions should be determined prior to mounting, the length (thickness of the jewel) mechanically and the diameter optically using a calibrated microscope eyepiece. Any orifice that is not a good circular cylinder should be rejected. A constant-current (rather than constant-voltage) power supply is recommended in order to decrease the effect of electrode polarization.

B. The Amplifiers

A low-input-impedance preamplifier will prevent amplifier distortion even with high-resistivity electrolytes and large parasitic capacitances. Any standard calibrated linear RC amplifier will do for the next stage, provided it has sufficient dynamic range. Adjustable integration and differentiation times are useful in improving the signal-to-noise ratio under a variety of experimental conditions (Grover *et al.*, 1969b).

C. The Selector

The electric field within the orifice is not uniform (Fig. 2) and so the signal produced by a cell will depend on the path it follows; this, in turn, is governed by the hydrodynamic field (Fig. 3). For a nonspherical cell, there is in addition the possibility that its shape factor will change due to particle rotation and distortion by the hydrodynamic field. Several techniques have been developed in an attempt to overcome these difficulties.

1. Gating

In the gating approach, every cell is measured, but only while it lies within the region of uniform electric field (Grover *et al.*, 1969b). There is one external control to set the delay in activating the peak detection circuit and another to fix the period during which it remains in the active state. (These controls can also be used to study the signal developed at different points along the length of the orifice.)

2. Form

There are several devices that select the signals to be processed on the basis of their shape; those not passing certain predetermined standards are rejected. Thus Waterman *et al.* (1975) reject all signals that they consider too long. Since the time it takes a cell to traverse the orifice increases the closer its trajectory lies to the walls of the orifice, such a selection favors cells that enter and exit near the orifice axis. Kachel (1973), on the other hand, rejects signals with short risetimes, the idea being that a short risetime usually reflects the presence of extraneous peaks in the generated signal. A commercial rejection circuit is on the market (Model C-1000, Coulter Electronics Ltd., Coldharbour Lane, Harpenden, Herts., England) that seems to operate along similar principles, but the complete absence of published specifications prevents us from making a proper evaluation.

3. Focusing

Hydrodynamic focusing was introduced independently by Spielman and Goren (1968) and by Thom *et al.* (1969) and later extended by Kachel *et al.* (1970) and by Schulz and Thom (1973); it is now available commercially (Model TF, Coulter Electronics Ltd., Coldharbour Lane, Harpenden, Herts., England). In this method, all cells are confined to a narrow band of trajectories about the orifice axis by means of an auxiliary capillary tube mounted in front of the main orifice opening and directed at its center. The cell suspension is sucked from the capillary through the orifice together with a surrounding sheath of particle-free medium (Fig. 6), the cross section of the core being determined by the magnitude of the pressure differences and by the geometry. Very narrow cores are expected to be unstable, and if so, this would impose a practical limitation on the size of the orifice and hence of the smallest cell that can be measured reliably. In addition, there is a problem of deformation: fresh human red blood cells, which retain their normal biconcave shape in conven-

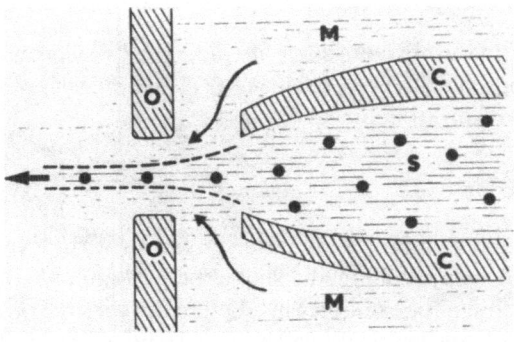

FIGURE 6. Schematic diagram of hydrodynamic focusing system showing particle-free medium (M), capillary tube (C), cell suspension (S), and orifice (O). Dashed lines: boundary of core region. Arrows: direction of fluid flow.

tional orifices (Grover *et al.,* 1972; Golibersuch, 1973), become long, thin rods following hydrodynamic focusing (Thom *et al.,* 1969; Kachel *et al.,* 1970).

D. Data Handling

The primitive discriminators in the original commercial instruments have been widely replaced by multichannel pulse-height analyzers. Standard stand-alone units can be used here, but care must be taken to feed them with pulses that are properly and uniformly shaped. Much more flexibility can be obtained with a simple analog-to-digital converter and a microcomputer; in kinetic studies in which experiments must be carried out in rapid succession, the latter is essential. In either case, live display of the distribution histogram is an asset, and provision should be made for hard-copy output and further processing off-line.

E. Auxiliary Units

The amplified signal should be accessible prior to shaping in order to permit oscilloscope monitoring of the input, amplification, and selection (where applicable) stages. We have found this to be a very sensitive means of detecting slight transient partial obstructions of the orifice in addition to aiding in the adjustment of the selector controls and in the evaluation of orifice geometry. An oscilloscope with storage facilities can prove particularly useful. A low-power ultrasonic cleaner should be kept handy to unclog plugged orifices or capillaries in those systems that do not include a built-in reverse-flow facility designed for this purpose.

IV. OPERATION

In this section we discuss some methodological aspects of sample preparation (including coincidence), choice of orifice dimensions, amplifier settings, and instrument calibration.

A. Cell Suspensions

The recommended use of long orifices carries with it an increased risk of clogging, and all suspending media should be well filtered (through a 0.22-μm Millipore membrane, for example) just before the cells are added. Coincidence, the simultaneous occurrence of two or more particles within the equivalent volume of the orifice, can always be reduced to negligible proportions by making the suspensions sufficiently dilute: about 10,000 cells/ml is usually adequate, depending on orifice volume (Mazumdar and Kussmaul, 1967). When that is not practicable, for whatever reason, a correction can be applied to the results, but not without some difficulty (Princen and Kwolek, 1965).

B. Orifice Dimensions

For a given current (or voltage), the signal is very nearly inversely proportional to orifice volume. There are two restrictions, however, that must be borne in mind. The first concerns orifice length and requires that it be sufficiently greater than orifice diameter to provide a region, no smaller than the size of a cell, within which the electric field is uniform. The second is less well-defined but states roughly that the diameter must be at least several times larger than that of the cells it purports to measure (Grover et al., 1969a, 1972). Within these limitations, therefore, the orifice chosen should be as small as possible, consistent with the necessity of keeping it reasonably free from blockage.

C. Amplification

The signal generated by a cell during its passage through the orifice is proportional to the current I, whereas the noise produced by the power supply is proportional to $\sqrt{I}$. Thus the higher the current, the greater the signal-to-noise ratio. Here, too, there are limitations, both physical and biological. Electrolysis increases with current density, and the bubbles generated produce spurious signals when they pass through or even near the orifice; the effect can be reduced by using large-surface electrodes (platinized platinum, for example) and keeping them well removed from the orifice. Electrode polarization also depends on the current density but is normally of no consequence with constant-current (as opposed to constant-voltage) power supplies.

An absolute limitation on the magnitude of the electric field within the orifice is imposed by the cells themselves: cell membranes undergo dielectric breakdown when exposed to sufficiently high fields (Grover et al., 1972; Zimmermann et al., 1973; Tsong et al., 1976). The precise level at which this breakdown occurs depends on the particular type of cell involved and is of the order of 1 kv/cm. Fortunately dielectric breakdown causes a marked drop in the electrical size and so is readily detected and avoided.

Having raised the current as much as permissible, one then increases the amplification of the RC amplifier until the output signals fall within the dynamic range of the pulse-height analyzer or microcomputer.

(The constant-current power supply should, of course, be able to provide sufficient voltage to remain below saturation even at the highest currents for all combinations of electrolyte conductivity and orifice dimensions likely to be encountered in practice.)

D. Calibration

There are no unknown parameters in the expression relating the volume of a nonconducting particle of given shape to the change in current produced during its passage through the orifice. Thus, all that is necessary for the absolute measurement of cell size are a calibrated power supply and amplifier and a precise determination of the effective orifice volume V. For accurate results, V should take account of the resistance contributed by the electrolyte suspension in the regions between the orifice and the electrodes (Grover *et al.*, 1969b).

It is advisable to check the entire measuring system routinely for overall stability and short-term drift with some sort of inert particle such as polystyrene latex spheres.

V. APPLICATIONS

The inherent reliability and sensitivity of electrical sizing, coupled with its high sampling rate, makes it a valuable tool in a variety of applications. We shall not review here the routine use of this technique to measure the size distributions of particular types of cells (Coulter Electronics, 1978) or groups of cells (England *et al.*, 1975), nor shall we discuss the data on cell volume following treatment with various reagents (Rosenberg and Gregg, 1969) or during the cell cycle (Steen and Lindmo, 1978). Instead, we consider three novel applications: fast kinetics of cell size changes, effects of high electric fields, and early alterations in membrane permeability.

A. Fast Kinetics

An obvious application of electrical sizing, one that follows directly from its original design as a measure of particle volume, is in osmosis. The monitoring of cell volume following exposure to hypotonic shock reveals an interesting pattern of osmotic regulation: after an initial swelling, which is complete within 1 or 2 min, nucleated cells shrink back to their physiological volumes during a period that depends on the type of cell involved and the condition of the donor (Buckhold *et al.*, 1965; Doljanski *et al.*, 1974). Thus peripheral blood lymphocytes from healthy donors take 3 min (at room temperatures), whereas those from chronic lymphocytic leukemia patients require about half an hour to regain their original volume (Ben-Sasson *et al.*, 1975).

A similar application concerns the monitoring of cell volume as an aid in understanding the processes that take place during complement-induced lysis (Valet and Opferkuch, 1975).

B. High Fields

The membranes of living cells undergo dielectric breakdown when the applied electric fields within the orifice are sufficiently high (Grover *et al.*, 1972, 1982; Zimmermann *et al.*, 1973, 1974; Tsong *et al.*, 1976; Kinosita and Tsong, 1977a). The minimum field at which this occurs, termed the critical or threshold level, is expected (Zimmermann *et al.*, 1974; Grover *et al.*, 1982) to be approximately proportional to the largest linear dimension of the cell; of more importance is the fact that it depends on cell type (Ben-Sasson *et al.*, 1974; Grover *et al.*, 1982) and so can be used to characterize different kinds of cells (Zimmermann *et al.*, 1974).

As the electric field is raised beyond its critical level, the signal produced by a cell begins to drop by an amount that depends on the internal resistivity of the cell: the lower the resistivity, the greater the drop. This phenomenon has been elegantly applied by Gear (1977) to measure the osmotic fragility of human erythrocytes. At high fields, resealed ghosts, their internal milieu having more or less equilibrated with the external medium during lysis, now produce a much smaller signal than do the intact red blood cells with their high internal resistance (Pauly and Schwan, 1966); the resulting bimodal distribution permits the ready determination of the proportion of lysed cells as a function of the tonicity of the lysing medium.

The damage caused to the membrane during dielectric breakdown can be interpreted in terms of the creation of pores (or the enlarging of existing ones), the number and size of which depend on the experimental conditions. Thus Kinosita and Tsong (1977b) were able to vary the effective pore size, albeit only between 30 and 60 nm, by changing the parameters of the applied field. Indeed, a certain amount of insight can be gained into the membrane structure itself by judicious manipulation of the conditions just prior to, during, and following the onset of dielectric breakdown (Ben-Sasson *et al.*, 1982).

C. Membrane Permeability

Even minor damage to the cell membrane can be expected (Grover *et al.*, 1982) to cause a substantial decrease in the electrical size of a cell, as illustrated in Fig. 7 for human red blood cells, rbc ghosts, and a lymphoblast line. This sensitivity can be exploited to monitor early changes in membrane permeability such as those involved in immune cell lysis (Reif *et al.*, 1977).

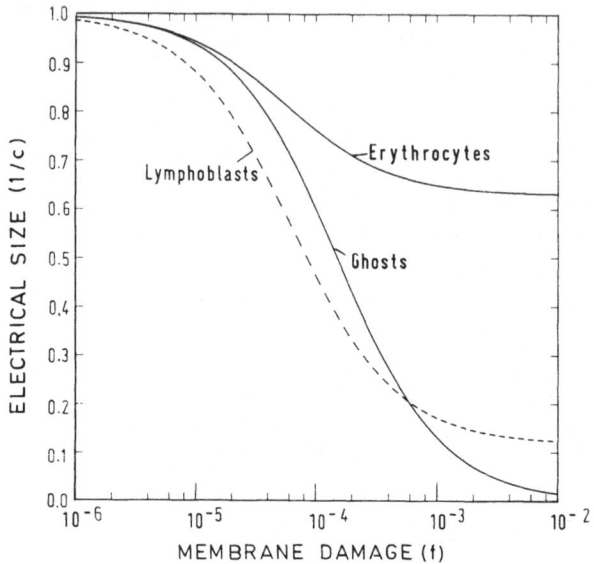

FIGURE 7. Relative electrical size $1/c$ as a function of proportion f of cell surface area permeable to electrolytes. Solid curves: human red blood cells ($\gamma = 1.18$: Grover *et al.*, 1972) intact ($\omega = 3$: Pauly and Schwan, 1966) and after gentle hemolysis and resealing ($\omega = 1$: Gear, 1977). Dashed curve: murine lymphoblast line ($\gamma = 1.5$, $\omega = 1.2$: Irimajiri *et al.*, 1978).

VI. THE FUTURE

Electrical cell sizing is over a quarter of a century old, yet interest in it continues unabated among scientists and commercial interests alike. Indeed, the lapse of the original Coulter patents (1953, 1955) has provided fresh impetus to the development of instruments with enhanced reliability and sensivitity, on the one hand, and, on the other, to the search for new and more sophisticated applications, especially among the latter community.

A. Sensitivity

Considerable improvement in the signal-to-noise ratio can be achieved by designing the orifice as part of a bridge circuit (Hoffman and Britt, 1979); the more suitable the matching between the components, the greater the rejection of common-mode noise (Leif *et al.*, 1979). Contamination by electrode products, including gas bubbles, can be eliminated by employing a four-electrode system to separate those providing the power from those detecting the signal. It would appear that even without such devices, however, a substantial reduction in the noise level can still be attained by proper attention to basic concepts of design and construction (Haynes and Shoor, 1978; Haynes, 1979).

B. Multiparameter Systems

By replacing the dc power supply with a high-frequency signal generator and making appropriate modifications in the detection system, Hoffman and Britt (1979) were able to measure several electrical properties of the cell in addition to its volume. A more ambitious setup, one that includes optical scanning, is now under development by Leif *et al.* (1979). But the ultimate in sophistication would seem to be an apparatus that not only measures electrical and optical parameters simultaneously but also sorts the individual cells accordingly, as first described by Fulwyler (1965). Such instruments, although relatively new on the market, are already beginning to provide quantitative information, and the interested reader is referred to a detailed review on the subject by Menke *et al.* (1977).

REFERENCES

Ben-Sasson, S., Patinkin, D., Grover, N. B., and Doljanski, F. (1974) Electrical sizing of particles in suspensions. IV. Lymphocytes, *J. Cell. Physiol.* **84**:205.

Ben-Sasson, S., Shaviv, R., Bentwich, Z., Slavin, S., and Doljanski, F. (1975) Osmotic behavior of normal and leukemic lymphocytes, *Blood* **46**:891.

Ben-Sasson, S. A., Naaman, J., and Grover, N. B. (1982) Membrane changes induced by high electric fields: Evidence for sulfhydryl group involvement, *Anal. Quant. Cytol.* **4**:000.

Buckhold, B., Adams, R. B., and Gregg, E. C. (1965) Osmotic adaptation of mouse lymphoblasts, *Biochim. Biophys. Acta* **102**:600.

Cole, K. C. (1928) Electric impedence of suspensions of spheres, *J. Gen. Physiol.* **12**:29.

Coulter Electronics (1978) *Coulter Counter Medical and Biological Bibliography,* Coulter Electronics Ltd., Coldharbour Lane, Harpenden, Herts., England.

Coulter, W. H. (1953) Means for counting particles suspended in a fluid, U.S. Pat. No. 2,656,508.

Coulter, W. H. (1955) Apparatus for studying the physical properties of a suspension of particles in a liquid medium, Br. Pat. No. 722,418.

Doljanski, F., Ben-Sasson, S., Reich, M., and Grover, N. B. (1974) Dynamic osmotic behavior of chick blood lymphocytes, *J. Cell. Physiol.* **84**:215.

England, J. M., Bashford, C. C., Hewer, M. G., Hughes-Jones, N. C., and Down, M. C. (1975) Simple method for automating the differential leucocyte-count, *Lancet* **1**:492.

Fulwyler, M. J. (1965) Electronic separation of biological cells by volume, *Science* **150**:910.

Gear, A. R. L. (1977) Erythrocyte osmotic fragility: Micromethod based on resistive-particle counting, *J. Lab. Clin. Med.* **90**:914.

Golibersuch, D. C. (1973) Observation of aspherical particle rotation in Poiseuille flow via the resistance pulse technique, *Biophys. J.* **13**:265.

Grover, N. B., Naaman, J., Ben-Sasson, S., and Doljanski, F. (1969a) Electrical sizing of particles in suspensions. I. Theory, *Biophys. J.* **9**:1398.

Grover, N. B., Naaman, J., Ben-Sasson, S., Doljanski, F., and Nadav, E. (1969b) Electrical sizing of particles in suspensions. II. Experiments with rigid spheres, *Biophys. J.* **9**:1415.

Grover, N. B., Naaman, J., Ben-Sasson, S., and Doljanski, F. (1972) Electrical sizing of particles ⁻ in suspensions. III. Rigid spheroids and red blood cells, *Biophys. J.* **12**:1099.

Grover, N. B., Ben-Sasson, S. A., and Naaman, J. (1982) Electrical sizing of particles in suspensions. V. High electric fields, *Anal. Quant. Cytol.* (in press).

Haynes, J. L. (1979) Particle counting apparatus utilizing various fluid resistors to maintain proper pressure differentials, U.S. Pat. No. 4,165,484.

Haynes, J. L., and Shoor, B. A. (1978) Particle density measuring system, U.S. Pat. No. 4,110,604.

Hoffman, R. A., and Britt, W. B. (1979) Flow-system measurements of cell impedence properties, *J. Histochem. Cytochem.* **27**:234.

Irimajiri, A., Doida, Y., Hanai, T., and Inouye, A. (1978) Passive electrical properties of cultured murine lymphoblast (L5178Y) with reference to its cytoplasmic membrane, nuclear envelope, and intracellular phases, *J. Membr. Biol.* **38**:209.

Kachel, V. (1973) Eine elektronische Methode zur Verbesserung der Volumenauflösung des Coulter-Partikelvolumenmessverfahrens, *Blut* **27**:270.

Kachel, V., Metzger, H., and Ruhenstroth-Bauer, G. (1970) Der Einfluss der Partikeldurchtrittsbahn auf die Volumenverteilungskurven nach dem Coulter Verfahren, *Z. Gesamte Exp. Med.* **153**:331.

Kinosita, K., Jr., and Tsong, T. Y. (1977a) Hemolysis of human erythrocytes by a transient electric field, *Proc. Nat. Acad. Sci. U.S.A.* **74**:1923.

Kinosita, K., Jr., and Tsong, T. Y. (1977b) Formation and resealing of pores of controlled sizes in human erythrocyte membrane, *Nature (London)* **268**:438.

Leif, R. C., Guarino, V., and Lefkove, N. (1979) The automated multiparameter analyzer for cells (AMAC) IIA, a true bridge circuit Coulter-type electronic cell volume transducer, *J. Histochem. Cytochem.* **27**:225.

Mazumdar, M., and Kussmaul, K. L. (1967) A study of the variability due to coincident passage in an electronic blood cell counter, *Biometrics* **23**:671.

Menke, E., Kordwig, E., Stuhlmüller, P., Kachel, V., and Ruhenstroth-Bauer, G. (1977) A volume activated cell sorter, *J. Histochem. Cytochem.* **25**:796.

Pauly, H., and Schwan, H. P. (1966) Dielectric properties and ion mobility in erythrocytes, *Biophys. J.* **6**:621.

Princen, L. H., and Kwolek, W. F. (1965) Coincidence corrections for particle size determinations with the Coulter Counter, *Rev. Sci. Instr.* **36**:646.

Reif, A. E., Robinson, C. M., and Incze, J. S. (1977) Assay of immune cytolysis of lymphocytes and tumour cells by automatic determination of cell volume distribution, *Immunology* **33**:69.

Rosenberg, H. M., and Gregg, E. C. (1969) Kinetics of cell volume changes of murine lymphoma cells subjected to different agents *in vitro, Biophys. J.* **9**:592.

Schulz, J., and Thom, R. (1973) Electrical sizing and counting of platelets in whole blood, *Med. Biol. Eng.* **11**:447.

Spielman, L., and Goren, S. L. (1968) Improving resolution in Coulter counting by hydrodynamic focusing, *J. Colloid Interface Sci.* **26**:175.

Steen, H. B., and Lindmo, T. (1978) Cellular and nuclear volume during the cell cycle of NHIK 3025 cells, *Cell Tissue Kinet.* **11**:69.

Thom, R., Hampe, A., and Sauerbrey, G. (1969) Die elektronische Volumenbestimmung von Blutkörperchen und ihre Fehlerquellen, *Z. Gesamte Exp. Med.* **151**:331.

Tsong, T. Y., Tsong, T., Kingsley, E., and Siliciano, R. (1976) Relaxation phenomena in human erythrocyte suspensions, *Biophys. J.* **16**:1091.

Valet, G., and Opferkuch, W. (1975) Mechanism of complement-induced cell lysis: Demonstration of a three-step mechanism of EAC1-8 cell lysis by C9 and of a non-osmotic swelling of erythrocytes, *J. Immunol.* **115**:1028.

Waterman, C. S., Atkinson, E. E., Jr., Wilkins, B., Jr., Fischer, C. L., and Kimzey, S. L. (1975) Improved measurements of erythrocyte volume distribution by aperture-counter signal analysis, *Clin. Chem.* **21**:1201.

Zimmermann, U., Schulz, J., and Pilwat, G. (1973) Transcellular ion flow in *Escherichia coli* B and electrical sizing of bacteria, *Biophys. J.* **13**:1005.

Zimmermann, U., Pilwat, G., and Riemann, F. (1974) Dielectric breakdown of cell membranes, *Biophys. J.* **14**:881.

5

Light Scattering Analysis of Single Cells

GARY C. SALZMAN

I. INTRODUCTION

Light scattering is an attractive tool for flow cytometric analysis because it is an inherently nondestructive probe and can be used for the analysis and sorting of unstained, viable cells. Since every cell passing through the laser beam scatters light, it can be used in conjunction with fluorescence probes to discriminate between stained and unstained fractions of a sample.

Light scattering is sensitive to cell size, shape, and refractive indices of the plasma membrane and internal structures within the cell relative to that of the surrounding medium. Fixatives and stains change refractive indices and thus affect light scattering. The sensitivity of light scattering to all these parameters is dependent on the scattering angle, the shape of the illuminating beam, and the solid angle over which the light is collected. The purpose of this chapter is first to show the sensitivity of several commonly used scattered light detector configurations to small changes in cell morphology and then to review recent applications of flow cytometric forward and 90° light scatter to cell analysis. Finally, we will discuss several new areas in which light scattering may contribute new information about cells.

Work on the use of light scattering to classify bacteria (Wyatt, 1968; Koch, 1968) led to its early use in flow cytometry for cell sizing (Mullaney *et al.*, 1969; Mullaney and Dean, 1969, 1970; Steinkamp *et al.*, 1973). This probe was first used by Mullaney and West (1973) to discriminate against debris in conjunction with fluorescence measurements. Julius *et al.* (1975) and Loken

GARY C. SALZMAN • Life Sciences Division, Los Alamos National Laboratory, University of California, Los Alamos, New Mexico 87545.

et al. (1976) first used light scattering to discriminate between live and dead cells in a flow cytometer. This latter discrimination is probably due to changes in refractive index.

Light scattering is a complex tool because its sensitivity to all the parameters mentioned above are folded together in the detected scattered light. Problems with cell orientation, fluid stream control, and the correct choice of scattering angles and angular resolution have hindered confident use of this powerful tool. The next section begins with a brief discussion of some of the theoretical models for the scattering from dielectric objects such as biological cells. In the remainder of the section, the coated-sphere cell model is used to determine expected detector responses for several commonly used scatter detector configurations.

II. THEORETICAL CONSIDERATIONS

A. Scattering Theory

The scattering of light from dielectric objects such as biological cells can be described by solutions to Maxwell's electromagnetic field equations (Stratton, 1941). Several investigators (Lorenz, 1890; Debye, 1909; Mie, 1908) have solved the scattering problem for a homogeneous sphere. More recent solutions have been presented by Stratton (1941), Born and Wolf (1975), van de Hulst (1957), and Kerker (1969). The exact solutions to the scattering from a dielectric object consisting of concentric spherical shells of different refractive indices has been given by Aden and Kerker (1951). The scattering from nonspherical objects has been considered by many investigators (Waterman, 1971; Purcell and Pennypacker, 1973; Barber and Yeh, 1975; Weil and Chu, 1976; Zerull *et al.*, 1977; Kerker *et al.*, 1978; Latimer *et al.*, 1978; Welch and Cox, 1978; Asano, 1979; Druger *et al.*, 1979; Fikioris and Uzunoglu, 1979; Fowler and Sung, 1979; Wang and Barber, 1979; Asano and Sato, 1980). Few of these models have been used to calculate the scattering from realistic models of single biological cells because of the difficult task of obtaining the distribution of refractive indices within a cell. Numerous approximation methods have been developed to deal with complex scattering objects. Some of these techniques have been reviewed earlier (Salzman *et al.*, 1979a) and will not be discussed here. The concentric spherical shell model (Aden and Kerker, 1951) is also referred to as the coated-sphere model. In the model, the inner sphere represents the nucleus, and the concentric outer shell represents the cytoplasm. This model was used for the experimental tests and for the predicted detector responses discussed below.

B. Coated Sphere Model Tests

Chinese hamster tissue culture cells are useful for modeling studies as they are spherical and have concentric nuclei. Brunsting and Mullaney (1974) have measured the refractive indices of the nucleus and cytoplasm for Chinese hamster tissue culture cells and have compared calculations based on the coated-sphere model of Aden and Kerker (1951) with measurements on suspensions of Chinese hamster cells blocked in various stages of the cell cycle. They were able to fit the data out to a scattering angle of about 15° with respect to the laser beam axis (0° is forward scattering). The concentric sphere model has also been used to estimate red blood cell membrane thickness by comparing calculations to scattering data from suspensions of red blood cell ghosts (Mullaney and Fiel, 1976).

C. Expected Detector Responses

In the following set of figures illustrating hypothetical scattered-light detector responses, the laser vacuum wavelength is 633 nm and refractive index of the medium is 1.3345, which is that for a 0.9% aqueous NaCl solution. The real part of the refractive index for a live cell nucleus (1.392) and cytoplasm (1.3703) are taken with respect to air and are from the measurements of Brunsting and Mullaney (1974). The real part of the refractive index for a fixed cell nucleus (1.4755) and cytoplasm (1.4525) are calculated based on an assumed 6% increase due to fixation. A change of this magnitude is approximately that required to explain the decrease in forward scatter intensity, assuming no decrease in cell diameter. No change of medium is assumed between the cell at the origin and the collection lens. The detector responses are normalized so that the relative response of the forward, 90°, and back-scatter detectors can be compared. Since diffraction-based calculations for forward scattering assume that a spherical object can be represented as an opaque disk with a cross-sectional area equal to that projected into a plane perpendicular to the illumination beam, the data are plotted as a function of diameter squared. Also, experimental comparisons for forward scatter geometries similar to that in Fig. 1 show an approximate linear increase as a function of cross-sectional area (proportional to diameter squared) (Visser *et al.*, 1978b). A cell is assumed to be uniformly illuminated with monochromatic light (633-nm wavelength). For a TEM_{00} (gaussian) laser beam with a diameter of 50 μm at the $1/e^2$ intensity points as in a typical fluorescence-activated cell sorter (FACS), this is a reasonable assumption for cells up to 20 μm in diameter.

Figure 1 shows the relationship between the cell stream axis, laser beam axis, and the light collection lens configuration for forward scatter. Figures 2

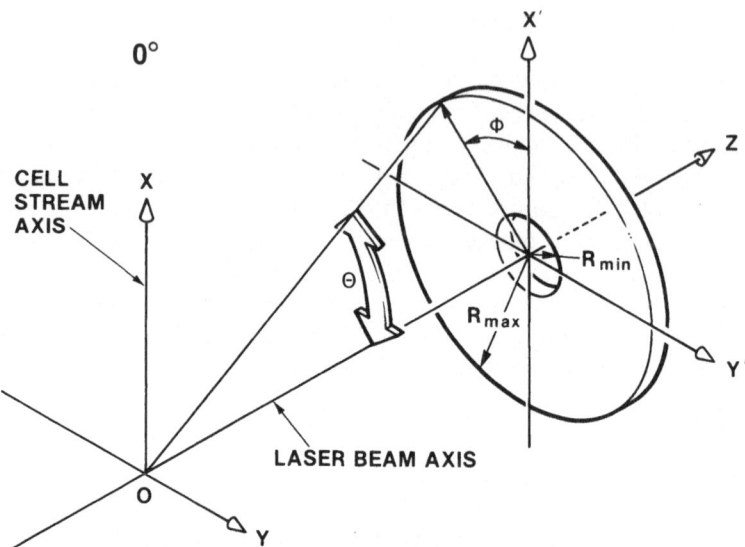

FIGURE 1. Typical forward scatter detector geometry for a flow cytometer. The cell stream is parallel to the x axis and intersects a focused laser beam (z axis) at the origin of coordinates 0. A collection lens of radius R_{max} is located in the x–y plane at a distance z from the origin. A beam stop of radius R_{min} is also located in this plane. The laser beam polarization is usually parallel to the x axis. The collection lens focuses the light scattered by a cell into the annular region between R_{min} and R_{max} onto a photodetector. Since the azimuthal angle ranges over a full circle, all scattering planes are represented. Hence, no polarization effects can be observed in the scattered light with this detector configuration. A negative value of the angle ϕ is shown. For the detector response calculations, $z = 23$ mm and the step sizes in x and y are 0.1 mm.

and 3 show the configuration for 90° and 180° scatter, respectively. In each figure, the circular disk represents a light-collection lens, which focuses the scattered light onto a photodiode or photomultiplier tube. A spherical, nucleated cell is assumed to be located at the origin of coordinates, 0. For each of these hypothetical detector configurations, the scattering plane is that determined by the laser beam axis and the vector from the origin defined by the angle pair (θ,ϕ). Since the forward and backscatter detectors allow the azimuthal angle to sweep through 360°, all scattering planes are allowed, and these detectors are insensitive to the plane of polarization of the laser beam. The 90° detector is somewhat sensitive to the polarization of the laser beam.

1. Forward Scatter Detector

For objects in the size range of most mammalian cells (3–20 μm in diameter), the scattered light intensity falls off rapidly with increasing scattering

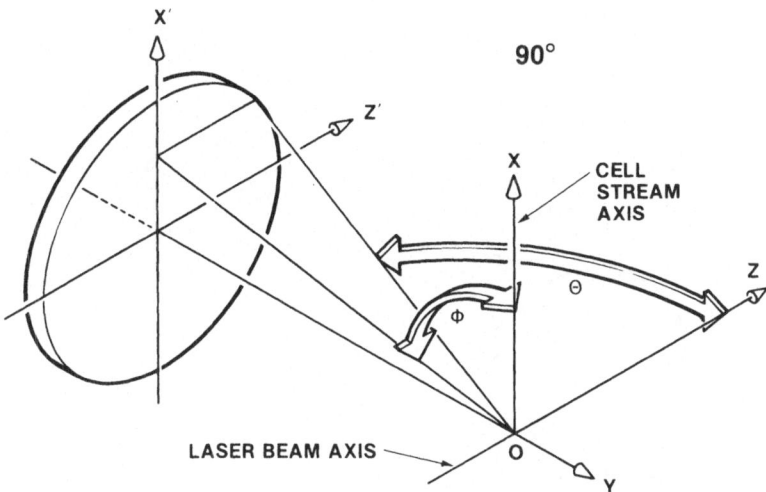

FIGURE 2. Typical 90° scatter detector geometry for a flow cytometer. The cell stream and laser beam polarization are parallel to the x axis. The laser beam is parallel to the z axis and intersects the cell stream at the origin 0. The 90° scatter collection lens is parallel to the $x-z$ plane and is located at a distance y from the origin. A negative value of the angle ϕ is shown. For the detector response calculations, the detector is at $y = -5.0$ mm and the step sizes in x and z are 0.1 mm.

angle. Figure 4 illustrates this effect for two beam-stop diameters and two maximum polar collection angles. For cells with diameters less than 10 μm (100 μm^2), the smaller beam stop (1°) is preferred because the response curve is steeper than for the 1.7° beam stop. However, the slope of the 1° beam-stop curve oscillates over a broader range than that for the 1.7° beam stop as a function of increasing cell diameter.

For the live cell coated sphere model, the forward-scatter detector response is relatively insensitive to changes in nuclear-to-cytoplasmic diameter ratio (N/C) until the cell diameter increases beyond about 14 μm (196 μm^2) (Fig. 5). For small cells the detector is quite insensitive to N/C ratio changes. In the case of fixed cells, the forward scatter detector response is also insensitive to N/C ratio changes (Fig. 6). In addition, the response function has several regions of nearly zero slope over which the detector is unresponsive to size changes. In the cell diameter region between 8 μm and 20 μm, the forward-scatter detector response is quite sensitive to changes in refractive index. The case for a 6% overall refractive index change is shown in Fig. 7.

a. FACS Forward Scatter. In the FACS (Bonner *et al.*, 1972), the laser beam intersects the 80-μm-diameter cell stream in air. The forward scatter detector configuration is similar to that shown in Fig. 1 except that the circular

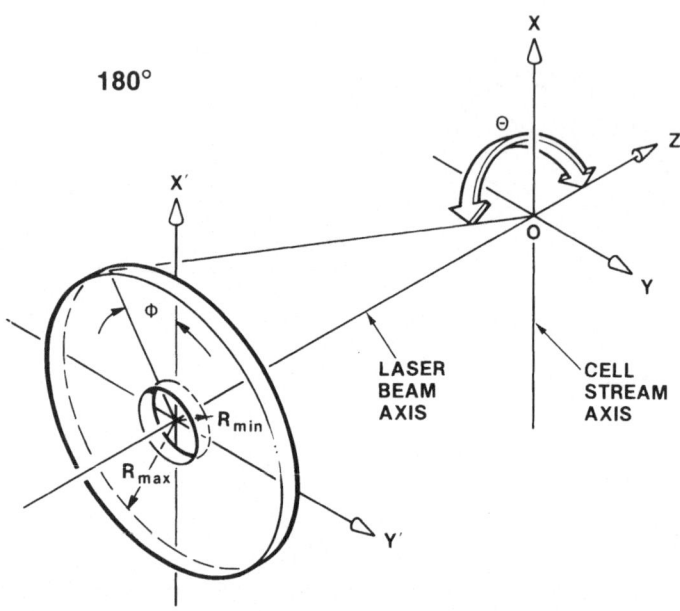

FIGURE 3. Possible backscatter detector configuration. Again, the laser beam axis (z) intersects the cell stream axis (x) at the origin 0. The laser beam polarization is parallel to the x axis. A collection lens of radius R_{max} is located at position z with respect to the origin. An aperture of diameter R_{min} allows for passage of the laser beam. A turning mirror or auxiliary optics would be required to focus the scattered light onto a photomultiplier tube. A negative value of the angle ϕ is shown. For the detector response calculations, the detector is at $z = -23.0$ mm and the step sizes in x and y are 0.1 mm.

beam stop is replaced by a horizontal bar, which is required to block the light refracted by the stream. In Fig. 8 (Salzman *et al.*, 1979b), the calculated FACS forward scatter detector response is compared to experimental data obtained with plastic microspheres. Visser *et al.*, (1978b), using fluorescein-ated aminoethyl-sephadex-G25 microspheres, collected bivariate histograms of fluorescence intensity (proportional to particle volume) vs. forward scattered light intensity at 488 nm and showed that the forward scatter intensity was approximately proportional to particle cross-sectional area.

    b. Forward Scatter in a Microscope-Based Flow Cytometer. A new flow cytometer (Steen and Lindmo, 1979; Lindmo and Steen, 1979) based on an inverted fluorescence microscope has been developed. A forward scatter detec-tor has been adapted to the instrument (Steen, 1980) so that light scattered by a cell between approximately 0° and nearly 90° is collected by the detector. Since the exciting light is from a broadband Xe or Hg arc source, the light

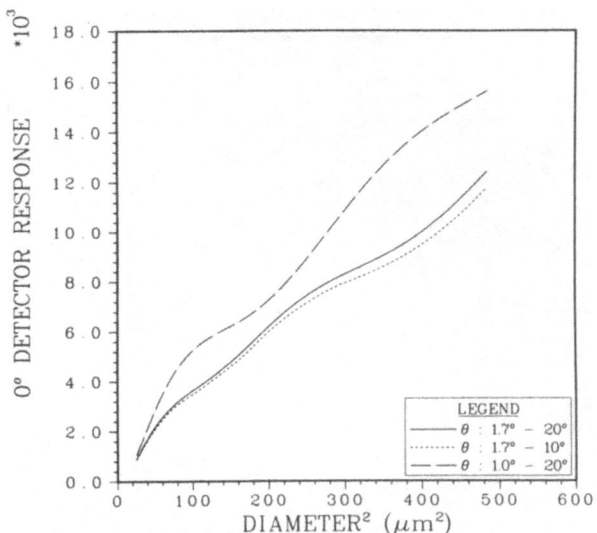

FIGURE 4. Forward scatter detector response (see Fig. 1) as a function of diameter squared for three different combinations of scattered light collection angles for the live cell coated sphere model using an N/C ratio of 0.72. The solid line shows the response for the configuration in which the polar collection angle, θ, varies from 1.7° to 20.3°. The dotted line shows the response for $1.7 < \theta < 10°$. The small difference between these curves indicates that most of the light is scattered within the first 10° of the forward direction. The dash–dot curve shows the response for the configuration with a smaller beam stop such that $1.0° < \theta < 20.3°$. This curve oscillates as a function of diameter squared more than the other two but has a steeper slope for smaller cells (diameters less than about 9 μm). The laser beam numerical aperture at the cell determines the minimum size of the beam stop.

scatter detector integrates with respect to wavelength as well as angle. Steen analyzed a sample of logarithmically growing tissue culture cells (NHIK 3025) which were fixed in 70% ethanol and the cytoplasmic and basic nuclear proteins stained with fluorescein isothiocyanate (FITC). He found that the scattered light intensity increased linearly as a function of FITC fluorescence. He concluded that the scattered light signal is proportional to cell dry mass. A more conservative conclusion would be that it is proportional to cellular protein content for 70% ethanol fixed cells. It is not clear what this relationship would be for viable cells.

2. 90° Scatter Detector

For objects in the size range of mammalian cells (3–30 μm in diameter), the 90° scatter detector response is roughly three orders of magnitude lower

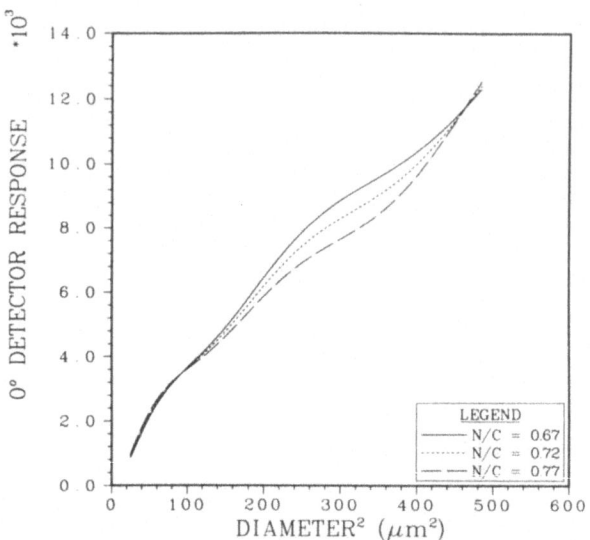

FIGURE 5. Forward scatter detector (Fig. 1) response as a function of diameter squared for the live cell coated sphere model with three different N/C ratios. The polar collection angle ranges from 1.7° to 20.3°. The forward scatter detector response is insensitive to N/C ratio changes over a range of 14% for cells with diameters less than about 14 μm (196 μm^2).

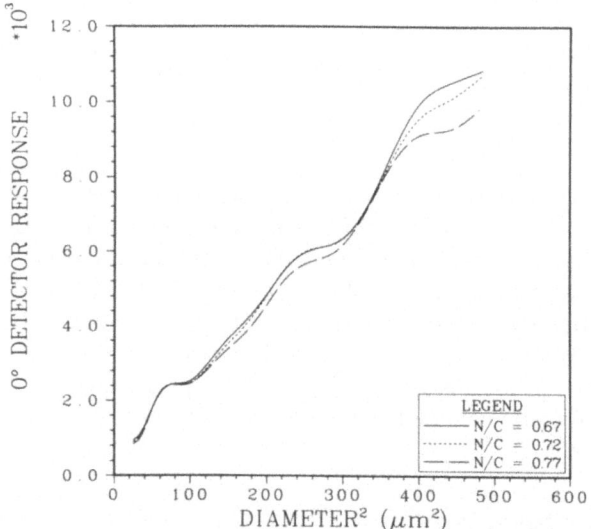

FIGURE 6. Forward scatter detector (Fig. 1) response as a function of diameter squared for the fixed cell coated sphere model with three different N/C ratios. The polar collection angle ranges from 1.7 to 20.3°. The forward scatter detector response is insensitive to N/C ratio changes of 14% for cells with diameters less than about 19 μm (361 μm^2). However, the response functions have two "knees" in this range over which the response is insensitive to changes in cell size.

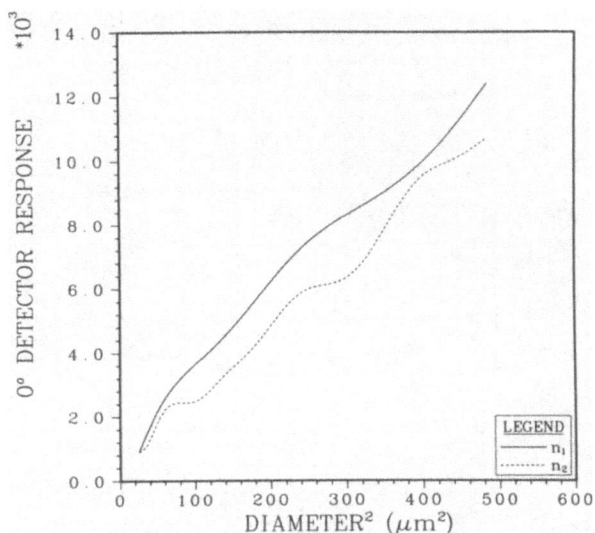

FIGURE 7. Forward scatter detector (Fig. 1) response as a function of diameter squared for the coated sphere model. The N/C ratio is 0.72 and n_1 represents the nucleus and cytoplasm real refractive index pair (1.392, 1.3703) while n_2 represents the pair (1.4755, 1.4525). The 6% increase of n_2 over n_1 produces a significant change in the forward scatter detector response. The detector spans the polar angular range from 1.7 to 20.3°. The imaginary part of the refractive index is zero in all cases.

than that for the forward-scatter detector. A photomultiplier tube is generally used as the detector instead of a photodiode because of the higher gain available with these tubes. To collect an adequate amount of light, a relatively large solid angle is used. Figure 9 illustrates the effect of changes in the collection angle on the 90° scatter detector response. The 90° scatter detector is not useful for detecting changes in N/C ratio for live (Fig. 10) or fixed (Fig. 11) cells.

For a 14-μm diameter cell (196 μm²), the 90° scatter detector response is roughly five times more sensitive to overall refractive index changes (Fig. 12) than is the forward scatter detector (Fig. 7). It may, therefore, be useful for detecting morphological changes which are manifest as changes in refractive index. The growth of lipid or micropolysaccharide storage vesicles in the cytoplasm could change the refractive index.

3. Backscatter Detector

Theoretical models for the scattering from biological cells with complex internal structures (Kerker, 1978; Meyer, 1979) predict that backscattering is the region of greatest sensitivity to small changes in morphology. Figure 13

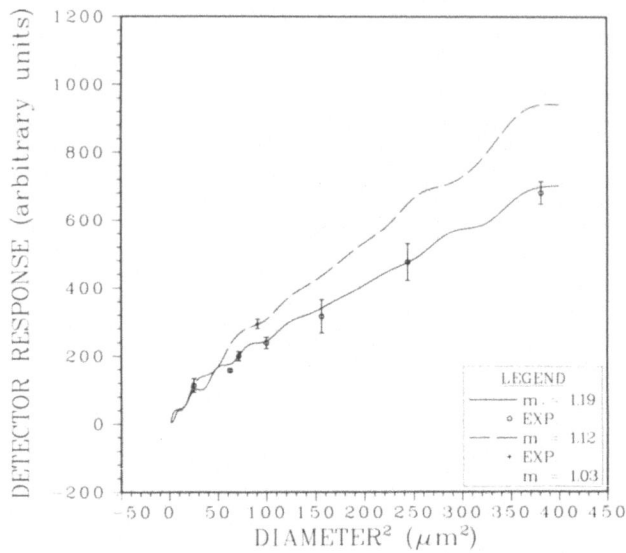

FIGURE 8. FACS forward scatter detector response as a function of homogeneous spherical particle diameter squared for three values of *m*, the particle refractive index relative to that of the medium (*n* = 1.3345). The open circles, associated with the solid line, are experimental data points for polystyrene latex microspheres with diameters 5.0, 7.9, 8.45, 10.0, 12.5, 15.6, and 19.5 μm. The other data point (+) (on the dash–dot curve) is for a 9.55-m-diameter polymethyl-methacrylate microsphere (*m* = 1.12). The curves are from homogeneous sphere Lorenz–Mie theory calculations (Dave, 1968) and are normalized to the data for the 10.0-m-diameter micro-sphere. The error bars on the experimental data points are two standard deviations high. From Salzman *et al.* (1979b).

shows the calculated backscatter detector response as a function of diameter squared for three N/C ratio values using the live-cell coated sphere model. The functions oscillate even more rapidly than those for the 90° detector. For the fixed cell model (Fig. 14), the oscillations disappear as does the sensitivity to changes in the N/C ratio. The backscatter detector is even more sensitive to changes in refractive index (Fig. 15) than is the 90° detector. If the coated sphere is an adequate model for a nucleated biological cell, it is unlikely that backscattering will prove useful for cell discrimination with viable cells. Besides the problem of low signal levels, backscatter detection is subject to interference from reflections of the main beam at air–glass and glass–water interfaces in the flow chamber. If the beam reflected from the window in the forward scatter direction strikes the cell, the "forward" scatter from this reflected beam will impinge on the backscatter detector and swamp the weak backscatter signal (Kratohvil, 1966).

In this section we have presented model calculations to show the sensitivity of various scattered light detector configurations to small changes in important

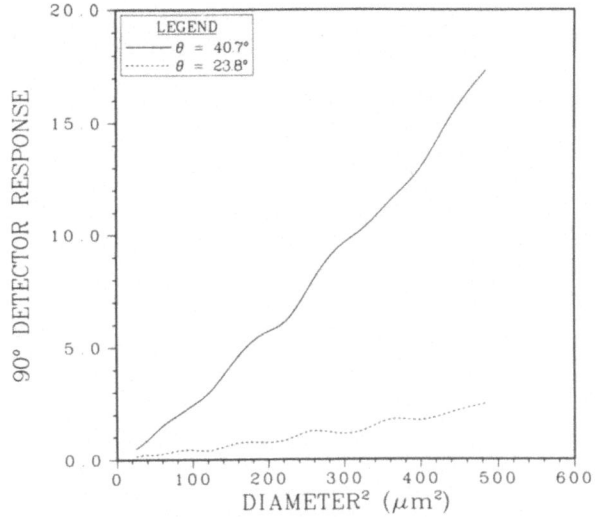

FIGURE 9. A 90° scatter detector (Fig. 2) response as a function of diameter squared for two values of the maximum collection angle using the coated sphere model with nucleus and cytoplasm refractive index values of 1.392 and 1.3703, respectively. Both functions are roughly proportional to diameter squared. The N/C ratio is 0.72. The angle θ here represents the collection angle subtended at the origin by the negative y axis and a ray to the rim of the detector. Elsewhere, θ is as defined in Fig. 1.

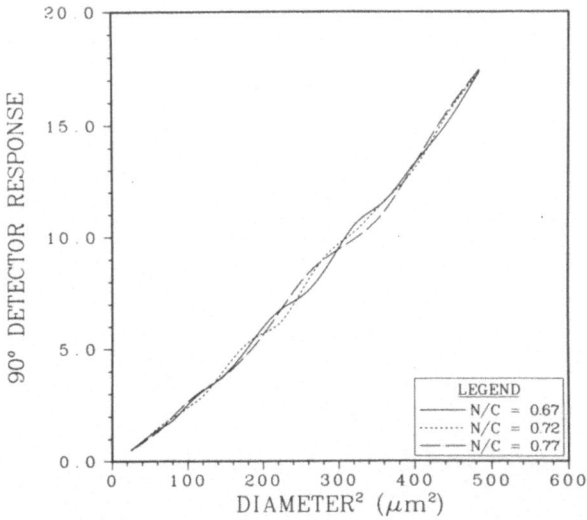

FIGURE 10. A 90° scatter detector (Fig. 2) response vs. diameter squared for the live cell coated sphere model with three different values of the N/C ratio. The collection angle is 40.7° as in Fig. 8. The nucleus and cytoplasm refractive indices are 1.392 and 1.3703, respectively. The 90° scatter detector response is somewhat sensitive to N/C ratio in certain cell size ranges but the functions overlap significantly in others.

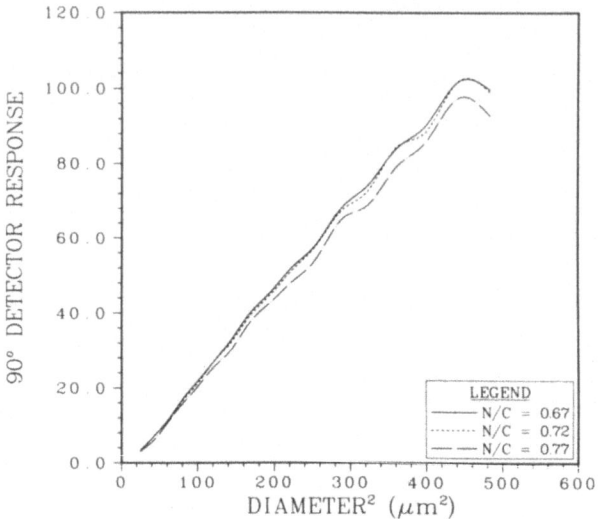

FIGURE 11. A 90° scatter detector (Fig. 2) response vs. diameter squared for the fixed cell coated sphere model with three different values of the N/C ratio. The collection angle is 40.7° as in Fig. 9 and the nucleus and cytoplasm refractive indices and 1.4755 and 1.4525, respectively. The sensitivity to changes in N/C ratio is quite small. Note that the response functions are multivalued over the diameter range 20–22 μm (400–484 μm²).

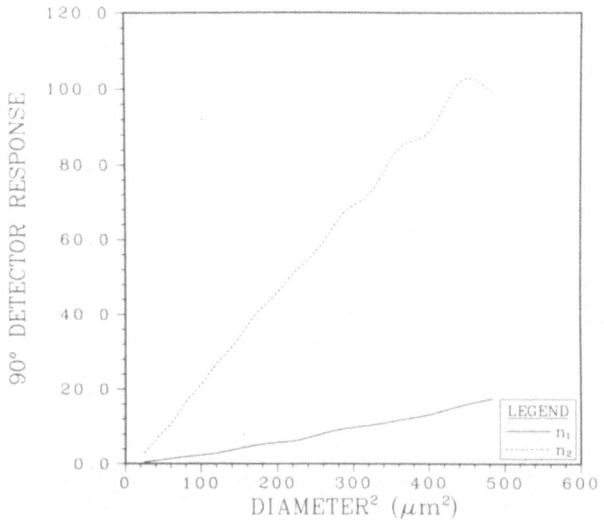

FIGURE 12. A 90° scatter detector (Fig. 2) response vs. diameter squared for the coated sphere model with an N/C ratio of 0.72. The refractive indices n_1 and n_2 correspond to those for live cells and fixed cells, respectively. Ninety-degree scattering is significantly more sensitive to refractive index changes than is forward scattering (see Fig. 7) for this detector configuration.

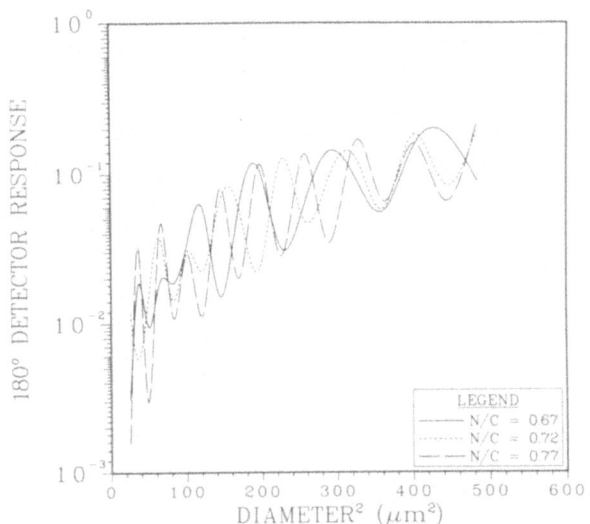

FIGURE 13. A 180° scatter detector (Fig. 3) response vs. diameter squared as a function of N/C ratio for the live cell coated sphere model. The detector includes a polar angular range of 20.3° about 180°. The detector response oscillates rapidly as a function of increasing cell diameter for each N/C ratio and the responses overlap significantly. Backscattering would be useful for N/C ratio discrimination only for carefully selected narrow size ranges. It is also a very-low-level signal.

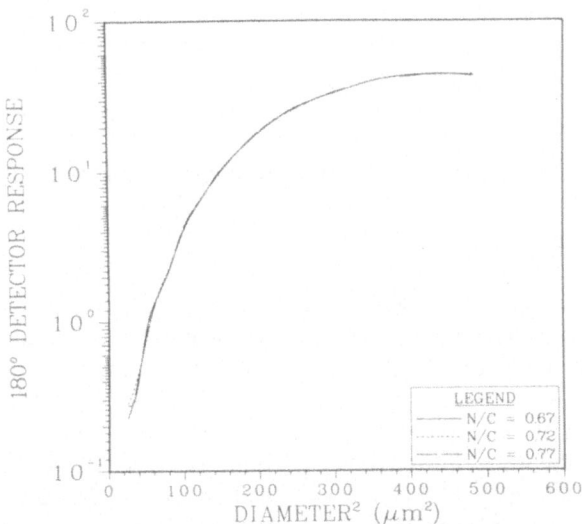

FIGURE 14. A 180° scatter detector (Fig. 3) response vs. diameter squared as a function of N/C ratio for the fixed cell coated sphere model. The detector configuration is the same as in Fig. 13. There is no sensitivity to N/C ratio over a wide range of cell diameters.

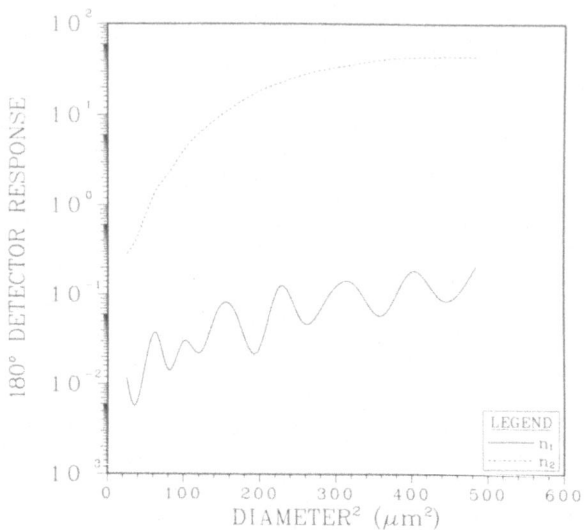

FIGURE 15. Comparison of the 180° scatter detector (Fig. 15) response vs. diameter squared for live (n_1) cell and fixed (n_2) cell coated sphere models. The detector configuration is the same as that in Fig. 14. The N/C ratio is 0.72. Backscattering is very sensitive to changes in refractive index.

cell parameters such as cell diameter, N/C ratio, and refractive index. In the next section we present a number of applications of forward and 90° scattered light detection to cell discrimination.

III. RECENT APPLICATIONS OF FORWARD AND 90° SCATTER

A. Human Peripheral Blood

An improvement in the separation of subsets of a heterogeneous population can frequently be obtained by collecting the data as a bivariate histogram. Figure 16 (Salzman *et al.*, 1975b) shows such a bivariate histogram for unstained human leukocytes. The monocytes (peak 2) cannot be clearly identified as a separate subpopulation with either dectector alone.

From the theoretical discussion in Section II, it may be inferred that forward scatter is dominated by a cross-sectional area dependent on cell size (Fig. 4). It may also be deduced that 90° scatter has both a cross-sectional area dependence and a strong dependence on refractive index (Fig. 11). As demonstrated below, the discrimination at 90° depends on refractive index differences and nuclear size and shape. Both forward and 90° scatter may also have a strong dependence on the shape of the cell.

B. Osmolarity Effects

Visser *et al.* (1980) used combined FACS forward and 90° scatter from mouse blood cells to show the sensitivity of both parameters to the osmolarity of the suspending medium (Fig. 17). The crenated blood cells at high osmolarity scattered significantly more light at 90° than did the spherocytes at low osmolarity. They attributed the effect to the shape of the cells. It is, however, possible that the 90° effect is partly due to refractive index changes. The high-osmolarity solution causes water to leave the cell, increasing the protein concentration, and thus, the refractive index within the cytoplasm. The low-osmolarity solution has the opposite effect and causes a reduction in the cytoplasm

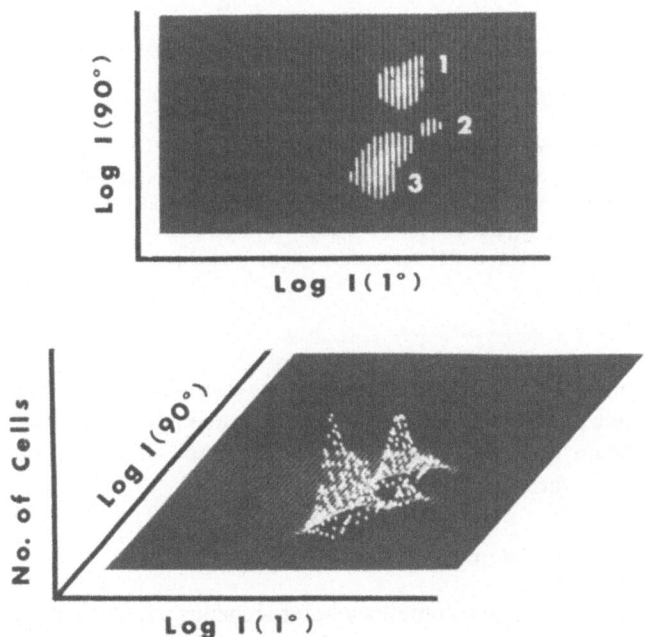

FIGURE 16. Bivariate histograms showing the number of unstained human leukocytes (z axis) as a function of the logarithms of the scattered light intensity at $1° \pm 0.1°$ (x axis) and at 90° $\pm 12.5°$ (y axis). The log scales span a three-decade range. The upper frame shows a contour map and the lower frame an isometric display. Sorting of the sample showed that peak 1 contained more than 80% neutrophils, peak 2 more than 77% monocytes, and peak 3 more than 93% lymphocytes. The excitation wavelength was 488 nm. From Salzmen *et al.* (1975b).

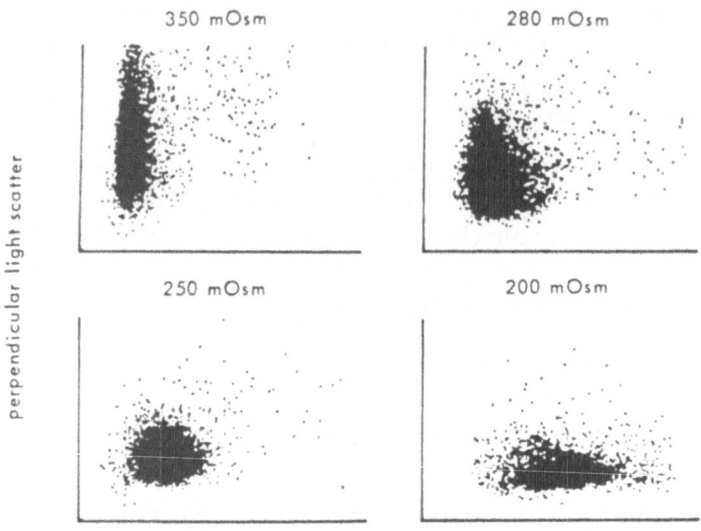

FIGURE 17. Bivariate scatter plot showing FACS forward (1.5° –13°) scatter vs. perpendicular (90° ± 25°) light scatter at 488 nm for live mouse blood cells suspended in NaCl solutions of various concentrations. The blood cells are swollen spherocytes at low osmolarity, discocytes at physiological osmolarity, and crenated shapes at high osmolarity. Note that the crenated red cells have significantly higher 90° scatter than do the spherocytes. Both axes are linear. From Visser *et al.* (1980).

refractive index. The forward scatter effect may, indeed, be due to the influence of shape. The crenated cells will present a smaller cross-sectional area to the laser beam than the swollen spherocytes and will, therefore, scatter less light in the forward direction.

C. Hematopoietic Stem Cells

Mouse bone marrow contains a heterogeneous population of cells with different morphological characteristics. Loken *et al.* (1976) used FACS forward scattering at 488 nm in the polar angular range from 2° to 8° to separate mouse bone marrow from a two-month-old C57BL/10 mouse into four subpopulations, which they identified by sorting and Wright-Giemsa staining of the sorted cells. In order of increasing forward scatter, the four groups contained (a) mature red cells, (b) small lymphocytes and a few normoblasts, (c) polymorphonuclear leukocytes, large lymphocytes and normoblasts, and (d) very large blast cells and contaminants from group (c). The discrimination amoung these subpopulations is probably dominated by differences in cell diameters (see Figs. 4 and 5). In this same paper, Loken and co-workers report setting up a helium neon laster (633 nm) so that they could observe the FACS forward scattering from each cell simultaneously at two wavelengths. They used a bivariate scatter plot of 488-nm scatter at 5° ± 3° vs. 633-nm scatter at 3° ± 1° to show some enhancement in the separation of these subpopulations. The enhanced separation may be explained by the wavelength-dependent differences in the cell refractive indices.

Using the same experimental arrangement as Salzman *et al.* (1975a), Visser *et al.* (1978a) separated NH$_4$Cl-treated mouse bone marrow (erythrocytes lysed) into four major subpopulations. Group I (lobocytes) had significantly less forward scatter (1.0 ± 0.1°) than groups II (normoblasts and lymphocytes), III (metamyelocytes and lymphocytes), and IV (granulocytes), which all scattered about the same amount of light in the forward direction. In order of increasing 90° ± 12.5° scatter, the subpopulations were group II, groups I and III, and group IV. By performing an *in vitro* colony-forming assay, they showed that group III contained the myleoid progenitor cell CFU-c, which in the presence of a colony-stimulating factor from pregnant mouse uteri (PMUE) gave rise to colonies of granulocytes and macrophages.

Van den Engh and Visser (1979) and van den Engh *et al.* (1979, 1980) used FACS forward (0.5°–13°) and perpendicular (90° ± 25°) scatter to extend this work to the separation and identification of the hematopoietic stem cell CFU-s (*in vivo* spleen-colony-forming unit) and three *in vitro* colony-forming cells (CFU-c1,2,3) leading to macrophages and granulocytes. On a bivariate scatter plot of forward vs. 90° scatter, the three major subpopulations contained (in order of increasing forward scatter) (I) erythrocytes, (II) lymphocytes, and (III) (in order of increasing 90° scatter) blast cells, CFU-s, CFU-c1, CFU-c2, CFU-c3, and granulocytes.

Goldschneider *et al.* (1978) used FACS forward light scatter at 488 nm to gate the fluorescence frequency histogram of fluoresceinated anti-Thy-1 sera to select subpopulations of rat hematopoietic cells expressing Thy-1 antigen on their surfaces. The scatter gating was used to exclude erythrocytes from analysis. Goldschneider *et al.* (1980a), using the experimental arrangement above, isolated pluripotent hematopoietic stem cells from rat bone marrow and identified three subsets of granulocyte-macrophage progenitor cells with an *in vitro* colony-forming assay. They used the *in vivo* colony-forming assay (CFU-s) to identify pluripotent stem cells. This same technique (Goldschneider *et al.,* (1980b) has been used to isolate rat hematopoietic cells which contain the intracellular enzyme terminal deoxynucleotidyl transferase (TdT). This enzyme catalyzes the polymerization of any 3'-OH-terminated segment of DNA without template direction. Ninety-seven percent of the TdT$^+$ cells occurred in the forward light scatter frequency histogram between peaks formed by lymphocytes and myeloid cells. The TdT$^+$ cells were identified by an indirect immunofluorescence assay. The fluoresceinated anti-Thy-1 method was used to show that TdT$^+$ cells were also Thy-1$^+$.

In a study of the separation of mouse bone marrow cells using wheat germ agglutinin affinity chromatography, Nicola *et al.* (1978) used combined FACS forward scatter and 90° ± 20° scatter to analyze the distribution of cell types as they were eluted from the column. They obtained roughly the same subpopulation distribution as Visser *et al.* (1978a). In seeking to understand the differential expression of lectin receptors during hematopoietic differentiation, Nicola *et al.* (1980) were able to obtain a 10- to 15-fold enrichment for *in vitro* colony-forming cells by FACS sorting using forward scatter, 90° scatter, and pokeweed mitogen fluorescence. They obtained bivariate scatter plots similar to those of Visser *et al.* (1978a). This use of three parameters for cell discrimination should be noted because it produced better discrimination than when two parameters were used alone. Few investigators have used more than two parameters for sorting because of the difficulty of visualizing a function of three or more independent variables. Three-parameter display techniques for flow cytometry have been investigated in several laboratories (Stohr and Futterman, 1979).

In the Cytograf (model 6301, Biophysics Systems, Inc., Boston, MA) flow cytometer, Kamentsky and Melamed (1965) and Adams and Kamentsky (1971) incorporated an extinction detector at 0° and used a helium neon laser (633 nm) as a light source. Extinction measures the sum of light absorbed by and scattered from a cell. Doukas *et al.* (1977) used this flow cytometer to follow lymphocyte stimulation with the mitogen concanavalin A.

D. T-Lymphocyte Subclasses

The light scattering techniques described above can be combined with fluorescence methods to obtain additional discrimination of subsets in a heterogeneous population of cells. Hoffman *et al.* (1980) used a modified Cytofluorograf (model FC200/4800A, Ortho Instruments, Westwood, Massachusetts 02090), to distinguish among the subclasses of human T lymphocytes. An NH$_4$Cl/EDTA-treated, buffy-coat preparation was incubated with various fluoresceinated monoclonal antibodies and analyzed in the flow cytometer. A bivariate histogram of forward-vs.-90° scatter at 488 nm was used to identify the lymphocyte subpopulation. Single-parameter fluorescence histograms, gated on the lymphocyte subpopulation, were then used to identify the T-lymphocyte subclasses by their affinity for the various fluoresceinated monoclonal antibodies.

E. Non-Hodgkin's Lymphomas

In a flow cytometric study using a FACS, Diamond and Braylan (1980) analyzed 30 cases of non-Hodgkin's lymphoma using nuclei isolated from chicken erythrocytes as an internal standard. The cell samples were fixed in 50% ethanol. Nonneoplastic controls were obtained from lymph nodes, spleens, tonsils, and peripheral blood. Separate Coulter volume analysis was performed for some of the cases using unfixed samples. Bivariate and single-parameter histograms were collected using forward light scatter at 488 nm and fluorescence from the propidium-iodide-stained nuclei. Whereas Coulter volume could be used for 7 of 8 high grade (severe) lymphomas to differentiate them from the nonneoplastic cases, light scatter could be used for this discrimination in only 5 of 12 high-grade lymphomas. This failure might be explained by the ethanol fixation procedure. Ethanol perforates the plasma membrane and coagulates the protein in the cytoplasm, raising the refractive index of the cell. As shown in Section II, the sensitivity to size changes is less for fixed cells than for live cells.

F. Cervical Cell Analysis with 90° Scatter

Barrett *et al.* (1978, 1979) have used a modified FACS to investigate the use of cellular DNA content and 90° light scatter for detection of abnormal cells in samples exfoliated from the female genital tract. The cellular DNA content was measured by staining with chromomycin A_3 and exciting with 457 nm argon laser light. To analyze the 90° light scatter signals, cells were sorted onto slides from eight regions with increasing 90° scattered light intensity. The areas of the cells on the slides were measured with a planimeter using a projection microscope at 800× magnification. The 90° light scatter response as a function of cell area had three different slopes as a function of increasing cell area. Although the scattering function increased monotonically with increasing cell area, the sensitivity to size changes was different for different size cells.

G. Sputum Samples

Frost *et al.* (1979) used a FACS to enrich sputum samples for neoplastic cells. The enriched samples were subsequently placed on slides and Papanicolaou-stained for digital image analysis. They used forward scatter between polar angles 0.5°–15° to gate the acridine orange green fluorescence in the cells. In three cases of squamous cell carcinoma of the lung, the investigators were able to enrich the samples in neoplastic cells more than 10-fold by using the two parameters together. In the broad scatter frequency histograms, the

malignant cells had scatter intensities (cross-sectional areas) similar to macrophages and squamous cells.

IV. MULTIANGLE AND MULTIWAVELENGTH SCATTERING

A. Multiangle Scattering

Forward and 90° scattering with large collection angles have contributed
significantly to the efforts to discriminate among classes of cells in a heterogeneous population. In an effort to extract more information from the forward
scattering region, Salzman *et al.* (1975a) and Mullaney *et al.* (1976) developed a flow cytometer based on static cell work by Meyer *et al.* (1974) with
which the scattering from a cell could be sampled simultaneously at up to 32
angles between 0° and 21° with respect to the laser beam axis. The detector
is a semicircular array of concentric ring photodiodes. Crowell *et al.* (1978)
used this system to demonstrate discrimination between viable human blood
lymphocytes and monocytes in the angular range from 0° to 1.6°. Price *et al.*
(1978) showed that the differences in the scatter patterns could be used to
distinguish among several kinds of microalgae with the above flow cytometer.
In a study of respiratory tract cells, Steinkamp *et al.* (1977) used the scattered
light patterns between 0.4° and 2° from hamster lung cells to discriminate
among three classes of cells. Schafer *et al.* (1979) used the multiangle scatter
flow cytometer above to distinguish between normal, cultured human fibroblasts and mutant cells containing cytoplasmic inclusions. In a study of murine
teratocarcinoma cells, Swartzendruber *et al.* (1979) showed that multiangle
forward scattering could be used to distinguish stem cells from parietal yolk
sac cells, visceral yolk sac cells, neuronal cells, and squamous cells. The largest
differences among the subpopulations occurred in the angular region between
0.4° and 2°. It is clear from the data in all of the above experiments that the
signals on many of the detector elements are highly correlated. It is perhaps
possible that a forward scatter detector such as that shown in Fig. 1 might have
been able to obtain the same discriminating power.

Loken *et al.* (1976) developed a multiangle sweep scatter system using a
FACS flow cytometer. As a cell passes through an expanded argon laser beam,
it subtends larger and larger scattering angles at an aperture which is imaged
onto a photomultiplier tube. The scatter pattern spans an angular range from
1° to about 49° with respect to the laser beam axis. A waveform recorder is
used to store the scatter pattern for each cell. A group of scatter patterns for
mouse thymocytes is presented. No further work has been published using this
instrument.

Ludlow and Kaye (1979) have developed a flow cytometer employing 174 fiber optic light guides, each of which subtends a polar angular range of 1° at the center of the flow chamber. As a particle moves slowly through the HeNe laser beam, the light scattered in the horizontal plane between 3° and 177° is scanned by a rotating disk onto a photomultiplier tube. The device has been used to determine size distributions and mean refractive indices for the spores *Penicillium chrysogenum* and *Aspergillus niger*. The refractive index and size information is extracted by using the homogeneous sphere Lorenz–Mie theory to fit the experimental data.

Morris *et al.* (1979) have developed a 360° scattering photometer, which they have used to measure the diameter of phospholipid vesicles and micelles in chromatographic column effluents. A turntable rotates a photomultiplier tube (PMT) telescope around a cylindrical curvette containing the flowing sample. The device scans from 20° to 160° on either side of the 2-W argon laser beam with a resolution of 0.25°. The PMT telescope rotates at a frequency of 1 Hz. The instrument has also been used to study the dynamics of polymerization of microtubular protein and the osmotic lysis of chromaffin granules. The system has a relatively large viewing volume as the laser beam is not focused on its passage through the cuvette.

Bartholdi *et al.* (1980) have investigated the scattering from single polystyrene latex microspheres over a polar angular range from 2.5° to 177.5° on either side of an argon laser beam. The flow cytometer can acquire 32 signals per particle at rates up to 1000 particles per second. Figure 18 shows the optical arrangement of this flow cytometer. The argon laser beam waist is 150 μm in diameter at its intersection with the 20- to 30-μm-diameter cell stream. Fig-

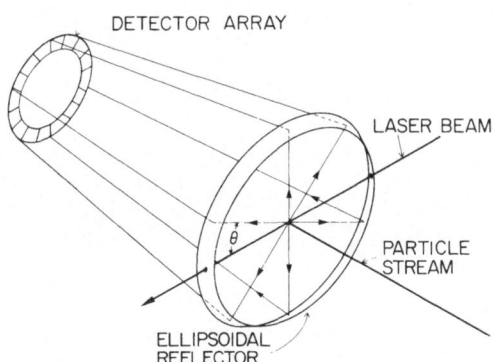

FIGURE 18. Optical basis for the 360° flow cytometer showing the intersection of the laser beam and particle stream at the primary focus of a circular strip of an ellipsoidal reflector and also the reflected scattered rays converging to the detector array (not to scale). From Bartholdi *et al.* (1980).

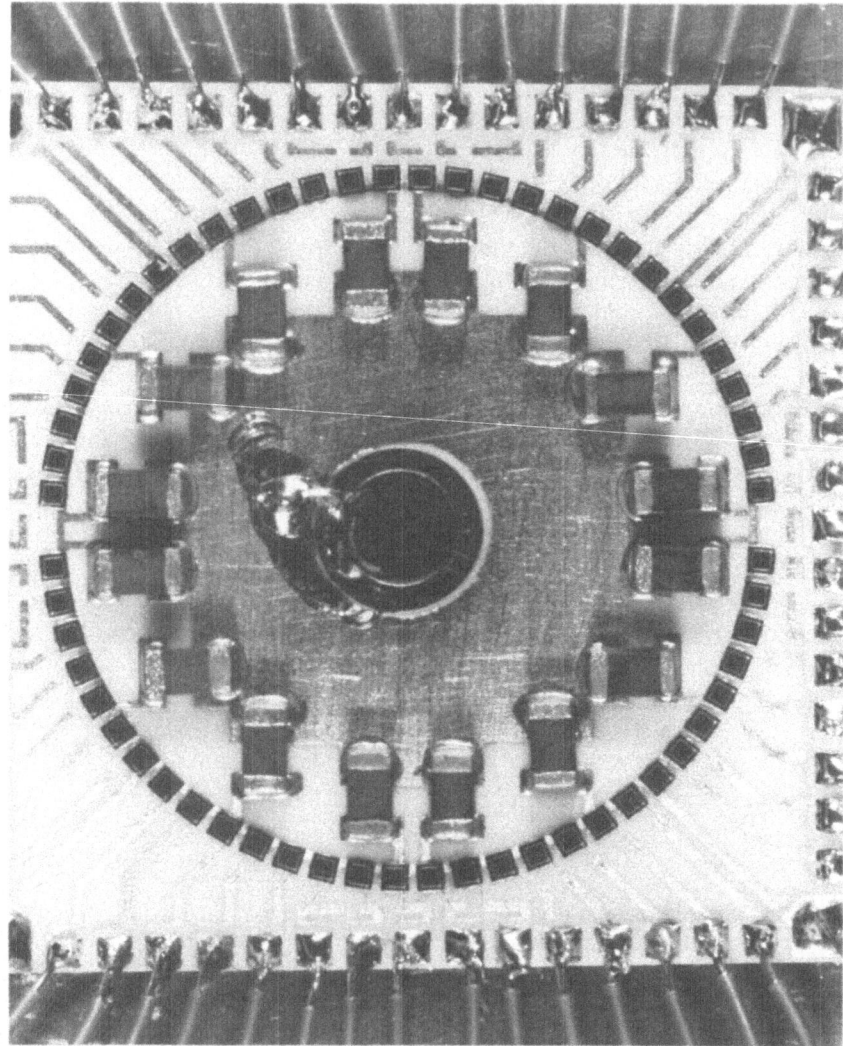

FIGURE 19. Photograph of the detector array consisting of 60 photodiodes arranged on a 2-cm-diameter circle. The photodiodes subtend a polar angular range of 3° at the flow stream–laser beam intersection. The signals from the photodiodes are fed to individual hybrid microcircuit preamplifiers on the rear of the substrate. The rectangular objects are decoupling capacitors to reduce cross talk between the photodiodes. From Bartholdi *et al.* (1980).

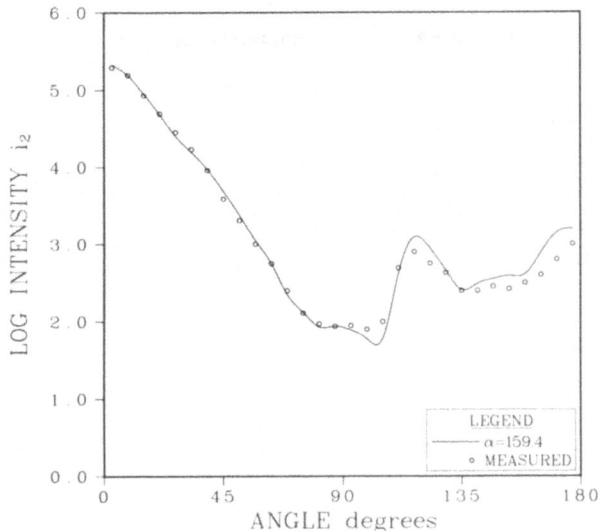

FIGURE 20. Measured angular scattering pattern (open circles) from a single particle in a polystyrene latex sample (19.5-μm nominal diameter) compared with a theoretical curve corresponding to size parameter (circumference/wavelength in the medium) α = 159.4 and relative refractive index m = 1.20. From Bartholdi *et al.* (1980). -

ure 19 shows the custom detector array used with this device. Figure 20 shows a comparison between the sampled scatter pattern for a 19.5-μm-diameter particle and the expected response according to the Lorenz–Mie theory. Figure 21 shows a set of eight detector frequency histograms for a mixture of five different polystyrene latex spheres. The smaller particles are much better resolved at the low detector numbers. When viable tissue culture cells were analyzed with this flow cytometer, reasonable signal-to-noise ratios were obtained out to scattering angles of only about 45° with a laser input power of 1.2 W at 514.5 nm. Since the instrument cannot be used for sorting, it has not been used for further biological investigations.

B. Two-Color Light Scattering

Loken and Houck (1981) used a FACS on which the argon laser contained mirrors permitting the output of all lines from 351 nm to 488 nm. A second forward scatter detector was added so that forward light scatter could be detected simultaneously in the ultraviolet (UV) and at 488 nm. Single-parameter scatter frequency histograms at the two wavelengths show that the

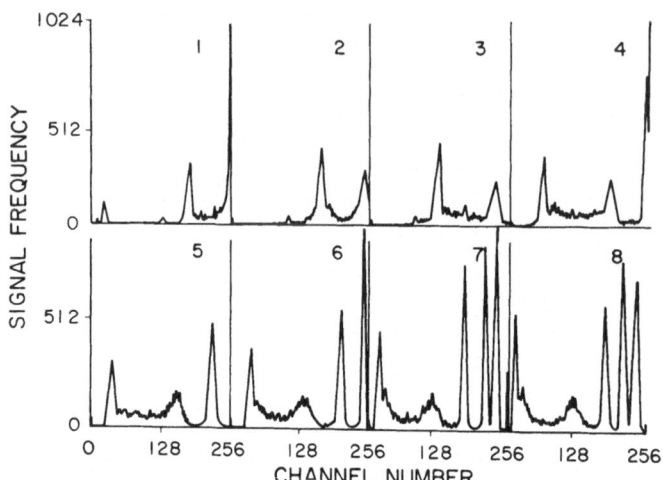

FIGURE 21. Scattered light frequency histograms from detectors 1 through 8 for a mixture of polystyrene latex spheres of nominal diameters 1.1, 5.0, 10.0, 15.6, and 19.5 μm. Detectors 1 through 8 are at mean polar angles 5.9°, 11.7°, 17.5°. 23.3°, 29.2°, 35.0°, 40.8°, and 46.6°, respectively. Channel number corresponds to log scattered light intensity over a 2.5-decade range. Detector 1 sees only the 1.1-μm diameter particles; the rest are off scale. Detectors 2, 3, and 4 see 1.1- and 5.0-μm-diameter particles. Detector 5 sees 1.1, 5.0, and 10.0; detector 6 sees 1.1, 5.0, 10.0, and 15.6; detectors 7 and 8 see all the particles. At detectors 6, 7, and 8, the 1.1-μm-diameter particles are not resolved from smaller debris. Samples of each particle size were analyzed separately to identify the peaks in each detector histogram. From Bartholdi *et al.* (1980).

resolution of subpopulations in a mouse bone marrow sample is significantly better at 488 nm than in the UV. This may be due to the refractive index dispersion relation. The refractive indices of the cell subpopulations may be different at 488 nm and nearly the same in the UV.

V. FOURIER TRANSFORM TECHNIQUES

If a forward scatter detector (Fig. 1) is sufficiently far from a scattering object such as a biological cell that the conditions for the Fraunhofer diffraction limit apply, the scattering of light in the forward direction can be described by a two-dimensional Fourier transform (Bracewell, 1965; Goodman, 1968). The cell and its internal structures are modeled as an *arbitrarily* complex two-dimensional distribution of optical density. The information about the cell is projected into a plane perpendicular to an incident laser beam and is contained in the optical density distribution. The nucleus is modeled as a region of higher optical density than the cytoplasm with variations in density to model chromatin condensation. Vacuoles in the cytoplasm can be modeled as regions of lower optical density than that of the cytoplasm. The Fourier transform converts spatial variations in the optical density distribution in the object plane into spatial frequency in the Fourier transform domain at the detector. Spatial frequency can be understood by considering a scattering object which has a sinusoidal variation in optical density as a function of increasing particle radius. If there are only a few periods of the sine wave across the object, only low spatial frequencies will be present at the detector and most of the scattered energy will be near the laser beam axis ($0°$). If there are many periods of the sine wave across the object, high spatial frequencies (larger scattering angles) will have significant energy content. Low spatial frequencies occur when the optical density distribution changes smoothly across the cell. High spatial frequencies appear as a result of sharp edges or discontinuities in the optical density distribution. Spatial frequency magnitude increases from zero at the center of the detector. The rim of the detector collection lens sets an upper limit on detected spatial frequency (scattering angle). In a static system arranged to optically compute the Fourier transform of enlarged transparencies of cervical cells, Kopp *et al.* (1974, 1976), Pernick *et al.* (1978a,b), and Wohlers *et al.* (1978) showed that the best discrimination between normal and abnormal cells occurred in the spatial frequency range between 20 and 40 cycles/mm, which corresponds to scattering angles of less than $3°$ with respect to the laser beam direction. The investigators used the same ring/wedge detector as Salzman *et al.* (1975a). The normal cells had a much more regular diffraction pattern than did the malignant cells. This effect was attributed to optical discontinuities in the nuclear image due to unusual chromatin condensation in the malignant cells.

Fourier transform measurements are difficult to perform in static systems because of potentially high background light levels resulting from dust on the lens and diffraction pattern modifications introduced by apertures. If the incident gaussian intensity distribution laser beam is cut off sharply near the particle in the object plane, the diffraction pattern at the detector will display large oscillations due to this aperture. These oscillations may mask the signal from the particle. Pernick *et al.* (1978c) examined the scattering geometry in an enclosed chamber flow cytometer and showed that Fourier transform techniques could be applied to the analysis of forward scatter in flow. They also showed that the forward scatter signature of a disk-shaped buccal cell is insensitive to rotation of the plane of the cell about the cell stream axis for angles up to nearly $45°$. Their model indicated that displacement of the cell from the center of the laser beam and displacement of the nucleus from the center of the cell had significant effects on the small-angle scattering.

A lens performs an optical Fourier transform of an object in the specimen plane. This transform can be viewed in a microscope by imaging the rear focal plane of the objective.

Seger *et al.* (1977) and Turke *et al.* (1978) developed a Fourier optical microscope with which they could scan the optical Fourier transforms of single cells on Papanicolaou-stained slides of cervical smears. They also found that abnormal cells had more energy scattered at higher spatial frequencies than did the normal cells. Genter and Salzman (1979) developed a graphic technique for displaying the complex data of Seger *et al.* (1977) and showed that two population discriminant analysis methods could be usefully applied to this kind of data.

Using phase contrast photographic images of pure populations of diatoms, Almeida and Fujii (1979) showed that the appropriately averaged transforms could be used to distinguish among several classes of diatoms. Fourier transform techniques are particularly useful for modeling the effects of beam shape, cell displacement from the center of the beam, and changes in complex cell shapes on the forward scatter detector response.

VI. POLARIZATION STUDIES—NEW DIRECTIONS

The light sources used in most of the studies described above have been lasers with linearly polarized beams. In most cases the scattered light detectors are insensitive to the polarization state of the scattered light. A biological cell is an optically active scattering object because it contains chiral (asymmetric) molecules. The polarization state of the scattered light is, in general, different from that of the incident beam.

A. Polarization Transformations

The most general polarization state for a beam of monochromatic light is elliptical. This beam can be regarded as a vector sum of two orthogonal coherent plane-polarized vibrations with different amplitudes and with a constant phase difference between them. This beam can be completely described by four variables called the Stokes parameters (Kerker, 1969). These variables are (1) the total intensity, (2) the degree of linear polarization at 0° or 90° with respect to the scattering plane, (3) the degree of linear polarization at ±45° to the scattering plane, and (4) the degree of left or right circular polarization.

The Stokes parameters describing the incident beam are modified by the beam's interaction with the cell and result in a set of Stokes parameters describing the beam scattered toward a detector. The Stokes parameters can be combined into a four-component vector and the interaction with the cell described by a four-by-four Mueller matrix (Mueller, 1948).

Hunt and Huffman (1973) developed a static light scattering photometer with which all the elements of the Mueller matrix can be determined for a bulk sample of scattering objects. The device uses a piezo-optical birefringence device to modulate the polarization state of the incident light beam. Phase-sensitive detection is used for the scattered light. Bickel *et al.* (1976) have used this tool to measure some of the transformation matrix elements for spores of two strains of *Bacillus subtilis.* One strain (Marburg 168) has normal spore structure. The other strain (UVS-42DPA) has spores that lack dipicolinic acid. They found that one matrix element was particularly sensitive to this small structural modification in the spore. This matrix element is involved in the transformation of circularly polarized light into light which is linearly polarized at 45° from the vertical plane. It would, of course, be extremely difficult to infer the exact nature of the structural change from the Mueller matrix. However, this tool may find application in distinguishing between closely related structures or in following time-dependent structural changes. Other investigators (Perry *et al.,* 1978; Thompson *et al.,* 1980) have examined Mueller scattering matrices for nonspherical particles.

B. Circular Intensity Differential Scattering

Because of their asymmetric structures, chiral molecules preferentially absorb and scatter circularly polarized light. These molecules present different refractive indices for right and left circularly polarized light. The real part of the refractive index gives rise to the phenomenon of optical rotatory dispersion (ORD), which is measured by observing the rotation of the plane of polarization of linearly polarized light. The imaginary part of the refractive index is responsible for circular dichroism (CD), which is the differential absorption of left and right circularly polarized light.

The disymmetry of the medium produces the differential scattering of left and right circularly polarized light. This phenomenon is called circular intensity differential scattering (CIDS) (Atkins and Barron, 1969; Barron *et al.,* 1973). If $I_L(\theta)$ and $I_R(\theta)$ are the scattering intensities for incident left and right circularly polarized light, respectively, then

$$\text{CIDS } (\theta) = \frac{I_L(\theta) - I_R(\theta)}{I_L(\theta) + I_R(\theta)}$$

The periodicities in molecular aggregates are proportional to the wavelength of the incident light. Different orders of molecular regularity are sensed at different wavelengths. The CIDS signals are strongest at 90° and in the backward direction (Bustamente *et al.,* 1980a,b). Maestre and Reich (1980) examined the scattering contribution to the circular dichroism of DNA films with twisted structures. They showed that the primary interaction is light scattering via a resonance phenomenon similar to that produced in cholesteric liquid crystals. They were able to use the Bragg law for cholesteric liquid crystals to determine the periodicity of the long-range ordered structures.

Reich *et al.* (1980) investigated the CD of DNA in ethanolic solutions. They showed that polynucleotide particles are shown to exhibit behavior similar to that of cholesteric liquid crystals. They assert that the scattering patterns are sensitive to the tertiary structure of the condensed DNA particles.

Maestre has developed a microscope system for measuring the combined CD and CIDS from the nuclei of single, intact cells. Salzman and Maestre (1981, private communication) have used this instrument to show that the CD/ CIDS signal is sensitive to changes in the DNA conformation as a function of cell cycle position for one tissue culture cell line. This technique may provide a new, nondestructive parameter for flow cytometry.

VII. CONCLUSION

Light scattering can be used for cell sizing in flow cytometry if care is taken in the choice of scattering angles and angular resolution. The most effective use of scattering has been in discriminating among subclasses of cells in a heterogeneous population. Here the sensitivity of light scattering to size, shape, and refractive index is used to advantage even though the detailed modeling of real cells remains a formidable problem.

The exploitation of scattered light polarization phenomena to explore changes in intracellular molecular conformations may be the next important area for use of this powerful but complex tool.

ACKNOWLEDGMENTS. This work was performed under the auspices of the U.S. Department of Energy and was supported in part by National Institutes of Health grant No. GM 26857. The author would like particularly to thank Mrs. Valerie Hoover for invaluable assistance in preparing the manuscript. The author would also like to thank L. Scott Cram, Dale M. Holm, and John C. Martin for helpful comments, as well as Craig F. Bohren and Donald R. Huffman for the use of their computer programs for the calculations in this chapter. The programs will appear as an appendix in their forthcoming book (Bohren and Huffman, 1982).

REFERENCES

Adams, L. R., and Kamentsky, L. A. (1971) Machine characterization of human leukocytes by acridine orange fluorescence, *Acta Cytol.* **15**:289.

Aden, A. L., and Kerker, M. (1951) Scattering of electromagnetic waves from two concentric spheres, *J. Appl. Phys.* **22**:1242.

Almeida, S. P., and Fujii, H. (1979) Fourier transform differences and averaged similarities in diatoms, *Appl. Opt.* **18**:1663.

Asano, S. (1979) Light scattering properties of spheroidal particles, *Appl. Opt.* **18**:712.

Asano, S., and Sato, M. (1980) Light scattering by randomly oriented spheroidal particles, *Appl. Opt.* **19**:962.

Atkins, P. W., and Barron, L. D. (1969) Rayleigh scattering of polarized photons by molecules, *Mol. Physics* **16**:453.

Barber, P. W., and Yeh, C. (1975) Scattering of electromagnetic waves by arbitrarily shaped dielectric bodies, *Appl. Opt.* **14**:2864.

Barrett, D. L., King, E. B., Jensen, R. H., and Merrill, J. T. (1978) Cytomorphology of gynecologic specimens analyzed and sorted by two-parameter flow cytometry, *Acta Cytol.* **22**:7.

Barrett, D. L., Jensen, R. H., King, E. B., Dean, P. N., and Mayall, B. H. (1979) Flow cytometry of human gynecologic specimens using log chromomycin A_3 fluorescence and lot 90° light scatter, *J. Histochem. Cytochem.* **27**:573.

Barron, L. D., Bogaard, M. P., Buckingham, A. D. (1973) Raman scattering of circularly polarized light by optically active molecules, *J. Am. Chem. Soc.* **95**:603.

Bartholdi, M., Salzman, G. C., Hiebert, R. D., and Kerker, M. (1980) Differential light scattering photometer for rapid analysis of single particles in flow, *Appl. Opt.* **19**:1573.

Bickel, W. S., Davidson, J. F., Huffman, D. R., and Kilkson, R. (1976) Application of polarization effects in light scattering: A new biophysical tool, *Proc. Nat. Acad. Sci. U.S.A.* **73**:486.

Bohren, C. F., and Huffman, D. R. (1982) *Absorption and Scattering of Light by Small Particles,* Wiley, New York (in press).

Bonner, W. A., Hulett, H. R., Sweet, R. G., Herzenberg, L. A. (1972) Fluorescence activated cell sorting, *Rev. Sci. Instrum.* **43**:404.

Born, M., and Wolf, E. (1975) *Principles of Optics,* 5th ed., Pergamon Press, London.

Bracewell, B. M. (1965) *The Fourier Transform and Its Applications,* McGraw-Hill, San Francisco.

Brunsting, A., and Mullaney, P. F. (1974) Differential light scattering from spherical mammalian cells, *Biophy. J.* **14**:439.

Bustamente, C., Maestre, M. F., and Tinoco, I., Jr. (1980a) Circular intensity differential scattering of light by helical structures. I. Theory, *J. Chem. Phys.* **73**:4273.

Bustamente, C., Maestre, M. F., and Tinoco, I., Jr. (1980b) Circular intensity differential scattering of light by helical structures. II. Applications, *J. Chem. Phys.* **73**:6046.

Crowell, J. M., Hiebert, R. D., Salzman, G. C., Price, B. J., Cram, L. S., and Mullaney, P. F. (1978) A light-scattering system for high-speed cell analysis, IEEE Transactions on Biomedical Engineering, **BME-25**:519.

Dave, J. V. (1968) Subroutine for Computing the Parameters of the Electromagnetic Radiation Scattered by a Sphere, International Business Machines Scientific Center, Palo Alto, CA, Report 320-337.

Debye, P. (1909) Der Lichtdruck anf Kugeln von beliebigem Material, *Ann. Physik* **30**:57.

Diamond, L. W., and Braylan, R. C. (1980) Flow analysis of DNA content and cell size in non-Hodkin's lymphoma, *Cancer Res.* **40**:703.

Doukas, J. D., Ruckdeschel, J. C., and Mardiney, M. R., Jr. (1977) Quantitative and qualitative analysis of human lymphocyte proliferation to specific antigen *in vitro* by use of the helium neon laser, *J. Immunol. Meth.* **15**:229.

Druger, S. D., Kerker, M., Wang, D. S., and Cooke, D. D. (1979) Light scattering by inhomogeneous particles, *Appl. Opt.* **18**:3888.

Fikioris, J. G., and Uzunoglu, N. K. (1979) Scattering from an eccentrically stratified dielectric sphere, *J. Opt. Soc. Am.* **69**:1359.

Fowler, B. W., and Sung, C. C. (1979) Scattering of an electromagnetic wave from dielectric bodies of irregular shape, *J. Opt. Soc. Am.* **69**:756.

Frost, J. K., Tyrer, H. W., Pressman, N. J., Albright, C. D., Vansickel, M. H., and Gill, G. W. (1979) Automatic cell identification and enrichment in lung cancer. I. Light scatter and fluorescence parameters, *J. Histochem. Cytochem.* **27**:545.

Genter, F. C., and Salzman, G. C. (1979) A statistical approach to the classification of biological cells from their diffraction patterns, *J. Histochem. Cytochem.* **27**:268.

Goldschneider, I., Gordon, L. K., and Morris, R. J. (1978) Demonstration of Thy-1 antigen on pluripotent hemopoietic stem cells in the rat, *J. Exp. Med.* **148**:1351.

Goldschneider, I., Metcalf, D., Battye, F., and Mandel, T. (1980a) Analysis of rat hemopoietic cells on the fluorescence activated cell sorter, *J. Exp. Med.* **152**:419.

Goldschneider, I., Metcalf, D., Mandel, T., and Bollum, F. J. (1980b) Analysis of rat hemopoietic cells on the fluorescence-activated cell sorter. II. Isolation of terminal deoxynucleotidyl transferase-positive cells, *J. Exp. Med.* **152**:438.

Goodman, J. W. (1968) *Introduction to Fourier Optics,* 1st ed., McGraw-Hill, San Francisco.

Hoffman, R. A., Kung, P. C., Hansen, W. P., Goldstein, G. (1980) Simple and rapid measurement of human T lymphocytes and their subclasses in peripheral blood, *Proc. Nat. Acad. Sci. U.S.A.* **77**:4914.

Hunt, A. J., and Huffman, D. R. (1973) A new polarization-modulated light-scattering instrument, *Rev. Sci. Instrum.* **44**:1753.

Julius, M. H., Sweet, R. G., Fatham, C. G., and Herzenberg, L. A. (1975) Fluorescence activated cell sorting and its application, in: *Mammalian Cells: Probes and Problems* (C. R. Richmond, D. F. Petersen, P. F. Mullaney, and E. C. Anderson, eds.), ERDA Symposium Series CONF-731007, Tech. Information Center, Oak Ridge, TN, p. 107.

Kamentsky, L. A., and Melamed, M. R. (1965) Spectrophotometer: New instrument for ultrarapid cell analysis, *Science* **150**:630.

Kerker, M. (1969) *The Scattering of Light and Other Electromagnetic Radiation,* 1st ed., Academic Press, New York.

Kerker, M., Cooke, D. D., Chew, H., and McNulty, P. J. (1978) Light scattering by structured spheres, *J. Opt. Soc. Am.* **68**:592.

Koch, A. L. (1968) Theory of the angular dependence of light scattered by bacteria and similar-sized biological objects, *J. Theor. Biol.* **18**:133.

Kopp, R. E., Lisa, J., Mendelsohn, J., Pernick, B., Stone, H., and Wohlers, R. (1974) The use of coherent optical processing techniques for automatic screening of cervical cytologic samples, *J. Histochem. Cytochem.* **22**:598.

Kopp, R. E., Lisa, J., Mendelsohn, J., Pernick, B., Stone, H., and Wohlers, R. (1976) Coherent optical processing of cervical cytologic samples, *J. Histochem. Cytochem.* **24**:122.

Kratohvil, J. P. (1966) Calibration of light scattering instruments. IV. Corrections for reflection effects, *J. Colloid Interface Sci.* **21**:498.

Latimer, P., Brunsting, A., Pyle, B. E., and Moore, C. (1978) Effects of asphericity on single particle scattering, *Appl. Opt.* **17**:3152.

Lindmo, T., and Steen, H. B. (1979) Characteristics of a simple, high-resolution flow cytometer based on a new flow configuration, *Biophys. J.* **28**:33.

Loken, M. R., and Herzenberg, L. A. (1975) Analysis of cell populations with a fluorescence activated cell sorter, *Ann. N.Y. Acad. Sci.* **254**:163.

Loken, M. R., and Houck, D. W. (1981) Light scattered at two wavelengths can discriminate viable lymphoid cell populations on a fluorescence activated cell sorter, *J. Histochem. Cytochem.* **29**:609.

Loken, M. R., Sweet, R. G., and Herzenberg, L. A. (1976) Cell discrimination by multiangle light scattering, *J. Histochem. Cytochem.* **24**:284.

Lorenz, L. (1890) Lysbevaegelse i uden for en auf plane lysbelger belyst Kugle, *Vidensk. Selsk. Skrifter* **6**:63.

Ludlow, I. K., and Kaye, P. H. (1979) A scanning diffractometer for rapid analysis of microparticles and biological cells, *J. Colloid Interface Sci.* **69**:571.

Maestre, M. F., and Reich, C. (1980) Contribution of light scattering to the circular dichroism of deoxyribonucleic acid films, deoxyribonucleic acid-polylysine complexes, and deoxyribonucleic acid particles in ethanolic buffers, *Biochemistry* **19**:5214.

Meyer, R. A. (1979) Light scattering from biological cells: Dependence of backscattering radiation on membrane thickness and refractive index, *Appl. Opt.* **18**:585.

Meyer, R. A., Haase, S. F., Podulso, S. W., and McKhan, G. M. (1974) Light scattering patterns of isolated oligodendroglia, *J. Histochem. Cytochem.* **22**:594.

Mie, G. (1908) Beiträge zur Optik trüber Medien, speziell kolloidaler Metallösungen. *Ann. Physik* **25**:377.

Morris, S. J., Schultens, H. A., Hellweg, M. A., Striker, G., and Jovin, T. M. (1979) Dynamics of structural changes in biological particles from rapid light scattering measurements, *Appl. Opt.* **18**:303.

Mueller, H. (1948) The foundations of optics, *J. Opt. Soc. Am.* **38**:661.

Mullaney, P. F., and Dean, P. N. (1969) Cell sizing: A small-angle light scattering method for sizing particles of low relative refractive index, *Appl. Opt.* **8**:2361.

Mullaney, P. F., and Dean, P. N. (1970) The small angle light scattering of biological cells. Theoretical considerations, *Biophys. J.* **10**:764.

Mullaney, P. F., and Fiel, R. J. (1976) Cellular structure as revealed by visible light scattering: Studies on suspensions of red blood cell ghosts, *Appl. Opt.* **15**:310.

Mullaney, P. F., and West, W. T. (1973) A dual-parameter flow microfluorometer for rapid cell analysis, *J. Phys. E.* **6**:1006.

Mullaney, P. F., Van Dilla, M. A., Coulter, J. R., and Dean, P. N. (1969) Cell sizing: A light scattering photometer for rapid volume determinations, *Rev. Sci. Instrum.* **40**:1029.

Mullaney, P. F., Crowell, J. M., Salzman, G. C., Martin, J. C., Hiebert, R. D., and Goad, C. A. (1976) Pulse-height light scatter distributions using flow-systems instrumentation, *J. Histochem. Cytochem.* **24**:298.

Nicola, N. A., Burgess, A. W., Metcalf, D., and Battye, F. L. (1978) Separation of mouse bone marrow cells using wheat germ agglutinin affinity chromatography, *Austr. J. Exp. Biol. Med. Sci.* **56**:663.

Nicola, N. A., Burgess, A. W., Staber, F. G., Johnson, G. R., Metcalf, D., and Battye, F. L. (1980) Differential expression of lectin receptors during hemopoietic differentiation: Enrichment for granulocyte-macrophage progenitor cells, *J. Cell. Physiol.* **103**:217.

Pernick, B., Jost, S., Herold, R., Kopp, R. E., Mendelsohn, J., and Wohlers, R. (1978a) Screening of cervical cytological samples using coherent optical processing. Part 3. *Appl. Opt.* **17**:43.

Pernick, B., Kopp, R. E., Lisa, J., Mendelsohn, J., Stone, H., and Wohlers, R. (1978b) Screening of cervical cytological samples using coherent optical processing. Part 1, *Appl. Opt.* **17**:21.

Pernick, B., Wohlers, M. R., and Mendelsohn, J. (1978c) Paraxial analysis of light scattering by biological cells in a flow system. *Appl. Opt.* **17**:3205.

Perry, R. J., Hunt, A. J., and Huffman, D. R. (1978) Experimental determinations of Mueller scattering matrices for nonspherical particles, *Appl. Opt.* **17**:2700.

Price, B. J., Kollman, V. H., and Salzman, G. C. (1978) Light-scatter analysis of microalgae. Correlation of scatter patterns from pure and mixed asychronous cultures, *Biophys. J.* **22**:29.

Purcell, E. M., and Pennypacker, C. R. (1973) Scattering and absorption of light by nonspherical dielectric grams, *Astrophys. J.* **186**:705.

Reich, C., Maestre, M. F., Edmondson, S., and Gray, D. M. (1980) Circular dichroism and fluorescence-detected circular dichroism of deoxyribonucleic acid and poly[d(A-C).d(G-T)] in ethanolic solutions: A new method for estimating circular dichroic differential scattering, *Biochemistry* **19**:5208.

Salzman, G. C., Crowell, J. M., Goad, C. A., Hansen, K. M., Hiebert, R. D., LaBauve, P. M., Martin, J. C., Ingram, M., and Mullaney, P. F. (1975a) A flow system multiangle light-scattering instrument for cell characterization, *Clin. Chem.* **21**:1297.

Salzman, G. C., Crowell, J. M., Martin, J. C., Trujillo, T. T., Romero, A., Mullaney, P. F., and LaBauve, P. M. (1975b) Cell classification by laser light scattering: Identification and separation of unstained leukocytes, *Acta Cytol.* **19**:374.

Salzman, G. C., Mullaney, P. F., and Price, B. J. (1979a) Light-scattering approaches to cell characterization, in: *Flow Cytometry and Sorting,* 1st ed. (M. R. Melamed, P. F. Mullaney, and M. R. Mendelsohn, eds.), John Wiley and Sons, New York.

Salzman, G. C., Wilder, M. E., and Jett, J. H. (1979b) Light scattering with stream-in-air flow systems, *J. Histochem. Cytochem.* **27**:264.

Schafer, I. A., Jamieson, A. M., Petrelli, M., Price, B. J., and Salzman, G. C. (1979) Multiangle light scattering flow photometry of cultured human fibroblasts: Comparison of normal cells with a mutant line containing cytoplasmic inclusions, *J. Histochem. Cytochem.* **27**:359.

Seger, G., Achatz, M., Heinze, W., and Sinsel, F. (1977) Quantitative extraction of morphologic cell parameters from the diffraction pattern, *J. Histochem. Cytochem.* **25**:707.

Steen, H. B. (1980) Further developments of a microscope based flow cytometer: Light scatter detection and excitation intensity compensation, *Cytometry* **1**:26.

Steen, H. B., and Lindmo, T. (1979) Flow cytometry: A high resolution instrument for everyone, *Science* **204**:403.

Steinkamp, J. A., Fulwyler, M. J., Coulter, J. R., Hiebert, R. D., Horney, J. L., and Mullaney, P. F. (1973) A new multiparameter separator for microscopic particles and biological cells, *Rev. Sci. Instrum.* **44**:1301.

Steinkamp, J. A., Hansen, K. M., Wilson, J. S., and Salzman, G. C. (1977) Automated analysis and separation of cells from the respiratory tract: Preliminary characterization studies in hamsters, *J. Histochem. Cytochem.* **25**:892.

Stohr, M., and Futterman, G. (1979) Visualization of multidimensional spectra in flow cytometry, *J. Histochem. Cytochem.* **27**:560.

Stratton, J. A. (1941) *Electromagnetic Theory,* 1st ed., McGraw-Hill, New York.

Swartzendruber, D. E., Price, B. J., and Rall, L. B. (1979) Multiangle light-scattering analysis of murine teratocarcinoma cells, *J. Histochem. Cytochem.* **27**:366.

Thompson, R. C., Bottinger, J. R., and Fry, E. S. (1980) Measurement of polarized light interactions via the Mueller matrix, *Appl. Opt.* **19**:1323.

Turke, B., Seger, G., Achatz, M., Scelan, W. V. (1978) Fourier optical approach to the extraction of morphological parameters from the diffraction pattern of biological cells, *Appl. Opt.* **17**:2754.

van de Hulst, H. C. (1957) *Light Scattering by Small Particles,* 1st ed., John Wiley and Sons, New York.

van den Engh, G., and Visser, J. (1979) Light scattering properties of pluripotent and committed haemopoietic stem cells, *Acta Haematol.* **62**:289.

van den Engh, G., Visser, J., and Trask, B. (1979) Identification of CFU-s by scatter measurements on a light activated cell sorter, in: *Experimental Hematology Today 1979* (S. J. Baum and G. D. Ledney, eds.), Springer-Verlag, New York, p. 19.

van den Engh, G., Visser, J., Bol, S., and Trask, B. (1980) Concentration of hemopoietic stem cells using a light-activated cell sorter, *Blood Cells* **6**:1.

Visser, J. W. M., Cram, L. S., Martin, J. C., Salzman, G. C., and Price, B. J. (1978a) Sorting of a murine granulocytic progenitor cell by use of laser light scattering measurements, in: *Pulse-Cytophotometry, Part III* (D. Lutz, ed.), European Press, Ghent, Belgium, p. 187.

Visser, J., Haaijman, J., and Trask, B. (1978b) Quantitative immunofluorescence in flow cytometry, in: *Immunofluorescence Related Staining Techniques* (W. Knapp, K. Holubar, and G. Wick, eds.), North-Holland Biomedical Press, Amsterdam.

Visser, J. W. M., van den Engh, G. J., and van Bekkum, D. W. (1980) Light scattering properties of murine hemopoietic cells, *Blood Cells* 6:391.

Wang, D. S., and Barber, P. W. (1979) Scattering by inhomogeneous nonspherical objects, *Appl. Opt.* 18:1190.

Waterman, P. C. (1971) Symmetry, unitarity, and geometry in electromagnetic scattering, *Phys. Rev.* D3:825.

Weil, H., and Chu, C. M. (1976) Scattering and absorption of electromagnetic radiation by thin dielectric disks, *Appl. Opt.* 15:1832.

Welch, R. M., and Cox, S. K. (1978) Nonspherical extinction and absorption efficiencies, *Appl. Opt.* 17:3159.

Wohlers, R., Mendelsohn, J., Kopp, R. E., and Pernick, B. (1978) Screening of cervical cytological samples using coherent optical processing. Part 2, *Appl. Opt.* 17:35.

Wyatt, P. F. (1968) Differential light scattering: A physical method for identifying living bacterial cells, *Appl. Opt.* 7:1879.

Zerull, R. H., Giese, R. H., and Weiss, K. (1977) Scattering functions of nonspherical dielectric and absorbing particles vs Mie theory, *Appl. Opt.* 16:777.

6

Methods for Measuring Leukocyte Locomotion

P. C. WILKINSON, J. M. LACKIE, AND R. B. ALLAN

I. INTRODUCTION

The most direct way to study the movement of any object or living thing would seem to be to watch it moving. Strangely, where leukocytes are concerned, watching cells move has been something of a minority interest. The reasons for this are partly historical, partly that more people have been interested in the way that chemical substances modify leukocyte locomotion than in the locomotion itself, and the direct approach has not always been the most helpful for answering questions about such modifications. In this methodological review, we have deliberately placed much emphasis on visual assays. We feel that an understanding of locomotory behavior is fundamental to an understanding of cellular reactions such as chemotaxis and chemokinesis, and that some of the conceptual confusions which arise in the study of these reactions can be avoided if results obtained with filter assays or agarose assays are interpreted in the context of a detailed knowledge of the way leukocytes actually move. The use of a variety of assays, each of which provides information which is not fully provided by the others, seems increasingly important if balance is to be maintained in a field which is becoming highly popular. The study of leukocyte chemotaxis is attracting workers from a range of disciplines—both basic and clinical—some of whom may be fairly unfamiliar with the locomotory behavior of cells. We hope that this review will give them a feeling for the possible exper-

P. C. WILKINSON • Department of Bacteriology and Immunology, Western Infirmary, University of Glasgow, Glasgow G11 6NT, U.K.   J. M. LACKIE AND R. B. ALLAN • Department of Cell Biology, University of Glasgow, Glasgow G12 8QQ, U.K.

imental approaches that can be used to answer questions about leukocyte locomotion.

Leukocytes are of three major classes. Neutrophil leukocytes (polymorphonuclear leukocytes, PMNs) are the major representatives of the myeloid
cell series, the mature forms of which are frequently known as granulocytes.
They are typically present in large numbers in acute inflammatory lesions, are
actively phagocytic and microbicidal, and form the first line of defense in
pyogenic bacterial infections. They also have a scavenger function in removing
damaged tissue. Almost all the information about locomotion and chemotaxis
of leukocytes has been obtained using neutrophils. They are easy to purify in
large numbers and their locomotion is more vigorous than that of any other
leukocyte type. Most of the assays described here were developed and are
chiefly used for studies of neutrophils. These assays can also be used for the
study of other leukocyte types.

The eosinophil leukocyte is another type of granulocyte, though it now
appears that it may develop from a stem cell line different from the neutrophil.
It probably has a special function in killing large metazoan parasites by secre-
tion of damaging material rather than by phagocytosis. Eosinophil chemotaxis
has been studied chiefly using the filter assay and there is little visual information on eosinophil locomotion. The mononuclear phagocyte line includes a
wide variety of cells—including motile blood monocytes and macrophages,
about the locomotion of which there is still a great deal to be learned—as well
as fixed macrophages that are a permanent part of the architecture of organs
such as the liver and spleen and others and that may serve as phagocytic filters
to remove microorganisms and unwanted particles circulating in blood or
lymph. The motile macrophages are prominent in many types of chronic
inflammation. Finally, there are the lymphocytes, the cells of central interest
in immunology. Lymphocytes are highly motile cells, though little is known
and much is to be learned of how their locomotor properties are related to their
complex roles as initiators, effectors, and modulators of specific immune reactions. Quite a number of visual studies have been made of lymphocyte or lymphoblast locomotion, though we know rather little about their responses to
chemical attractants.

The movement of leukocytes, like that of other cells, is modified by the
environment of the cell (Fig. 1). There is good experimental evidence that the
direction of locomotion is determined by such substances *(chemotaxis)*. Chemical substances may also determine the rate of leukocyte locomotion. This is a
form of *chemokinesis,* known as *orthokinesis.* Thus a major aim of assays of
leukocyte locomotion is to test this locomotion under conditions where it can
be modified by chemical substances, e.g., either under conditions of uniform
concentration, in which chemokinesis (orthokinesis) can be studied, or by
exposure of the cells to concentration gradients in which chemotaxis can be

studied. Other reactions that determine cell locomotion may be physical rather than chemical. Thus locomotion of cells may be modified by the spatial organization of tissues through which they move, e.g., alignment and patterning of tissues, tissue contours, three-dimensional matrices *vis-à-vis* planar surfaces. These reactions come under the heading of *contact guidance*. They have been little studied in leukocytes. Cells may also show *contact inhibition of locomotion* on contact with other cells of the same or of other types.

Assays of leukocyte locomotion can be divided into two types. First, there are those visual assays in which the locomotory behavior of a small number of cells is observed in detail by sequential photography, by time-lapse cinematography, or simply by sitting and watching. It is almost impossible to quantify cell locomotion by just watching, and a photographic record of the experiment is usually required. Second, there are those assays in which a large population of cells is allowed to move in the presence or absence of substances that modify their locomotion. After appropriate time, the experiment is stopped, usually by fixing the cells, and the distribution of the population in relation to the attractant is measured by any appropriate means. In practice, the population assays tend to reveal rather less of the diversity of behavior that the cells exhibit since the end point scored (the leading front, for example) is more a function of the behavior of the most active cells than of the general population capability. The small population studied in direct visual assays may well be a more representative random sample.

Of the assay systems to be described here, the visual assays were the first to be developed. Comandon (1917, 1919) used time-lapse cinematography to study the locomotion of blood leukocytes toward red cells infected with plasmodia and described a directional locomotion of the leukocytes toward such infected cells. Several groups made studies of the morphology, speed, and response to gradients, of leukocytes during the succeeding years (see, for example, Lewis, 1931, 1934; McCutcheon, 1946; H. Harris, 1954), but little progress was made during this time in understanding the biochemistry and pharmacology. These assays established that leukocytes migrated on surfaces and that they adopted a polarized morphology as they moved, with a broad hyaline, organelle-free leading edge with ruffles, and with a tapered shape and frequently with a tail and posterior retraction fibers. The speed of movement was variable, up to 30 μm per minute. Cells moving at random took sinuous paths, but, in chemotactic gradients, they turned less frequently and moved in fairly straight paths toward the gradient source (McCutcheon, 1946; Zigmond, 1974; Allan and Wilkinson, 1978). Dixon and McCutcheon (1935) devised a "chemotactic index"—originally called a "chemotropism ratio" for assessing the chemotactic response of cells to gradients. If the ratio *displacement/total distance traveled* is expressed in relation to a gradient source as *straight-line distance from start to gradient source/total distance traveled by cell from start to*

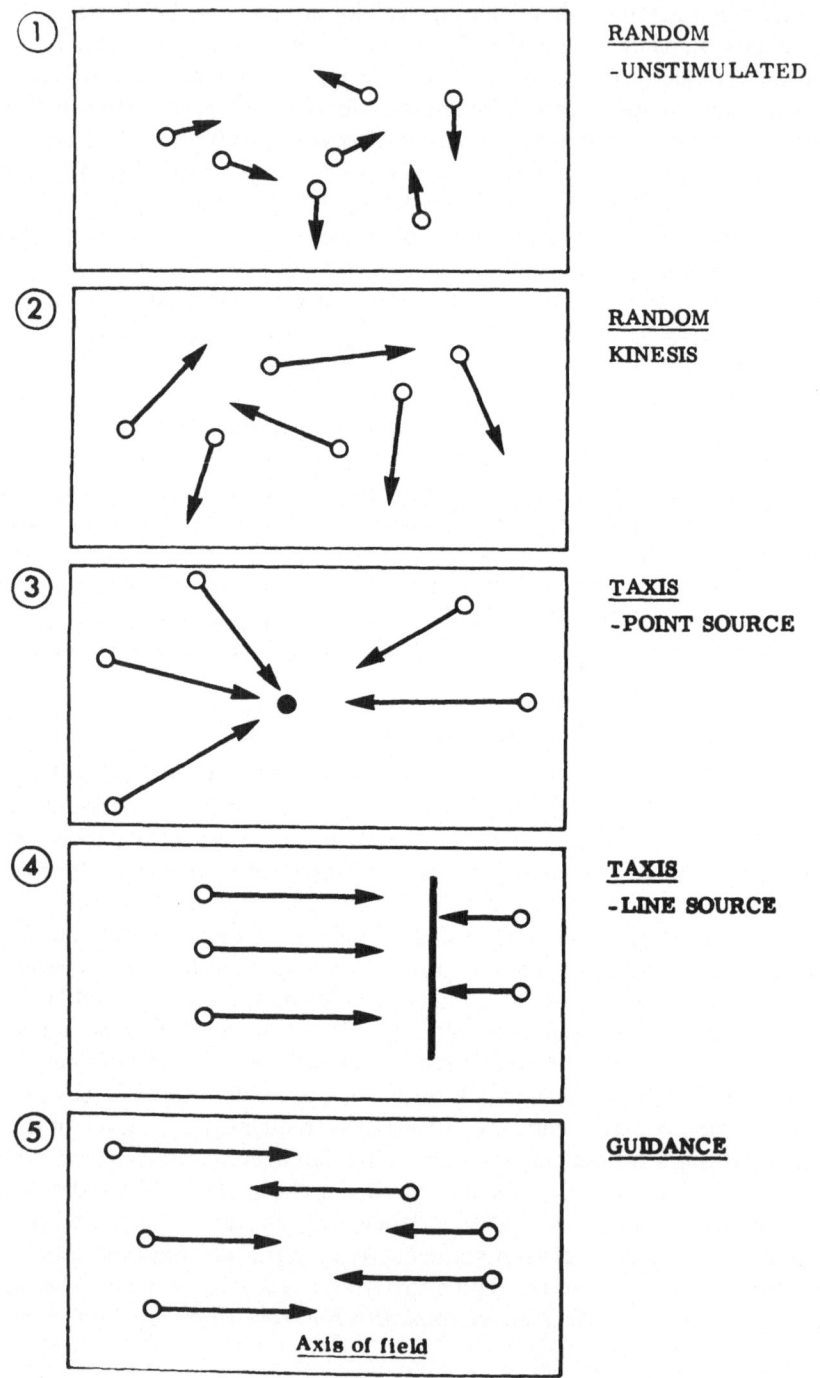

1. RANDOM
   -UNSTIMULATED

2. RANDOM
   KINESIS

3. TAXIS
   -POINT SOURCE

4. TAXIS
   -LINE SOURCE

5. GUIDANCE

Axis of field

gradient source, the limiting values are $+1.0$ for a cell traveling straight to the source and -1.0 for a cell traveling in a straight line away from it. Neutrophils in good gradients can show chemotactic ratios of $+0.8$ to $+0.95$.

In 1962, Boyden introduced the micropore filter assay. This was an assay of quite a different type, being based on measurement of the locomotion of a cell population rather than on the study of individual cells. A two-compartment chamber was used in which cells in the upper chamber were placed on top of a porous filter whose pore size was chosen to allow cells to squeeze through but not to drop through. Below the filter was an attractant. Cells migrated through the filter toward the attractant. After a given time, the experiment was stopped and the cells were fixed and stained. The number of cells reaching a given point (in early versions, the lower surface of the filter) or the distance traveled by the leading front of cells (Zigmond and Hirsch, 1973) was then estimated. This assay gives no information about how cells move; in fact, they are not examined while they are moving, only as a fixed preparation. However, it had the advantage that tests could be set up with large series of chambers with varying concentrations of chemotactic factors, inhibitors, etc., so that quantitative data about the effects of such factors could be obtained much more rapidly than with visual assays. It soon became very popular, and for some 10 years was the only assay in use. Later it was realized that it was a very imperfect way of distinguishing chemotactic from chemokinetic reactions (Zigmond and Hirsch, 1973), and a modification—the checkerboard assay—was introduced to achieve this distinction. During the last decade the agarose assay has been introduced (Cutler, 1974; Nelson *et al.,* 1975). This is a very simple assay in which wells are cut into agarose and appropriate wells can be filled with cells or with chemoattractant solutions. Cell migration toward wells with and without attractants can be measured. Like the filter assay, this is an "end-point" assay, as usually employed, since the experiment is usually stopped, the cells fixed, and their distribution examined. However, it can be adapted as a visual assay, since cells moving under agarose can be filmed.

Since there are a number of different assay techniques available, the investigator wishing to study leukocyte locomotion might ask which is appro-

---

FIGURE 1. A series of schematic diagrams to illustrate the differences between cells moving under various conditions. The lengths and orientations of the lines are intended to represent the extent and direction of displacement of cells. Positive kinesis, as in (2), enhances the extent but does not affect the direction of displacement. Guidance, as in (5), may also enhance displacement, but does so by restricting the angles of turn rather than by increasing speed. Chemotactic factors tend also to be chemokinetic, but the direction of movement is modified [(3) and (4)]. The nature of the source must be considered in analyzing displacement vectors; over the whole field the displacement vectors in example (3) may appear random although it is clear that the cells are moving directionally (see later).

priate to his/her particular problem. Some guidelines are therefore given below.

Visual assays are the best for studying the detailed behavior of moving cells. Details of the changes in shape that accompany movement on two-dimensional surfaces or three-dimensional matrices can be observed, and the paths taken by cells can be tracked and their turning behavior measured. Accurate measurements of cell velocity and cell speed can be obtained. The interaction of cells with other cells can be observed in detail, e.g., leukocytes with one another, or with vascular endothelium, fibroblasts, etc. Similarly, the interaction of cells with surfaces coated with various materials (e.g., serum, purified proteins) or with substrata of various shapes (e.g., planar, curved, grooved, ridged) can be observed. Direct observations can be made of the influence of sources of chemotactic gradients or of the presence of chemokinetic materials on cell morphology, speed, and turning. However, visual assays are time-consuming, and it is not easy to obtain information about the behavior of large populations of cells. It is not easy to obtain important information such as the dose–response of cells to a chemotactic factor, since the sampling error is high, apart from the fact that an impossibly large number of film sequences would need to be analyzed. Thus biochemical and pharmacological information is usually better acquired using a population assay.

End-point assays using large cell populations (filter, agarose assays) give poor information about the way locomotor cells behave. They have really come into their own, not as assays of locomotion *per se,* but as assays of how locomotion is influenced by the chemical environment of the cell. Thus a principle of these assays is that the cells are separated from a source of some chemical substance, and the behavior of the cell population in the presence or absence of that substance, or in different concentrations of that substance, is assayed. Almost all the leukocyte chemotactic factors were first identified and characterized using such assays. Similarly, substances that inhibit leukocyte chemotaxis have usually been identified in the same way. Such assays are preferable to visual assays for studying pharmacological and biochemical problems in which detailed dose–response data are required. They are easy to use, and many experiments can be set up on the same day. Two major drawbacks that have become apparent since they were introduced are, first, that it is possible to do many experiments on locomotion without ever looking at a living cell, a situation that can easily give rise to misconceptions about cell locomotion, and, second, that these assays do not give an unequivocal distinction of chemokinesis from chemotaxis.

Since both types of assay have their advantages and disadvantages, it would seem rational to use both side by side, and we devote considerable space to both in the remainder of this chapter.

II. VISUAL ASSAYS

A. Time-Lapse Filming

1. Principle

Even for rapidly moving cells such as leukocytes, direct observation of cells in real time is impractical; some means of speeding things up is essential. The most commonly adopted method is to take photographs at fixed time intervals and then to replay the film at cinematographic speed—a time-lapse movie, in fact. Since we are dealing with microscopic objects, the camera must be mounted on a microscope and the subjects of our film must be kept at 37°C. To avoid tedium it is normal to have an automatic timing system which opens the shutter at a preset interval and then moves the film on; but heroic films have been taken without these aids. Essentially the same thing can now be done with video-tape recording, and we will discuss this at length later.

The aim of taking a time-lapse film is to capture, at an appropriate magnification, a reasonably large number of cells moving about under conditions which can be defined and manipulated. The speed-up must not be so great as to sacrifice detail—cells should be seen to move continuously rather than to hop like fleas—but not so small as to require several reels of film per sequence. The purpose of the exercise should never be forgotten. These films are not home movies but a means of obtaining information, and the analysis of a film sequence is probably the most difficult part of the whole process. Anecdotal film footage exists almost everywhere and enlivens many a stuffy conference hall, but correctly analyzed, a film can tell us a great deal about cell locomotion. Thus the first question to be asked should always be, "What feature of the movement do we want to measure?" Once the film is taken, it is too late to alter the magnification, the lapse interval, and so on. These may seem obvious points, but they are frequently forgotten.

In the following pages we have attempted to indicate the basic methodology in sufficient detail to assist a complete novice in setting up visual assays, but we cannot give an exhaustive guide to commercially available equipment. An excellent recent monograph (Riddle, 1979) gives more detail in this respect. The equipment we use or know well we mention where appropriate; but offer no hostages to fortune in giving specific recommendations. The question of film analysis, despite the comments in the preceding paragraph, is left until the methods have been described.

2. Equipment

a. Microscopes and Accessories. An obvious requirement is a good microscopic system with, at the least, phase-contrast optics. Normal or inverted microscopes can be used; the particular application may dictate which one is most suitable. Unless very detailed observations of individual cells are important, fairly low-power objectives are suitable and have the advantage of greater focal depth and working distance. Most of our filming has been done with ×10, ×16, or ×25 objectives, and a range of objectives and eyepieces can be a help in achieving a compromise between detail and a representative sample of cells from the population. Long working distance objectives and condensers are necessary in some applications. Phase-contrast optics are generally quite satisfactory, but for analysis of collisions between cells, especially at high magnification, there are some advantages in using differential-interference-contrast optics (the Nomarski optics of Zeiss), which give sharply defined images of cell margins without the phase haloes which might obscure the contact zone (Fig. 2). These advantages must be balanced against cost, against the rather short working distance (even with the long-working-distance Leitz system), and

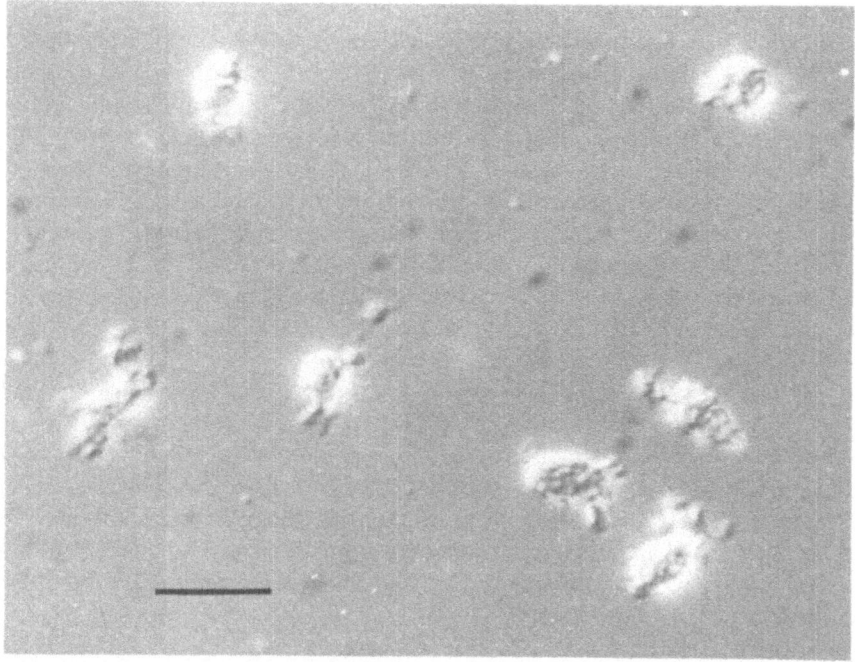

FIGURE 2. Human polymorphonuclear neutrophil leukocytes moving chemokinetically in casein (1mg/ml) as observed with Nomarski interference-contrast optics. This can be compared with the appearance of cells with conventional phase-contrast optics (Fig. 12). Bar = 20 μm.

against the major disadvantage that plastic substrata cannot be used because of strain-induced polarization in the plastic. Aesthetically, phase contrast is poorer. It should be self-evident that transposing an image to film will never improve the resolution, and the better the optical system the better the resulting film. Good optics depend to some extent upon the geometry and nature of the filming chamber. Possibly the ideal system is that in which the cells are moving over the ceiling of a chamber, the ceiling being a coverslip. Debris will fall off the substratum and the optical path is uncluttered. Nonadherent cells will also fall, however, and for this reason it may be better to use an inverted microscope. For many purposes we use simple chambers, shown in Fig. 3, made from a stainless steel slide with a "dry" coverslip sealed on with silicone grease and a "wet" coverslip held on with a hot wax–Vaseline (3:2) mixture. The "wet" coverslip can be one upon which cells have been grown, for example. These small chambers have a low thermal capacity, and the problem of con-

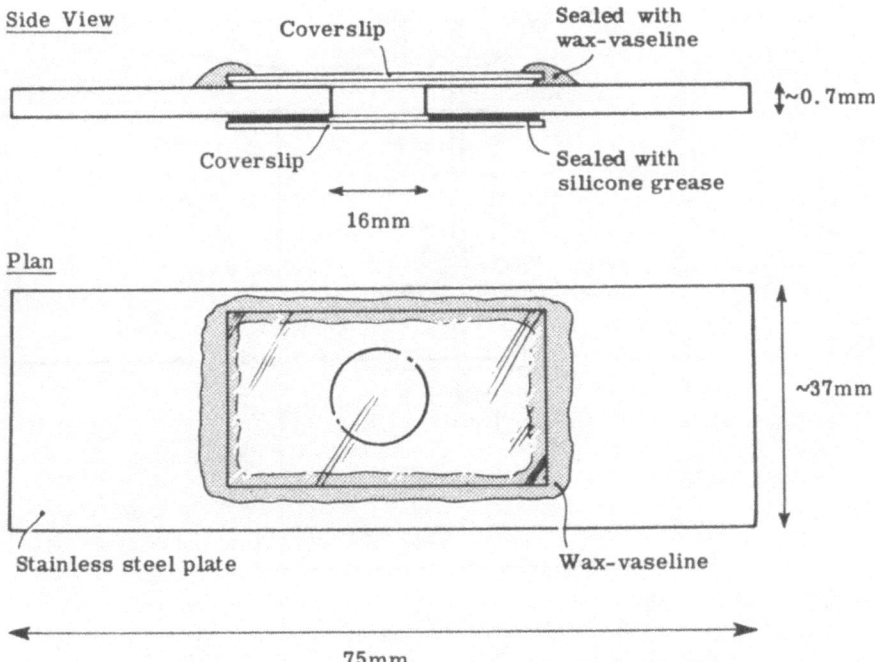

FIGURE 3. Diagram of the specimen chamber we normally use for high-resolution filming.

vection currents within the medium is minimized. Such chambers can be modified with inlet and outlet ports, can be partially filled with gel-like materials, or can have variously coated coverslips forming the substratum for movement. Other filming chambers can be devised for filming cells moving under agarose, moving in orientation chambers (see Sections II-D, II-E, and IV) or even in collagen gels. All these types of filming have been carried out successfully, and the only limit seems to be the ingenuity and dexterity of the experimenter.

Warm stages of various kinds are available to keep the chamber at 37°C; we have found it convenient (and preferable) to use an air-curtain incubator— basically a fan heater directed at the microscope stage, regulated via a proportional controller linked to a miniature-bead thermistor on the stage (Figs. 4 and 5). Proportional controllers are better than "on–off" switching of the heating element. We use a home-made device, based upon a design by G. A. Dunn, but commercial equivalents are available. Air-curtain systems seem to

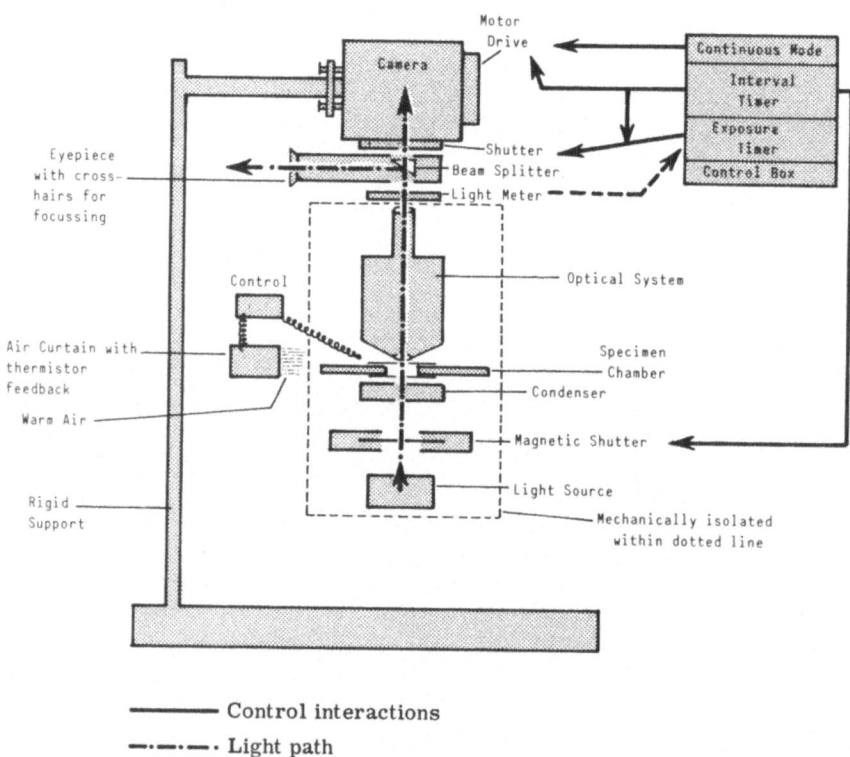

FIGURE 4. Schematic diagram of the time-lapse cinematography system, relating to the equipment shown in Fig. 5.

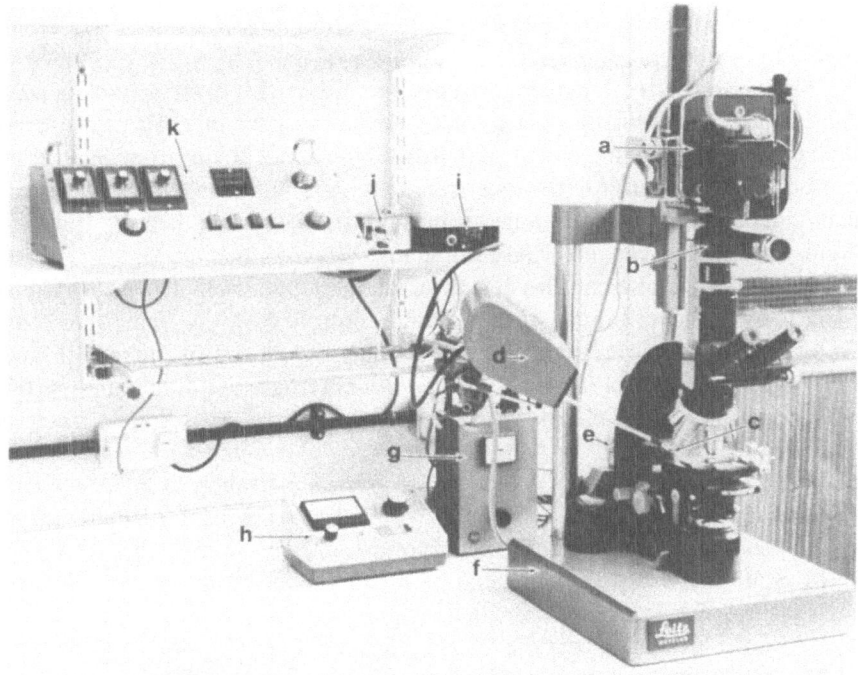

FIGURE 5. Time-lapse cinematography equipment. This system is built around a Leitz Ortholux microscope with phase- and interference-contrast optics. (a) Bolex camera with motor drive; (b) beam splitter and focusing eyepiece with integral light sensor; (c) thermistor taped to thermometer; (d) fan heater; (e) magnetic shutter; (f) stand; (g) light-source rheostat; (h) light meter; (i) magnetic shutter control (j) heater control; (k) intervalometer.

give a more uniform temperature field than warm stages, and the problems of convection are reduced. Control to within $\pm 0.1\,°C$ is easily achieved.

Local heating of the chamber by the illumination system of the microscope is liable to cause problems, and cells often appear to be less normal in morphology in the area that has been filmed. A heat filter in the light path may help, as will keeping the level of illumination as low as possible. For long-term filming it may be better to introduce a shutter into the light path, activated to open only when an exposure is to be made. A manual override is, of course, necessary for setting up.

 b. Camera and Accessories (Figs. 4 and 5). The microscope image must now be transferred to film. Various arrangements of beam splitter can be used and an eyepiece with cross hairs and an indication of frame size is essential. Without a correcting eyepiece in the optical plane of the film, focusing is impossible. Some beam splitters used for this purpose also incorporate an exposure meter which can be swung into position; this is the most convenient method but not the only one.

Two alternative linkages are possible: either the camera and its beam splitter are mechanically isolated from the microscope or the two are firmly linked. The latter arrangement may lead to mechanical vibration being transmitted from the shutter to the specimen, so the former arrangement of mechanical independence is preferable. Vibration from the fan heater, traffic, or clumsy colleagues can also cause problems.

Last, but by no means least, is the camera system itself. The basic requirement is for a movie camera with a motor, controlled, in the single-shot mode, by an external timer system. Various systems are marketed, some with automatic exposure control (Nikon), others with external control of exposure time and lapse interval (Paillard-Wild). Because the market is small, "intervalometers" (as the control systems are often known) are expensive. It should not be beyond the capability of a modest electronics workshop to make a control system that sends "open" and "shut" signals to the camera motor at set times—thus determining both the exposure and the interval between exposures. Such homemade systems can be much cheaper, and we have used two such systems for several years.

Almost without exception, 16-mm film has been the preferred medium; 16-mm movie cameras are not cheap, but they tend to be built to rather exacting professional specifications, and we have had trouble-free service from Bolex cameras. Rarely, 8-mm- and 35-mm-film formats have been used, but the availability of 16-mm camera systems and suitable projectors for analysis of film makes the choice obvious. For most purposes, using phase-contrast optics, monochrome film is quite adequate, and the use of a reversal film avoids transposition. We have generally chosen to use a rather slow film (Kodak Plus-X reversal) because it is possible to print from single frames. Processing of movie film is inconvenient unless the appropriate equipment is at hand and a few specialist firms still process reversal film. The alternative is color film.

c. Video Systems. The alternative to 16-mm film is to use a television camera and a video-tape recorder (Fig. 6). Monochrome television cameras are now much cheaper than 16-mm movie cameras and will work at comparable or even lower light levels without significant loss of resolution. The cheapest Vidicon-tube cameras are usable, although it is easy to damage the tube by mechanical jarring or by irreversibly bleaching the tube with excessive light. Chalnicon tube cameras will work at lower light levels, have a better spectral sensitivity than Newvicon tubes, give a lag-free picture, and are less easily damaged; they are (at present) approximately fivefold more expensive than the Vidicon tube but are still cheaper than a movie camera. The most sensitive low-light-level systems incorporating image intensification are much more expensive but will permit recording of fluorescent images without photobleach-

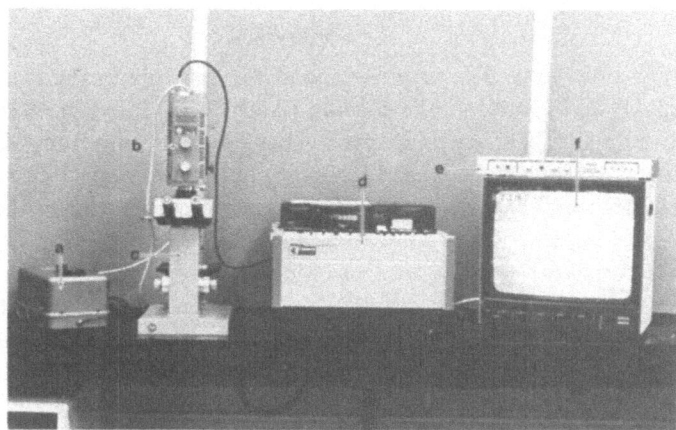

FIGURE 6. Video time-lapse equipment. This system is built around a Leitz Diavert microscope with long-working-distance phase- and interference-contrast optics. For clarity, the air-curtain incubator has been removed. (a) Light-source rheostat; (b) monochrome Chalnicon-tube television camera; (c) microscope; (d) time-lapse VTR (National VTR 8030); (e) time-date generator (For-A VTG-88); (f) video-monitor.

ing. Television monitors are more expensive than their domestic counterparts but are relatively cheap and allow several observers to argue about the image simultaneously, probably a good thing.

Color cameras and monitors are more expensive than monochrome and are unnecessary when phase-contrast optics are being used. The picture quality is good, however, and they are excellent for histological demonstration material.

Time-lapse videotape recorders are now available with a spot-frame facility on playback; although recording leads to a small loss of picture quality, the end result is still excellent.

It is almost essential to have a time-date generator (which displays the "real" time on the monitor) if speeds of movement are of interest; the automatic time base of film frames is absent and the frame counters on videotape recorders are nonlinear and unreliable for accurate timing. The capital outlay on a video system is approximately half that for a 16-mm movie system, the running costs are lower (videotapes are reusable—but not indefinitely), and the facility for immediate playback of a recorded sequence, while not essential, is enjoyable.

Our own experience of videotape recording, and that of colleagues elsewhere, suggests that this is probably the method of choice for a laboratory embarking upon time-lapse work from scratch; the only disadvantage is in analyzing the recorded data (see Section II-C).

3. Analysis and Measurement—Practice

Getting time-lapse film or videotape of moving cells is the least difficult aspect; having immortalized the moving image, how do we translate it into quantitative data? There are undoubtedly many miles of film from which nothing but anecdotal information has been extracted; there is no surer way of getting colleagues away from your time-lapse system than to make them try to analyze their first film.

For 16-mm film an analytical or "stop-action" projector is an essential (and fairly expensive) piece of equipment. Such a machine allows the film to be projected frame by frame or at 1, 2, 4, 8 frames/sec as well as at normal movie speed (16 or 24 frames/sec). We have found the L&W Photo-Optical Data Analyser (or Lafayette, Lafayette Instrument Co., Lafayette, Indiana) fairly satisfactory over several years of use. More expensive systems are of course available, although we have no experience of them. Essentially the film is projected frame by frame onto a paper screen and the position of the cell or cells marked at intervals. A mirror set at 45° to project the image down onto a flat surface makes this a marginally less painful task (Fig. 7). The usual strategy in deciding upon a tracking method is empirical. After the film is watched several times, the chosen method is attempted—and quite frequently modified. The actual analysis will inevitably take time, and the sequence must be repeated for every cell—tracking a lone cell is easy, but collisions can be

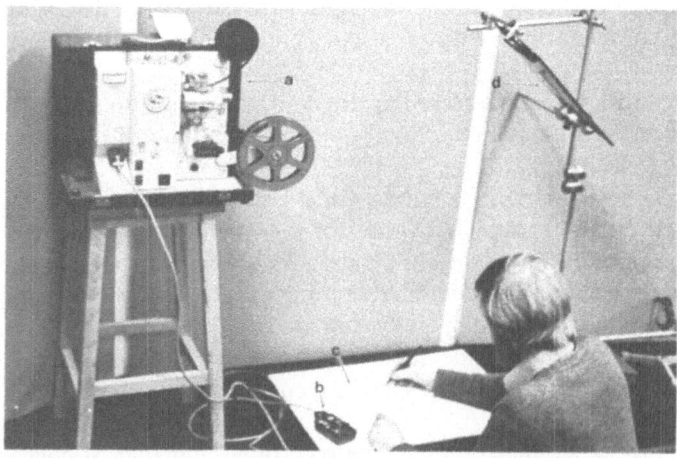

FIGURE 7. Analysis of 16-mm time-lapse film. (a) L-W Photo-optical Data Analyser; (b) remote control; (c) drawing pad; (d) mirror set at angle of 45°

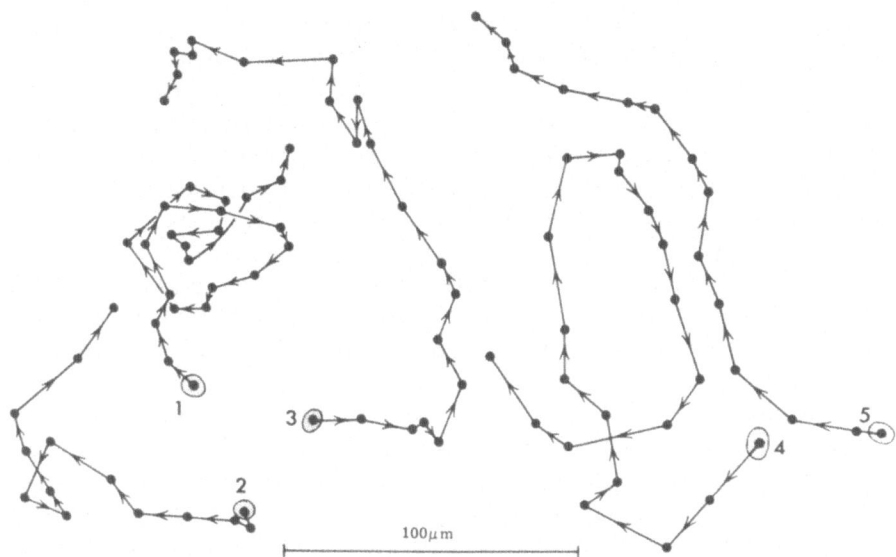

FIGURE 8. Five cell tracks from a film sequence of rabbit peritoneal exudate neutrophils moving in an isotropic environment with 50% exudate fluid in the medium. The glass surface will have been coated with protein from the medium. Single frames were exposed at 10-sec intervals, and the position of the cell every tenth frame was marked. As the paths of cells frequently cross, the tracks have been repositioned for clarity. The tracks are analyzed numerically in Table I. Track 1 shows a circuitous path; track 5 shows cells exhibiting considerable persistence.

very confusing. Various tricks make life easier—colored pens, transparent overlays, and simple codes for common events, for example. The cell track (for examples, see Fig. 8) may then be analyzed either with ruler and protractor, with a map-measurer, or, most conveniently, with a digitizer tablet linked to a computing system. In this last method, a stylus incorporating a microswitch, held over the tablet, is used to "dot" positions on the track and the (x, y) coordinates of each position are recorded automatically. Once one knows the (x, y) coordinates of each point, it is relatively easy to program the computer to calculate the step lengths, the angles of turn, or whatever parameter is of interest. We have used a 30-cm × 30-cm "bit pad" (Summagraphic Fairfield, C.T.) linked to a Commodore PET-2 desk-top computer and, with the assistance of Dr. M. Burns, Department of Zoology, Glasgow University, have devised programs which give mean step lengths, displacements, and so on (see section IIc). Although it would be possible to project the film directly onto the digitizer tablet for analysis, we have transposed tracks onto paper first to give us an easily accessible hard copy. By calibrating the digitizer tablet, the distances moved can be printed out directly in micrometers, and by specifying the scale line as "North–South," vectorial information can be obtained.

The use of this semiautomatic measuring device makes analysis much easier, although it is still necessary to follow individual cells frame by frame. With practice, it is possible to track 10 cells through 250 frames each in 30–60 min to get the track on paper, and 5–10 min to get measurements of the tracks. Operator eye strain vies with boredom in determining the number of sequences analyzed per session.

Fully automatic analysis is clearly an attractive concept, in view of the tedium of the method outlined above. The automatic analysis of static images is, however, a complex procedure and requires relatively sophisticated computing systems (e.g., the Quantimet Image Analysing Computer: Bradbury, 1977); introducing a time element to analyze moving images poses even greater problems. No doubt such automatic systems will eventually become available to research workers in small laboratories, but that time has not yet come. No doubt, too, there will always be those who argue, as we would, that interposing an intrinsically unimaginative interface between reality and data output means that novel observations of cell behavior will be missed. The problem, and the doubt expressed above, is that a relatively short time-lapse sequence encapsulates an enormous amount of information. The trick in analyzing a film is in deciding what pieces of information it is worth attempting to extract—and we will consider this under a separate heading (Section II-C).

An alternative approach to transposing cell tracks from movie film to paper is that used by Ramsey and Harris (1973), who projected their film onto large-format photographic paper. By having the cells bright against a dark background, the moving cells show up as bright, although rather smudgy, tracks on the print. Not having tried this method, we cannot really comment upon its practicality except to note that few people seem to have tried it.

The analysis of videotape is straightforward if, for example, the analysis is of the outcome of collisions, where nonparametric scoring will be used. Measuring tracks or turning angles is more difficult because transposing from screen to paper is not easy. The difficulties are trivial in origin—television screens are not flat but curved in two orthogonal axes, and the magnification at the edges is not the same as in the center. There are also parallax problems. Possibly a projected television image could be used, but we have not tried this. Since the information on a television screen is already digitally coded, it should be possible to gain access to this by interfacing with a computer (this is effectively what the Quantimet does for a direct camera input). Systems that have this capability are probably being developed—the Joyce-Loebl Magiscan Mk.II, for example, may have the capacity to handle videotape input, but the cost of such equipment is likely to be prohibitive.

4. Common Hazards

Time-lapse filming, although basically simple, has various pitfalls. No instruction schedule, however detailed, can prevent some lessons from being learned only by experience. Having permitted undergraduates to use our time-lapse system, we can identify several common problems, and these are listed below with comments.

1. *Sequence too short:* Probably a minimum of 200 frames at a 10-sec lapse interval for neutrophils and 30-sec lapse for fibroblasts. Gaps between sequences are a help.
2. *Lapse interval too long:* Ten seconds is a long time for a neutrophil; 2 sec is better at higher magnifications.
3. *Exposure time too long:* Images are blurred. The minimum on many cameras is 0.2 sec, but this permits adequate exposure at reasonable light levels. It is easier to adjust the light than the exposure time.
4. *Blurred film:* Frequent refocusing is necessary, especially when using high-numerical-aperture objectives. If this fails, the focusing beam splitter is incorrectly positioned.
5. *Unknown magnification factor:* Take a few shots of a stage scale on every film—otherwise, complex calculations must be undertaken every time the projector is moved.
6. *Unknown lapse interval:* Check with stopwatch; record interval used.
7. *Failure to record details of procedure in such a way that sequences can be identified when film finally returns from developer.*
8. *Warm-up time not accounted for.*
9. *Inadequate controls:* Variation between batches of cells is enormous, and a control sequence is essential every time.

B. Indirect (Non-Time-Lapse) Methods

Two other methods for assessing the movement of individual cells have been developed: one—the optoelectronic method—has been used for neutrophils (Dahlgren, 1979), and the other—Albrecht-Buehler's phagokinetic track method—is probably inapplicable (Albrecht-Buehler, 1977). The latter method relies on cells clearing a track through colloidal gold carpeting the substratum. Since phagocytosis affects the adhesive and probably the locomotory properties of neutrophils, it is unlikely that this method would give an adequate measurement of "normal" movement. Preliminary trials (P. Sheterline, personal communication) have been unsuccessful.

The optoelectronic method (Dahlgren, 1979) substitutes, for one eyepiece of a binocular microscope, a grid of 32 × 32 photodiodes, and the output from each of these light-sensitive elements is used to detect movement, the output from each element being compared at 11-sec intervals. A change in light intensity, caused by the image of a cell on the substratum, is detected, and the digitally encoded information is processed by a small computer. Up to six cells could be handled simultaneously. Although this does constitute a fully automatic system, the quality of an image composed of only 1024 dots is inevitably low, and details of cell behavior are lost. Changes in light intensity might also arise because of cells spreading and becoming "phase-dark" as is frequently observed when cells are placed on very adhesive substrata.

C. Analysis of Cell Movement

The preceding sections have concentrated on methods for capturing the image of a moving cell, converting this into a crooked line upon a piece of paper, and measuring the length of the line.

What information should one try to obtain from a film (or videotape) sequence? To some extent this depends on the particular problem being investigated, and there can be no definitive answers. Frequently, however, the information required concerns the rate of movement, the frequency and the angle of turning, and, in an anisotropic environment, whether there is any external bias in the locomotion.

An excellent discussion of the locomotory behavior of cells and forms of directed behavior that might be observed is by Dunn (1981).

1. Movement in Uniform Environments

In an isotropic environment in which leukocytes (or any other cell types) are moving over a plane surface, the locomotion is best described as a random walk with internal bias. That is to say, cells set off with equal probability in any direction, and the path followed owes nothing to external cues. Once having committed itself to a particular direction, however, the cell may well tend to continue in that general direction, and the angles (of recognizable "turns") are normally (for neutrophils) of the order of 50°–60°; both these factors contribute to the pattern of tracks observed. Whether tracks can in fact be considered a series of straight-line segments is not clear, and few workers would follow Peterson and Noble (1972) in attempting to define "real" turns. Behavior is continuous even though our analysis must, perforce, be discontinuous. Near-continuous tracking is possible, especially with videotape recordings, but the whole point of time-lapse filming is lost. For convenience, most tracks are based on "steps" of fixed times—usually 1–2 min (5–10× frames taken at 5–10-sec intervals) for neutrophils, but the choice of time base is a balance between convenience and accuracy. If the position of the cell is marked at 10-frame

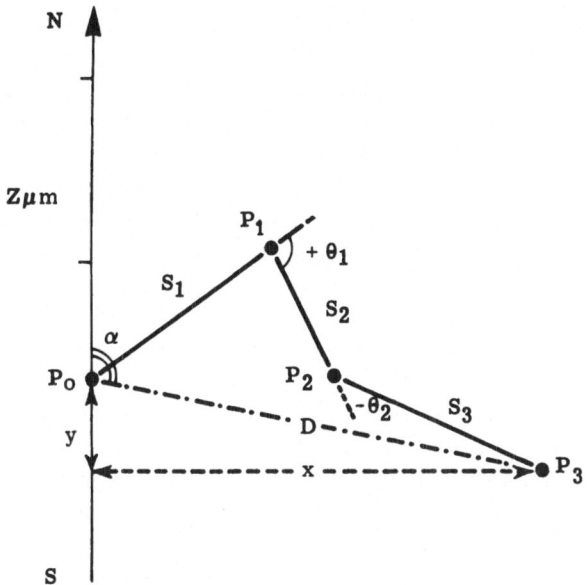

FIGURE 9. Diagram illustrating the information routinely obtained from cell tracks analyzed with the bit-pad/computer system described in the text. An axis N–S is defined with reference to the film frame (arbitrary with respect to the filming chamber unless it is appropriate to define an orientation) and is also the scale bar. Positions at equal time intervals (e.g., every tenth frame) are marked (P_o, P_1–P_n), and each step (S_1, S_2, etc.) is measured; the total step length ($\Sigma_1^n S$) is divided by n, the number of steps, to give the mean step length. Displacement (D) is computed from the coordinates of P_o and P_n, and the ratio $\Sigma_1^n S/D$ is the persistence index. Angles of turn (θ) are defined as positive when clockwise and negative when counterclockwise. The rectangular coordinates of P_n (x,y) can be computed easily from the polar coordinates (D, α). Data printout can either be complete—all steps and angles—or just the mean values. Other parameters can easily be measured, but the ones mentioned here are those most likely to be useful in defining locomotory behavior.

intervals—and a visual estimation of the center of the cell will usually suffice—then the step length gives a measure of speed and can be used as a direct indicator of locomotory capacity (Fig. 9). Alternatively, the initial and final positions can be marked and the displacement in known time used. This method is based on the assumption that particles (cells) executing a random walk will diffuse outward and that the mean square displacement will increase linearly with time. This method has been used for fibroblasts (Gail and Boone, 1972) and for neutrophils (Lackie and Smith, 1980).

Turning angle does not seem to correlate with speed, and since klinokinetic agents have not been described for neutrophils the measurement of angles may not be worthwhile unless an automated measurement system is available.

For a single track, step length, displacement, and mean angle of turn are obviously interesting parameters. Probably the most useful, although one of the most tedious to obtain, is mean step length (Fig. 9). Having made a series of measurements on, say, a control and experimental sequence with comparable cell densities and comparable step numbers, one encounters problems in comparing the populations. This is more difficult than might be expected. Some cells in a film remain effectively anchored by their tails—they may manage excursions of as much as two or three cell diameters, but invariably these excursions end by the cell's snapping back to its original position (not necessarily its position at frame 1, of course). It is worth recording the proportion of the cell population in this category. If the tail does detach, the cell may set off at relatively high speed. Only by watching the film with a critical eye over a reasonably long period is this behavior observed. Other cells achieve similar low net displacements through locomotory incompetence (insufficient traction, for example), through excessive adhesiveness, or through death. Even if we consider only the motile cells, the range of speed is very large—in a group with a mean speed of 15 μm/min, the slowest cells may be traveling at 2–3 μm/min, while the fastest may be traveling at 30 μm (Table I). Altering the environment by adding inhibitors or by making the substratum more adhesive may alter the speeds of all cells or may draw a new population of cells into action. The cells that are anchored under control conditions may be the fast-moving cells under experimental treatment, and only if all cells are moving in all sequences can we eliminate this—something that has never happened in any of our films.

Comparison of groups of cells exhibiting such high variability cannot be

TABLE I
Cell Tracks

Cell No.	Number of steps[a]	Mean step length (μm)	Displacement (μm)	Persistence index[b]	Mean turning angle
1	22	15.2	88.2	0.26	61°
2	13	16.4	84.2	0.39	70°
3	20	15.0	122.9	0.41	65°
4	19	21.7	96.1	0.23	46°
5	13	17.6	201.4	0.87	28°

[a]Each step represents 10 frames with 10-sec intervals, i.e., 100 sec total. The other measurements are defined in Fig. 9. The cell tracks shown in Fig. 8 are analyzed here in terms of mean step length, displacement, persistence, and turning angles. Further details are given in the legend to Fig. 8. In this series (12 cells in total), the mean speed was 10.2 ($\pm$12.4) μm/min, and the mean turning angle was 62° ($\pm$ 80°); all cells were moving. Standard deviations (shown in parentheses) were computed for all steps and all angles rather than from the 12 mean step lengths and angles.
[b]Persistence index = Displacement/total path length.

done simply by looking at means and standard deviations—nothing is ever significant except total cell death in the experimental population!

We have used Mann–Whitney U-tests for comparing groups, and although such nonparametric rank-order methods are rather more troublesome, they are more appropriate for this sort of data.

2. Nonuniform Environments: Taxes and Guidance

Most filming has been carried out in isotropic environments largely because of the problems of setting up stable or even metastable gradients in fluid phase; the *Candida* assay is an exception (see Section II-E). Occasionally, a sequence of film will show a general bias of displacement, presumably because some local chemotactic source is operating, and in such sequences a vector scatter diagram (e.g., Fig. 18) for the net displacements of cells will show a nonrandom pattern. The polar coordinates of the end point can easily be obtained, either by transformation of the rectangular (x, y) coordinates or by direct measurement. A more detailed discussion of this sort of analytical approach to movement in environments with gradients (tactic responses) or axial differences (guidance responses) is given by Dunn (1981). It will be obvious that relatively sophisticated analysis of a set of tracks can be done fairly easily once the coordinates of the points on various tracks are on record; it becomes a problem in programming once the questions have been formulated.

Filming under gradient conditions is possible with the *Candida* assay (Section II-E) and with two new methods devised in the first instance for other purposes—the orientation chamber described by Zigmond (1977) and discussed in Section II-D, and the under-agarose assay described in Section IV. Although we have filmed cells under these conditions, we have not done so systematically as yet.

Analytical methods for movement under gradient conditions must take account of the form of the gradient; if a point source is centrally located, then over the whole field the net displacement will be random, whereas if the source is large and the gradient parallel to the long axis of the field, then the net displacement will be distinctly nonrandom. Clearly, some common sense in the application of analytical methods is essential, and further detailed description is probably pointless since methods tend to be rather specific for particular problems.

D. The Orientation Assay

Use of visual assays for the study of chemotaxis has been hampered by lack of an assay that would allow cells to respond to gradients originating from a fluid source. Most of the early investigators used particles as point sources, e.g., parasitized cells or clumps of bacteria, and, though they did not know it, they were often measuring the response of leukocytes to gradients of C5a derived from complement activation at the particle surface. However, as progress in identifying and isolating chemotactic factors was made, it became important to study fluid-phase factors that had been prepared in advance rather than those generated during the experiment. Sally Zigmond devised a very simple chamber that allowed fluid-phase gradients to be set up. This is fully described by Zigmond (1977). It is a Perspex (or plexiglass) slide of dimensions 1 inch × 3 inches × ⅛ inch, cut so that a bridge 1 mm wide crosses the midline of the slide transversely and separates two identical wells 5 mm wide and 1 mm deep (Fig. 10). A coverslip is held in place over the bridge by two brass spring clips.

The test is carried out by spreading a drop (*ca.* 200 µl) of an appropriate cell suspension (1 to 2 × 10^6 cells/ml) down the middle of a clean coverslip (22 × 32 mm) and allowing the cells to settle on the glass, so that the cells will be aligned on the bridge once the coverslip is inverted over it. For cell orientation, protein is not required, but if the investigator wishes to observe locomotion, the cells must be in a protein-containing medium. After the cells have attached to the glass, the coverslip is inverted over the bridge and clipped

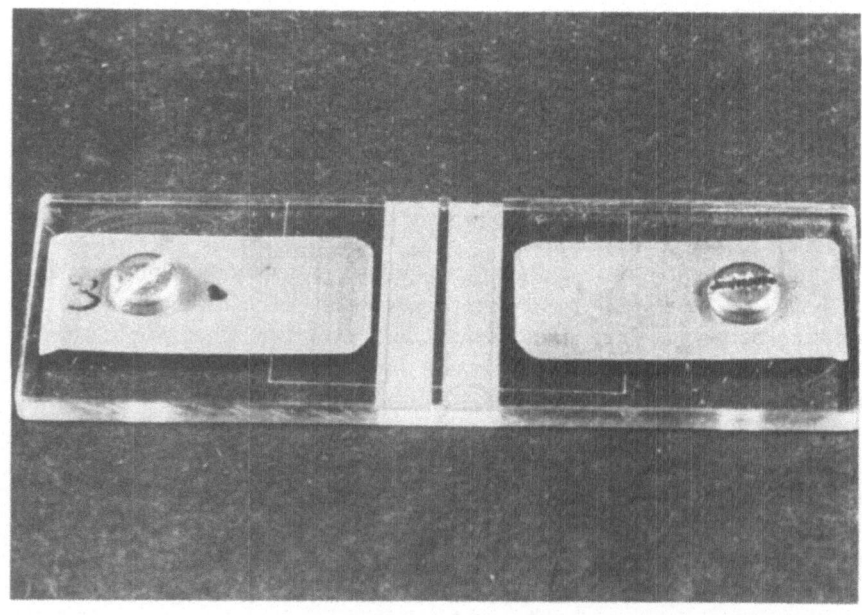

FIGURE 10. The orientation chamber (Zigmond, 1977). See text for description.

into place. One of the wells is filled with a control solution, the other with the chemotactic factor under test (or gradients of different amplitudes can be made by filling the wells with appropriate solutions). A gradient will form rapidly across the bridge (Lauffenburger and Zigmond, 1981). The apparatus can be left for 30–40 min in a damp chamber, by which time orientation should be apparent. Orientation is measured by counting 100–200 polarized cells and determining the percentage oriented into the 180° arc facing the gradient source and the 180° arc facing away from the source. Over 90% of cells should show orientation to the gradient source in optimal gradients of strong chemotactic factors such as formyl-Met-Leu-Phe or C5a (Fig. 11). With protein factors (casein, denatured proteins) it may be necessary to leave the slides longer, since gradients of such factors form more slowly. If it is wished to study locomotion in such chambers, it is simple to place them on a heated microscope stage and film the cells.

In the original description, Zigmond (1977) allowed human neutrophils to orient in gradients of f-Met-Leu-Phe without protein. Under these conditions, the cells were very adherent and formed long tails. In the presence of protein, orientation may not be quite so easy to score since the cells often do not show obvious tails and have more capacity for shape change (or for detachment from the substratum) than cells on uncoated glass.

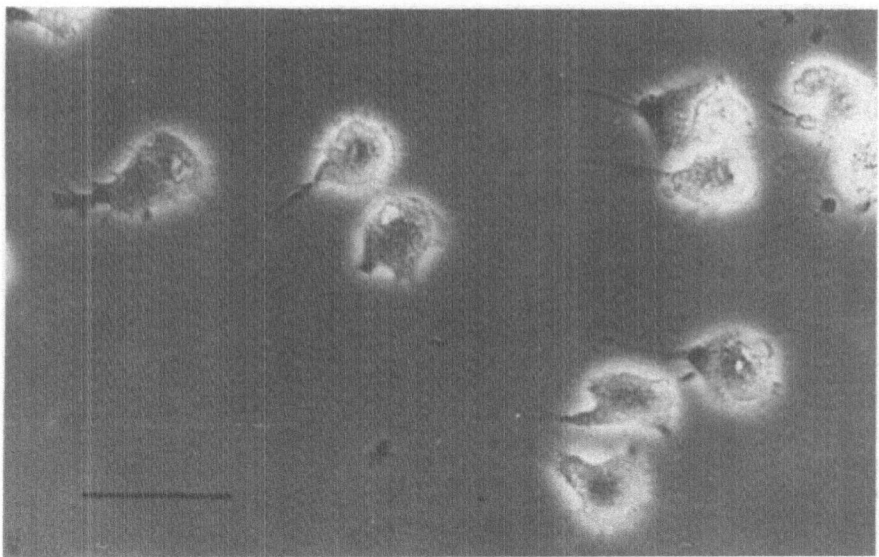

FIGURE 11. Human blood neutrophils showing morphologic orientation toward a source of f-Met-Leu-Phe (10^{-8} M) on the right. Note broad, anterior leading edges and posterior tails. Bar = 20 µm.

E. The *Candida* Assay

Most of the techniques available for the measurement of leukocyte locomotion do not distinguish between chemotactic and chemokinetic reactions of
the cells. In order to demonstrate unequivocal chemotaxis, it is necessary to
observe cells directly, either oriented in a fluid-phase gradient as in the Zigmond chamber (Section II-D) or by time-lapse filming of locomotion toward
a localized source of a chemotactic gradient. Clumps of bacteria have been
used as chemotactic foci in some studies (McCutcheon, 1946; Ramsey, 1972),
but the gradients produced from such large, irregularly shaped sources tend to
be erratic. More precise chemotactic gradients were made by Bessis and Burtè
(1965) by damaging individual erythrocytes in a microscopic field with a laser
beam; these damaged erythrocytes then acted as point sources to attract nearby
leukocytes. In a less elaborate system, we have used the blastospores of *Candida albicans* as very satisfactory point sources (Allan and Wilkinson, 1978).

Blastospores from the pathogenic yeast *C. albicans* in normal, human
plasma are potent sources for the continuous generation of complement-derived
chemotactic factors (probably C5a), and phagocytes will move directionally
toward and phagocytose the yeasts. *C. albicans* is grown on dextrose–peptone
agar (Oxoid Ltd., London), which allows the growth of discrete blastospores
about 3 μm in diameter but not the growth of pseudomycelial forms which
requires the presence of serum proteins. The yeast can be picked off the culture
plate with a loop and suspended in Hank's solution.

One drop of a suspension of leukocytes at a concentration of 2×10^6/ml
is placed on the bottom coverslip of a standard filming chamber (Fig. 3). After
5 min, nonadherent cells are removed by rinsing and the chamber is filled with
a suspension of *C. albicans* blastospores at approximately 2×10^5/ml in 50%
normal plasma. These proportions of leukocytes to spores are not critical, but
they are such that there is likely to be some distance between any neutrophil
and the nearest yeast (mean distance about 35 μm in the experiments of Allan
and Wilkinson, 1978), so that cell paths sufficiently long for filming and analysis are obtained. Since yeast spores adhere poorly to glass, the assay requires
the use of an inverted microscope so that the leukocytes are on the lower surface of the chamber. During the assay the blastospores settle gradually from
suspension onto the substratum around the leukocytes, and the subsequent chemotactic response of the leukocytes (Fig. 12) is recorded by time-lapse cinematography (Section II-A) with a magnification of $\times 100$ and a filming rate
of 15 frames/min. Normal neutrophils and monocytes respond rapidly by locomotion in straight paths toward the nearest yeast, which is phagocytosed on
arrival. The paths taken by the leukocytes can be tracked as described in Section II-A-3. Speeds, straightness of paths, and turning behavior can be
recorded and used to compare the behavior of different populations of cells. A

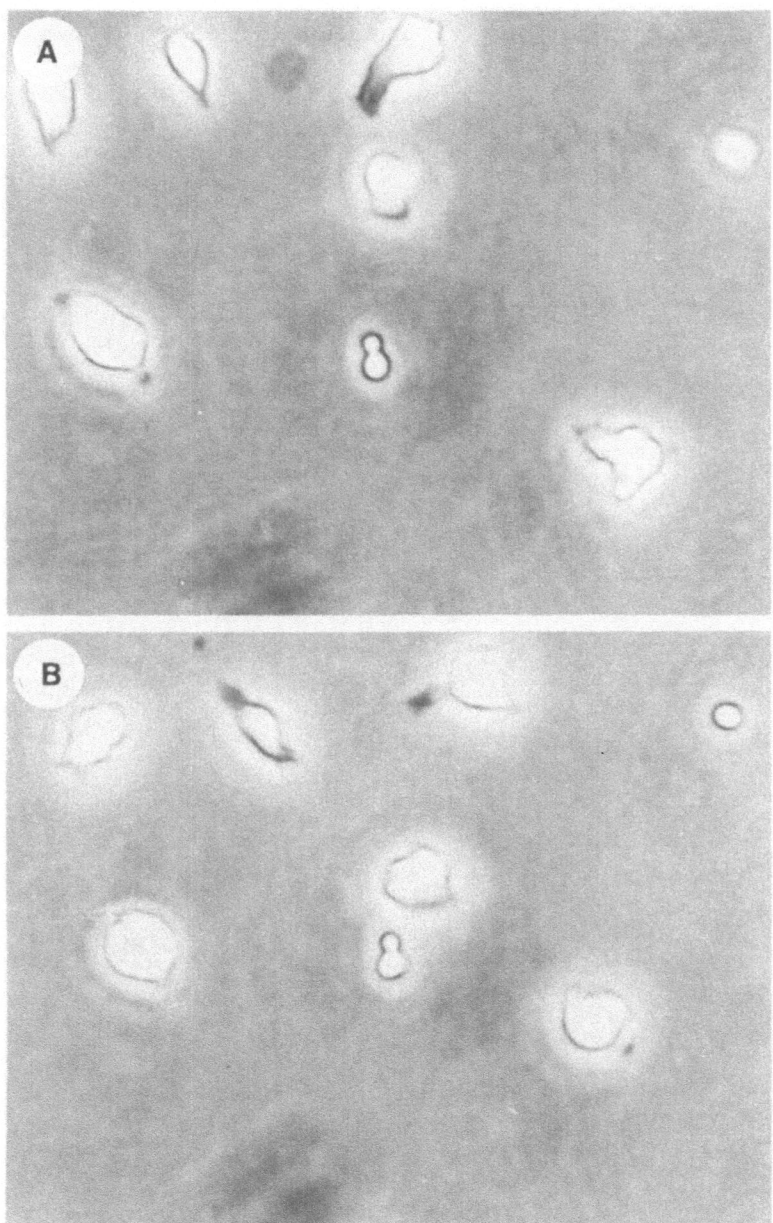

FIGURE 12. Human blood neutrophils responding to the presence of *Candida albicans* blasto-spores in 50% plasma. Note the formation of a broad leading edge facing the blastospore in the sequential photographs and the ultimate phagocytosis. Bar = 20 μm.

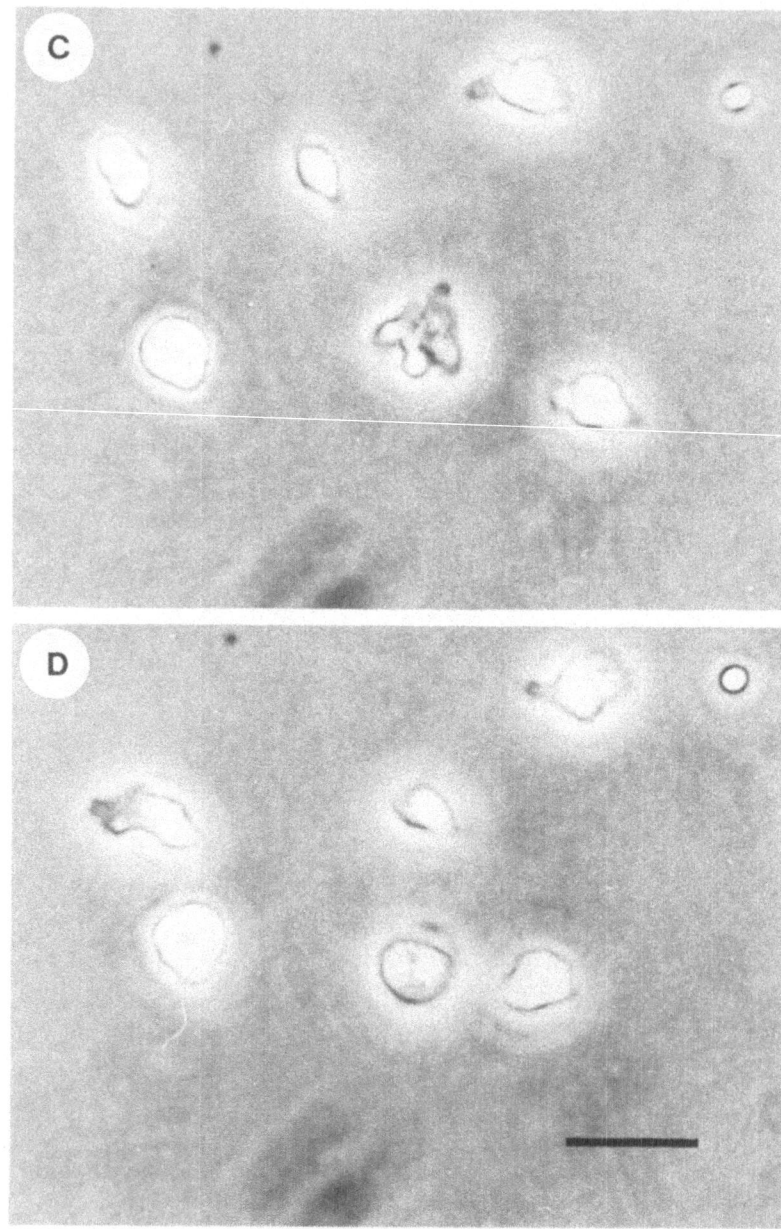

FIGURE 12. (Cont.)

detailed analysis of the behavior of human neutrophil leukocytes is given by Allan and Wilkinson (1978).

Probably the most useful information that can be obtained from the *Candida* assay is a McCutcheon "chemotactic index" (Section I) for each population of moving leukocytes; this is usually greater than +0.9 for neutrophils responding to blastospores in homologous plasma. More detailed information can be obtained from an analysis of turning behavior and calculation of mean speeds of cells responding to spores. The chemotactic gradients generated in this assay are short-range, intense, and continuously renewed at the gradient source, and they are probably stronger than in any other assay. The assay has obvious uses for determining whether previously unstudied cell types are capable of chemotactic reactions to plasma-derived factors, and we are using it in this way in a study of various populations of macrophages. It is also useful for discriminating the effects of drugs on locomotion from those on chemotaxis (see results of Allan and Wilkinson, 1978, who used colchicine). Its main limitation is that it measures only a single chemotactic system and other chemotactic factors cannot be studied.

III. THE MICROPORE FILTER ASSAY

A. Principle

The micropore filter assay is a way of measuring the locomotor response of cells to diffusible chemical substances. The essential requirement is to separate the cells from the substance under study. Thus all micropore filter assays employ a two-compartment chamber in which the cells are separated by a filter from the chemical substance. The filter is porous and the pores are of a mean diameter such that cells can squeeze through actively but not drop through. In the simplest form of the assay, cells are allowed to settle on the filter, and the substance being tested is placed underneath. A concentration gradient then forms as the substance diffuses up through the filter and cells on the upper surface of the filter detect the gradient and migrate into the filter.

B. Apparatus

1. Chambers

There are many forms of chambers available that fulfill the function outlined above. Boyden (1962) described the first, made of the material known in Great Britain as "Perspex" and in the United States as "Plexiglass." Chambers have been modified and simplified since Boyden's original description, chiefly to save reagents. In the original Boyden chamber, 3 ml of cell suspension needed to be placed above the filter and 3 ml of the attractant solution below. These are large volumes of materials that may be scarce, and most laboratories use scaled-down versions of the chamber in which the volume of cells and attractant required is less than a milliliter.Many plastic chambers are available. We use an apparatus shown in Figs. 13 and 14, in which the upper chamber is the barrel of a tuberculin syringe, sawn-off to an appropriate length, and to the lower end of which the filter is glued. "Uhu" glue (Lingner & Fischer, Bühl, West Germany) is appropriate and nontoxic, and the filters are readily detached since the glue is alcohol-soluble and dissolves when the filters are fixed. The lower chamber is a 5-ml beaker as shown in Fig. 14, but, for scarce chemoattractants, we use a narrow flat-bottomed glass tube (2 inches × ½ inch). When this type of chamber is used, it is important that the level of the fluid in the upper and lower compartments is the same, otherwise a good gradient across the filter will not be achieved. Other suitable chambers include variants of the Sykes–Moore chamber and a multiwell filter assay system described recently by Falk *et al.* (1980) and marketed by Neuro Probe, Bethesda, Maryland, U.S.A.

2. Filters

The filter is the essential component of the apparatus. All the action—both gradient formation and cell locomotion—takes place across it, and careful consideration should therefore be given to how filters are used.

The most popular type of filter is made of cellulose esters or cellulose nitrate. These filters are available from several sources (Sartorius, Göttingen,

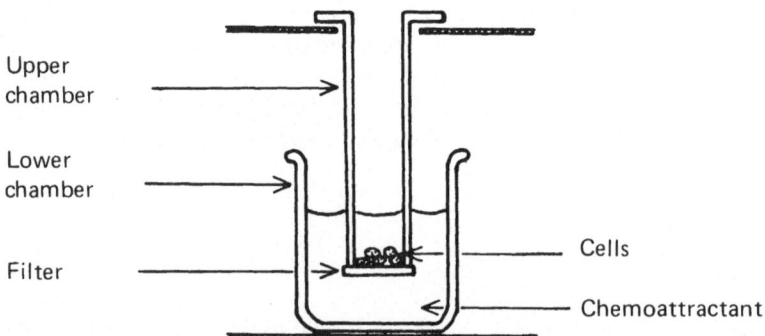

FIGURE 13. Schematic diagram of the two-compartment chamber used for the filter assay in which a filter separates cells in the upper compartment from the chemoattractant in the lower compartment.

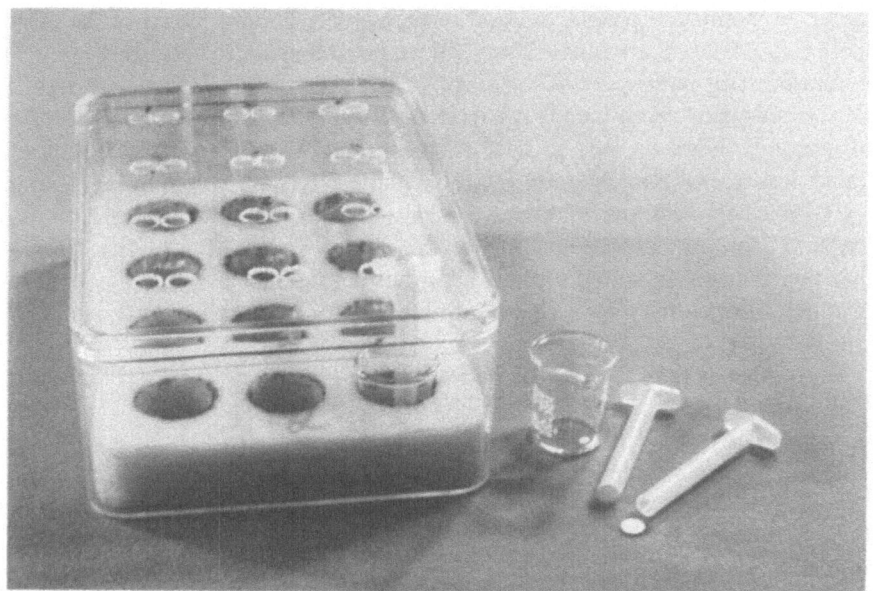

FIGURE 14. A simple apparatus for the filter assay consisting of a 5-ml glass beaker (as lower chamber) and a sawn-off tuberculin syringe barrel (as upper chamber). The filter is glued to the lower end of the syringe barrel. The whole is contained in a sandwich box with upper cell-containing chambers being suspended through the lid. The sandwich box contains foam rubber and acts as a damp chamber.

Germany; Schleicher & Scheull,Keene, New Hampshire, U.S.A.; Millipore, Bedford, Massachusetts, U.S.A.) and are about 100–150 μm in thickness, so the cells have to traverse the pores to this depth to reach the lower surface. It is essential that the filter be wettable. Cells will not move on, nor will gradients diffuse through, hydrophobic filters (Keller and Sorkin, 1967). It is important that there be protein in the medium. In the absence of protein, cells do not migrate on cellulose filters, and the cellulose must be coated with protein to act as a suitable substratum.

Cellulose filters are available in a variety of pore sizes. Those most suitable for studying cell locomotion have pore sizes varying between 3 μm and 12 μm. Neutrophil leukocytes are very deformable cells, and they, but not any other type of leukocyte (except eosinophils, which are usually an easily distinguishable minority), can crawl through 3-μm filters. This is convenient because, with blood leukocytes, if neutrophil locomotion is to be studied, the neutrophils do not have to be separated from other leukocytes; rather, the locomotor neutrophil population will separate itself by moving into pores that other cells cannot enter. Mononuclear phagocytes and lymphocytes will enter filters of the larger pore sizes (5 μm, 8 μm, and 12 μm). Since neutrophils also enter these, a prior separation (for instance on Ficoll-Hypaque, Pharmacia, or Percoll, Pharmacia) of the cell-types (from blood) is required. Blood lymphocytes are smaller than neutrophils, however, due to their large nuclear : cytoplasmic ratio, they are much less deformable. We use 8-μm-pore-size filters for studying both monocytes–macrophages and lymphocytes. However, for lymphoblasts, which may be very large, we prefer 12-μm-pore-size filters. Orientation of the filters— whether they are mounted top side up or bottom side up as they come from the manufacturers—may make a perceptible difference to cell migration (Nind, 1981).

Another type of filter is the polycarbonate filter (Nucleopore, Pleasanton, California, U.S.A.) (Horwitz and Garrett, 1971).This filter, only 13 μm thick, is not a matrix like the cellulose filter but an impermeable membrane with holes at intervals. Since the thickness of the filter is only one cell diameter, there is little room for formation of a gradient or for cell movement across the pore, and probably the important events take place on top of the filter by radial diffusion of a gradient from each pore. Use of filters as thin as these limits the ways in which locomotion can be measured and precludes use of checkerboard assays (see below). For these reasons we do not recommend them.

3. Media

Media for chemotaxis assays must be physiological and contain divalent cations and protein. Serum albumin is a suitable protein for coating the surfaces of filters. The pH should be in the physiological range. Leukocytes tolerate drops in pH fairly well but are less tolerant of rises in pH such as might be seen when using an inadequately buffered medium from which CO_2 is lost. Leukocytes move better in hypoosmolar solutions (Rabinovitch *et al.,* 1980) than in hyperosmolar solutions (Bryant *et al.,*1972). The optimal temperature for leukocyte locomotion is slightly above physiological (37–40°C) (Bryant *et al.,* 1966; Wilkinson, 1982). Leukocytes are not fastidious about slight or even quite marked changes in ionic composition, glucose concentration, etc., in the medium. They are cells that may migrate into fairly unphysiological surroundings in diseased tissue *in vivo*.

C. Measurement of Leukocyte Locomotion in Filter Assays

This is the aspect of the filter assay which has proved most contentious and which has frequently given rise to discrepancies between results obtained in different laboratories. After time has been allowed for cells to respond to an attractant by migrating into a filter, the filters are fixed, stained, cleared with xylene, and mounted for examination as stained preparations. Since it is impossible to plot the position of every cell that has entered the filter, some method of sampling is required. It is the way this sampling is done that has given rise to debate. Two methods can be used:

1. To count cells (a) at a given plane, (b) which have passed a given plane, or, (c) at all planes in the filter.
2. To measure the distance traveled by a sample of the population.

1. Counting Methods

The method used in the earliest studies was to allow cells to migrate through filters to the lower surface. The filters were mounted inverted and the number of cells on the lower surface in selected fields was counted. This "lower-surface" count has been shown to be inaccurate, since cells may drop off the lower surface (Keller *et al.,* 1972), and most groups have abandoned it. Another way of achieving the same result more accurately is to select a plane *in* the filter (e.g., 60 μm below the upper surface) and to count cells at that plane (Swanson and Becker, 1976).

A more popular counting method is to count cells which have *passed* a given plane. Two variants of this involve the use of two filters. In the method of Keller *et al.* (1972), a cell-impermeable filter (0.45 μm pore size) is placed below the filter being used for the experiment. All cells that reach or pass the plane of the lower surface of the top filter are counted by measuring the number of cells on the lower surface of the top filter plus the cells on the upper surface of the lower filter. Gallin *et al.* (1973) used chromium-labeled cells and a second cell-*permeable* filter below the top filter. They allowed cells to migrate into this second filter, then measured radioactivity in it. The two methods may give different results (Keller *et al.,* 1980). Alternatively, a given plane could be selected in the top filter and all cells below it counted.

2. Distance Methods

The simplest distance method is the "leading-front" assay of Zigmond and Hirsch (1973). The cells are allowed to migrate into the filter but not to reach the lower surface. When the stained preparation is examined, the filters are mounted top side up and the fine-adjustment micrometer of the microscope is racked down to a focal plane beyond the leading front of cells. It is then racked back up to the first point where two or more cells are in focus in a field. A reading is taken from the micrometer. The micrometer is racked back up to the top surface and a reading taken. The difference between the two readings gives the distance migrated by the leading front.

3. Comparison of Methods

Scatter is lower with the leading front method than with most counting methods, partly because counts at any level vary directly with number of cells placed on the upper surface of the filter for cells studied under standardized conditions, whereas for distance measurements, variation with cell numbers is much lower (for cells distributed randomly, the square of the distance migrated is directly proportional to the logarithm of the number of motile cells used).

The leading front method is mathematically valid for normal cell populations (i.e., the leading front distance accurately reflects the distance migrated by the whole population) (Knight, 1974—appendix to paper by Zigmond,

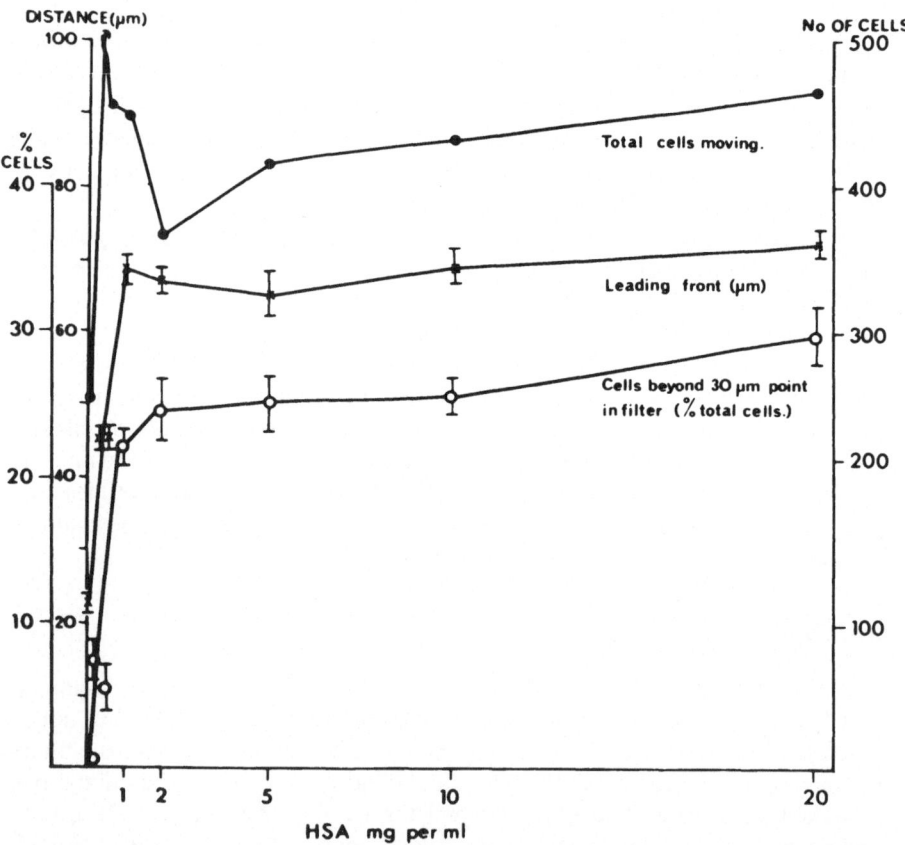

FIGURE 15. Chemokinetic dose–response curves of human neutrophils to human serum albumin (HSA) measured by the leading-front method, by counting cells at all levels in the filter and by counting cells that have passed a fixed point (30 μm from the top) in the filter. Note that the curves for all three methods are similar (apart from the dip in the curve for total cells moving, an explanation for which is not apparent). From Wilkinson (1982).

1974). However, it is possible that two cell populations showing heterogeneous behavior might not be distinguished with this method. Because cell counts in filters vary so much with the number of cells used, Keller's group (Keller *et al.,* 1975) express their counts as percentages of the cells originally placed on the filter.

Cells migrating at random into filters would be expected to show—and do show experimentally—a normal distribution with the top of the distribution curve at the top of the filter. If the number of cells (*n*) at different levels is plotted, and a curve is drawn of log *n* against d^2 (where *d* is distance), the plot is a straight line. Under such conditions, the leading front distance, the total number of cells moving into the filter, and the number of cells that have passed any given plane would be expected to show a good correlation, and this seems to be true experimentally (Fig. 15). However, the filter assay is most frequently used to study locomotion of cells in response to chemotactic gradients, and under such conditions the cell distribution would be expected to show a flux such that the peak of the distribution curve was not at the top of the filter but at some point in the filter (Fig. 16). Under such conditions, a plot of log *n*

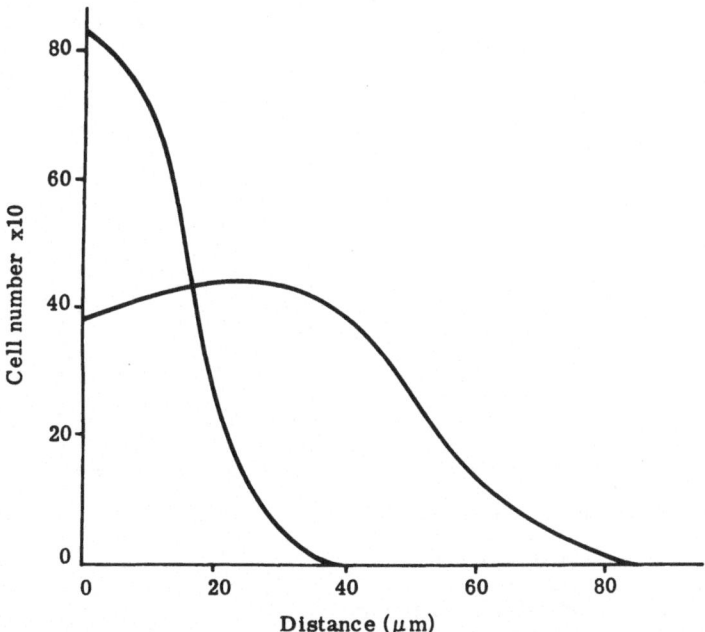

FIGURE 16. Theoretical curves for distribution of cells through a filter in response to a chemokinetic stimulus and to a chemotactic stimulus. In practice, chemotactic factors do not produce the kind of distribution shown since they may simultaneously stimulate both chemotaxis and chemokinesis.

against d^2 would not give a straight line. The number of cells beyond a given plane would not be directly proportional to the number of cells entering the filter, nor would the leading front measurement relate directly to the number of cells entering the filter. However, the leading front measurement would correlate with the number of cells beyond a given plane, provided that the plane chosen was beyond the peak of the distribution curve. In practice, these predictions seem to be borne out (Table II). It can be proposed, therefore, that measuring the total number of cells entering the filter or the number of cells *at* a given plane within the filter is not an accurate way of sampling distributions of cells responding to gradients. For such cells, measuring the leading front or counting cells beyond a given plane give better information.

It might be expected that measuring the distribution of the cell population

TABLE II
Comparison of Various Methods of Assessing Locomotion of Normal Human Blood Neutrophils into Filters toward Chemotactic Factors (3-μm Pore Size) after a 75-min Incubation[a]

Material below filter	Distance reached by leading front (μm)	Percent cells which had penetrated deeper than 40 μm into filter	Total number of cells which had entered filters	Cell counts at 40-μm depth in filter
HSA 1 mg/ml	54	6.2	198	15
f-tri-Tyr				
10^{-9} M	72	13.9	221	18
f-tri-Phe				
10^{-8} m	70	9.4	228	30
10^{-9} M	76	9.7	268	30
10^{-10} M	80	24.7	847	99
10^{-11} M	77	11.7	703	121
f-Met-Leu-Phe				
10^{-9} M	107	34.4	885	127
f-Met-Met-Phe				
10^{-8} M	114	50.5	902	98
10^{-9} M	103	44.0	497	63
Alkali-denatured HSA				
1 mg/ml	109	27.5	994	125

Correlations	r	p
Leading front and cells beyond 40 μm	0.90	0.0002
Leading front and total cells in filter	0.78	0.0037
Cells beyond 40 μm and total cells in filter	0.65	0.0196
Leading front and cell count at 40 μm	0.71	0.0112
Cells beyond 40 μm and cell count at 40 μm	0.53	0.058
Total cells in filter and cell count at 40 μm	0.95	<0.0001

[a]Cells in HSA (1 mg/ml).

at different points in the filter would be a useful assay of chemotaxis, since a chemotactic flux would give rise to a peak within the filter as has been observed by some authors (Todd and Dowdle, 1978). However, other authors (Schreiner, 1978; Schreiner and Vaula, 1978) have failed to observe such peaks, and it is our experience that they are frequently not obvious. This is partly because most leukocyte chemotactic factors are also chemokinetic, partly possibly because the tortuous matrix of the filter dampens the strong directional response that free cells would show.

D. The Checkerboard Assay

Cells moving into a filter in response to a gradient of attractant diffusing from underneath may move either because they detect and respond to the gradient or because they detect and respond to the absolute concentration of the attractant even when they are incapable of detecting the gradient. In both cases, cells will move further into the filter than when unstimulated, but in the first case their migration will be chemotactic and in the second, chemokinetic. None of the assays described in this section up to now make an accurate discrimination between the two, and for this reason the *checkerboard* filter assay was introduced by Zigmond and Hirsch (1973). In this assay, a series of chambers is set up in which cells are exposed to attractant at a range of concentrations both above and below the filter. In some, the concentration above and below will be identical and there will be no gradient. A dose–response curve can thus be drawn for the response of the cells to a range of absolute concentrations of attractant. In other chambers, cells are exposed to positive gradients of different amplitudes, and in still others to negative gradients of different amplitudes. The setup of such an experiment is shown in Table III. All results in the table are leading front measurements. The key to this table is the diagonal, from upper left to lower right, along which the results for cells responding to various absolute concentrations of attractant are displayed. Each point on this diagonal gives the velocity of cells in the presence of a given concentration of attractant. Acceleration (or deceleration) of cells traveling from a point at which the concentration is x to a second point at which the concentration is y can easily be calculated since the velocities at x and y are known. Using a straightforward calculation outlined as an appendix in the paper of Zigmond and Hirsch (1973), one can easily calculate the distances traveled between any two attractant concentrations if the velocities at those concentrations are known, assuming that acceleration is linear between the points. The more points there are on the diagonal, the more justified is this assumption. The figures obtained give the expected distance migrated in any given gradient *provided* the cells only respond to *absolute* concentration. The necessary calculations can be programmed into a desk calculator. In Table III, these calculated

TABLE III

Migration of Human Blood Neutrophils in Various Absolute Concentrations and Concentration Gradients of fMLP[a]

		Distance traveled (μm $\pm$ S.E.M.) by the leading front fMLP concentration below filter M/L			
		0	10^{-10}	10^{-9}	10^{-8}
fMLP concentration above filter M/L	0	**62** $\pm$ 2.0	104 $\pm$ 3.0(70)	118 $\pm$ 2.0(97)	127 $\pm$ 1.3(111)
	10^{-10}	63 $\pm$ 1.6(81)	**89** $\pm$ 1.3	122 $\pm$ 1.7(102)	129 $\pm$ 2.4(111)
	10^{-9}	47 $\pm$ 1.3(107)	75 $\pm$ 1.9(108)	**121** $\pm$ 1.6	131 $\pm$ 1.5(112)
	10^{-8}	41 $\pm$ 1.6(109)	53 $\pm$ 1.5(109)	74 $\pm$ 2.1(108)	**99** $\pm$ 4.4

[a]One milligram per milliliter of HSA is present throughout.
[b]The figures not in parentheses indicate the distances migrated by cells in a series of chambers, each containing a different concentration of peptide above and below the filter. Down the diagonal from upper left to lower right, the peptide concentration was the same above and below the filter. Above the diagonal, cells were moving in positive gradients; below it, in negative gradients. The figures in parentheses are estimates of what migration would have been in any chamber if cells only responded to the absolute concentration, and not to the gradient, of peptide. Note that cells show both chemokinesis (figures along diagonal) and chemotaxis (difference between figures in parentheses and figures not in parentheses) to the peptide.

figures are given in parentheses for each of the gradients studied. Chemotactic responses to the gradients cause a deviation from these calculated figures. It can be seen from Table III that the cells have migrated further into the filter in positive gradients than would be expected if they were responding only to absolute concentration; conversely, they have migrated a shorter distance in negative gradients than expected. This is suggestive of a chemotactic response to the gradient of f-Met-Leu-Phe. A good test of the validity of the checkerboard assay is to take a purely chemokinetic material (such as serum albumin) and see if the migration obtained experimentally is close to that calculated on the basis of a response to the absolute concentration only. The results of such an experiment are shown in Table IV. The experimental values are very close to those calculated.

Since it is based on a calculation of acceleration, the checkerboard assay can *only* be used with a method that measures the *distance migrated by cells,* not with counting methods. The assay is based on several assumptions, i.e., that acceleration between two known velocities is linear, that a linear gradient exists across the filter, and that the migrating cells are behaving as a homogeneous population (the distance migrated, e.g., leading front, being representative of the whole population). These assumptions are not completely justified; nevertheless, they approximate closely enough to the behavior of neutrophils in filters to give good experimental results.

The checkerboard assay has frequently been used in ways that are not justified. Checkerboards have been published in which cell counts were taken at the lower surface or other points in the filter; in which polycarbonate filters,

TABLE IV
Migration of Human Blood Neutrophils in Various Absolute Concentrations and Concentration Gradients of Serum Albumin[a,b]

| | | Distance traveled (μm $\pm$ S.E.M.) by the leading front in a 3-μm-pore-size filter in 75 min BSA concentration (μg/ml) below filter | | | | |
		0	50	200	350	500
BSA concentration (μg/ ml) above filter	0	19 $\pm$ 1.1				
	50		20 $\pm$ 1.1	20 $\pm$ 1.3(20)	21 $\pm$ 1.3(21)	23 $\pm$ 1.0(21)
	200		23 $\pm$ 0.9(24)	24 $\pm$ 1.0	27 $\pm$ 1.1(25)	27 $\pm$ 1.7(26)
	350		33 $\pm$ 1.2(31)	33 $\pm$ 1.6(32)	33 $\pm$ 1.1	35 $\pm$ 1.4(34)
	500		41 $\pm$ 1.7(37)	38 $\pm$ 1.9(37)	38 $\pm$ 1.6(38)	39 $\pm$ 1.9

[a]BSA: Sigma.
[b]Experimental conditions and layout as in Table III. From Wilkinson and Allan (1978). Note that cells show chemokinesis in response to different concentrations of albumin but no chemotaxis (figures in parentheses and figures not in parentheses are very close to one another).

whose thinness excludes velocity determinations, were used; or in which experimental values were quoted but acceleration calculations were omitted. The last assay may give some information; however, since most chemotactic factors are inhibitory at high concentrations and cause deceleration, the rule of thumb that cells always move further in positive gradients than in analogous negative gradients, or in even attractant concentrations on both sides of the filter, is not borne out in practice.

E. Automated Methods

Probably the best way to obtain the fullest information about the distribution of cells responding to attractants in filters is to use several different measures simultaneously, i.e., leading-front distance and counts at many different planes (e.g., 10 μm apart) in the filter. This can be done visually but is extremely tedious and boring except for those exceptional persons endowed with a Job-like patience and an ability to maintain concentration for hours at a time. The same information can be obtained by automated counts. Basically what is required is an image analyzer connected to the microscope that counts the number of objects in a field, together with a device for automatically racking the fine-adjustment micrometer up and down at 10-μm intervals. The whole assembly is beyond the pockets of the present authors (who have preferred to invest in equipment for visual assays), but a number of groups use such equipment. Descriptions are given by Valerius (1977), van Dyke et al. (1979), Moss et al. (1979), and Turner (1979). All the information from the printout can be used for plotting cell distributions in filters, cell counts at any desired level, leading-front measurements, etc.

IV. THE UNDER-AGAROSE ASSAY

The under-agarose technique was used originally with macrophages as an alternative to the capillary-tube method (Carpenter *et al.,* 1968), but it has since gained more popularity for use with neutrophil leukocytes, and the majority of published work now relates to these cells. Leukocytes migrate quite well on glass or plastic surfaces that have been coated with agarose since they are able to crawl along between the substratum and the agarose layer. This has the advantage that a fluid phase for the cells to move in does not have to be set up, and the physical conditions remain relatively stable during the assay.

In the simplest (and original) form of this method (Carpenter *et al.,* 1968) a suspension of cells was mixed with a test substance and placed in a small hole cut in a layer of agarose on a petri dish. In the presence of suitable proteins, phagocytes could migrate outward from the well between the agarose and the plastic and either the radial distance traveled or the area covered by the population could be used as an index of motility. This approach is essentially similar to the agarose microdroplet assay for chemokinesis (Section V) but requires less test material. It is necessary to use a concentration of agarose sufficient to provide the necessary rigidity for cutting holes. One percent agarose is most often used, but it is possible to reduce this to 0.5%.

More detailed information about the effects of attractants on cell locomotion can be obtained from the under-agarose assay by separating the cells and the test material into adjacent wells in the agarose (Cutler, 1974) (Fig. 17). The attractant diffuses out and can modify the locomotor behavior of the cells migrating out of the other well. This is now the most commonly used approach and, as the cells are migrating in gradients of substances, it allows the possibility of chemotactic reactions by phagocytes; by comparing the relative distances migrated toward wells containing a putative chemotactic factor with wells containing an appropriate control solution, one can calculate a "chemotactic index."

It is necessary to devise a method for measuring the effects of attractants on cell locomotion, and here the problems are essentially similar to those discussed earlier for the filter assay. In the absence of an attractant, cells leaving the well will migrate at random and be distributed as a circular halo around the well. As in the filter assay, the logarithm of cell numbers is proportional to the square of the distance migrated. The presence of a nearby attractant, whether chemotactic or chemokinetic, causes the distribution of cells around the well to assume an egg shape, the point of the egg pointing toward the source

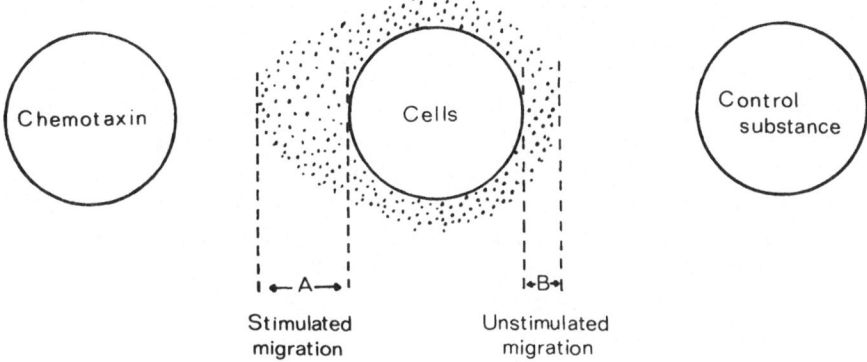

FIGURE 17. Schematic diagram of the under-agarose technique. See text for detailed description. The ratio A/B can be used as an index of locomotory activity.

of the attractant. It should be possible to draw a straight line through the center of the cell well, the point of the egg, and the center of the attractant well. A leading-front distance measurement can then be obtained at the point of the egg, or, by using an eyepiece graticule, cell counts at different distances along the line from the cell well to the gradient source can be obtained.

The under-agarose technique is eminently suitable for miniaturization. Chenoweth *et al.* (1979) have recently described a modification whereby six series of three wells can be cut with a template on a standard microscope slide. Wells cut with 12-gauge stainless steel tubing (2.4 mm outside diameter) in 5 ml of agarose on a microscope slide will hold 10 μl of fluid. The three wells are spaced 2.4 mm apart for each test, and each row is separated by 7.5 mm to prevent diffusion between tests.

To obtain good cell locomotion in the under-agarose assay, it is necessary to provide suitable proteins in the agarose medium. These may be supplied by adding a final concentration of 10% serum (Cutler, 1974; Nelson *et al.*, 1975). Serum has unpredictable effects on cell locomotion since it contains both stimulators and inhibitors, and, if desired, it is possible to replace the serum with albumin (Nelson *et al.*, 1975) or gelatin (Chenoweth *et al.*, 1979). Incubation of 2.5×10^5 neutrophils per well for 2 hr will produce good migration toward an active chemotactic factor.

One major disadvantage of the under-agarose technique is the low rate of diffusion of large molecules through agarose. This makes the assay unsuitable for the measurement of the activity of protein chemotactic factors, for example, and good migration will only be seen if relatively low-molecular-weight attractants such as C5a or formyl-methionyl peptides are used.

A higher concentration of attractant is required in the under-agarose assay than in the filter or Zigmond chamber assays. In the last two assays, the cells are exposed from the start to the mean of two attractant concentrations (e.g., above and below the filter), but the concentration of attractant reaching the cell-containing well in the agarose assay never exceeds 10% of the starting concentration (Lauffenburger and Zigmond, 1981). Another factor that must be borne in mind is that, as with the filter assay, the observation of increased migration toward a substance is not adequate evidence for the existence of a chemotactic effect; increased chemokinesis alone could be responsible. It is possible to distinguish between chemotaxis and chemokinesis in the under-agarose assay by setting up a checkerboard analysis as described by Zigmond and Hirsch (1973) for the filter assay. Orr and Ward (1978) have done this, and the results did correspond to a checkerboard pattern. However, their measure of locomotion was to count the numbers of migrating cells. The checkerboard assay depends on a calculation of cell velocity and acceleration, and these calculations can only be applied to migration-distance methods and not to cell counts (Zigmond and Hirsch, 1973). The Orr and Ward assay could be adapted easily to a leading-front or other migration-distance measurement.

The under-agarose techinque does have the advantage of allowing direct microscopic observation of the movement of the cells during the assay. It is thus possible to analyze the displacement of the cells with respect to the direction of the gradient to establish if the cells are, in fact, responding chemotactically. Our colleague, J. M. Shields, has filmed human blood neutrophils moving under agarose (0.5%) toward f-Met-Leu-Phe (10^{-7} M) in the presence of human serum albumin. The direction of cell locomotion can be analyzed in detail from such films, and convincing evidence for chemotaxis can be obtained. A vector scatter diagram showing displacements of cells in such an assay is shown in Fig. 18.

V. MISCELLANEOUS TECHNIQUES

A. The Capillary-Tube Assay for Macrophage Locomotion

The migration of macrophages from a capillary tube was introduced by George and Vaughan (1962) as an *in vitro* model for the study of delayed hypersensitivity. It has since gained wide popularity as a result of its simplicity. Capillary tubes (100-μl disposable micropipettes) are filled with a suspension of macrophages, sealed at one end, and centrifuged to form a dense pellet of cells. The capillary tubes are cut at the cell-supernatant interface and fixed horizontally on culture plates. The capillary tubes are then flooded with medium containing the substances to be assayed (usually inhibitory lympho-

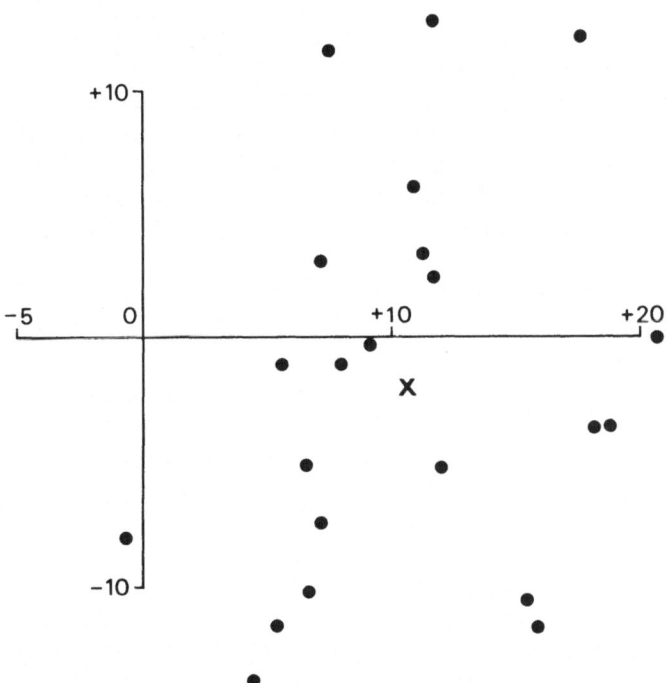

FIGURE 18. Vector scatter diagram of displacement of human neutrophils under agarose toward a source of f-Met-Leu-Phe (10^{-8} M). The point at which analysis of each cell track was begun has been taken as point zero. The x axis represents the perpendicular to the gradient source with positive values representing displacement toward the source (in micrometers per minute) and negative values representing displacement away from it. Displacement in the y axis is parallel to the source (values in micrometers per minute). The x axis crosses the center of both the cell well and the attractant well, and the attractant is diffusing radially from a point on the x axis to the right of the picture. It is very important, when plotting cell paths to obtain evidence for chemotaxis, that cells randomly distributed at the start of the experiment are tracked. Obviously, if cells are tracked as they leave the cell well, they will show net displacement away from the well whether or not a gradient is present. To obtain the data in this figure, cells were allowed to migrate some distance out of the well before tracking was begun. Cells starting in an arbitrarily chosen rectangular segment 30 μm wide and at a distance of about 100 μm from the cell well were then tracked. The tracks of these cells were plotted within an area which extended 60 μm from the starting point away from the well. This was to allow the cells a chance of movement toward the cell well (down gradient) equal to that of movement away from the cell well (up gradient). The diagram shows a marked displacement of cells to the right (up gradient toward the source). Mean displacement in the x axis was $+10.6$ μm/min, and in the y axis -2.1 μm/min ($p < 0.0001$). From Wilkinson (1982).

kines) and incubated for 24–48 hr. The macrophages migrate from the capillary tubes and spread out on the culture plate in a fanlike distribution. The leading front of the cells can be measured directly or the area covered by the cells can be assessed by projection and planimetry; these measurements are taken to be proportional to the locomotor capability of the cells.

The usefulness of the capillary-tube method as an assay for locomotion is limited by several factors: in the way it is usually done, no check is made that it is really measuring effects on locomotion at all; large numbers of cells are required, e.g., with blood monocytes 40–50 ml of blood is needed for a fairly small test; the effects of cell aggregation within the capillary tube cannot be assessed; and leukocytes outside the capillary tube may be spread by convection currents rather than locomotion. The assay is also poor at revealing subtle effects on movement.

Lymphokines such as macrophage-migration-inhibition factor (MIF) inhibit macrophage migration out of such capillary tubes. These lymphokines also increase the adhesiveness of macrophages for coated glass or plastic (Folger *et al.,* 1978; David and Remold, 1976); thus the migration-inhibition effect may be due to a MIF-induced increase in macrophage stickiness.

B. The Agarose Microdroplet Assay

The agarose microdroplet assay was developed by Harrington and Stastny (1973) as an alternative to the capillary-tube method for measuring lymphokine activity. They found that results from the two assays were comparable but that many fewer cells were needed for the microdroplet assay. Improvements to the assay have been described recently by Weese *et al.* (1978). A suspension of macrophages is mixed with an equal volume of 0.5% agarose (Indubiose Agarose A37, L'Industrie Biologique Francaise, Clichy, France), and a 2-μl droplet is placed on the bottom of each well of a microtiter plate. The well is then filled with 100 μl of test medium. After incubation for 24–48 hr, either the leading front or the area of distribution of the cells is measured.

The agarose microdroplet assay has been adopted recently by workers interested in acute inflammation for measuring the effects of antirheumatic drugs (Smith and Walker, 1980) and inflammatory mediators (Palmer *et al.,* 1980) on the locomotion of neutrophil leukocytes. With these cells an incubation of 4 hr is adequate to achieve measurable migration. It must be stressed that the agarose microdroplet assay is suitable only for measuring the chemokinetic effect of substances and provides no information about chemotactic activity. It does, however, have the advantages that only very small amounts of material are needed and it is a simple matter to set up several hundred tests rapidly in microtiter plates.

VI. FACTORS INFLUENCING LOCOMOTION

In this section we propose to discuss, very briefly, the factors that may affect the locomotion of cells in order to indicate the problems upon which the various assay methods may be used. Much of what we have to say applies to cells in general, not just to leukocytes, although the methods may have to be adapted in minor ways.

Three major factors influence the locomotion of cells: the nature of the contact of the cell with the substratum, the composition of the fluid phase, and interactions with other cells. There is some overlap between these since other cells may constitute a substratum, and modifying the fluid phase may affect substratum contact.

A. Cell–Substratum Interactions

Several features of the substratum may affect cell movement—in particular, adhesiveness, topography, and rigidity. Three-dimensional matrices have properties which are self-evidently different from two-dimensional surfaces: although we know that neutrophils migrate through tissue *in vivo,* and will migrate through micropore filters and hydrated gels *in vitro,* we know almost nothing about the essential determinants of matrix suitability. Circumstantial evidence suggests that the rules governing movement of neutrophils *in* filters and *on* protein-coated glass are rather different (Lackie and Smith, 1980; Lackie, 1982). Much more is known about two-dimensional substrata, partly because the other attributes with which we are concerned can more conveniently be estimated on these.

1. Cell–Substratum Adhesion

An optimum level of cell–substratum adhesion for efficient locomotion seems a reasonable expectation since the extremes of high and low adhesion do seem to block movement. Both fibroblasts and neutrophils seem to have adhesion optima for movement (Gail and Boone, 1972; Lackie and Smith, 1980), and it seems likely that this will be generally true. That locomotion studies are possible at all presupposes that we can find a substratum of suitable adhesiveness, something that has been very difficult for lymphocytes and for monocytes. One problem with a heterogeneous cell population may be that different subsets of cells move at different adhesion levels. Clean glass, i.e., well-washed glass without protein in the medium, is an unsuitable substratum for cells to move on. Orientation of cells can occur without net movement, and thus a clean glass substratum is used in the orientation-chamber assay (Section II-D) and cell locomotion does not complicate the picture. Gradients of adhesiveness may lead to directional locomotion (Haptotaxis: see Carter, 1967).

2. Substratum Shape

The shape of a plane substratum—whether convex, concave, grooved, or of complex topography—may well influence the direction of cell movement, the phenomenon known as contact guidance. This has been known for fibroblasts for many years, and various explanations have been proposed (Dunn and Heath, 1976; Dunn and Ebendal, 1978). Guidance reactions differ from taxes in that movement is restricted but not directed; thus cells may move only parallel to the long axis of a cylinder without being directed up or down the axis. A detailed discussion is inappropriate here, but the possibility of such a guidance phenomenon is worth bearing in mind when looking at cell tracks. Again, Dunn (1981) should be consulted for a more detailed review.

3. Substratum Rigidity

Intuitively, it seems likely that substratum rigidity is important since forward movement of the cell produces an equal and opposite force on the substratum. Little is known about this, although Harris (1973) showed clearly that fibroblasts required a much more rigid substratum than leukocytes. The viscoelastic properties of three-dimensional matrices are likely to modify movement, but almost nothing is known about this topic.

B. Composition of the Fluid Phase

Composition of the fluid phase is the most easily manipulated variable and little needs to be said. By appropriate choice of medium, the adhesion of cells can be altered and the addition of various agents may modify cellular function. The interpretation of the effects of drugs on a complex process such as cell movement is difficult, and it should be emphasized that inhibitors which are effective and specific *in vitro* on isolated enzyme systems may have multiple sites of action in whole cells—or be unable to reach their targets. Conversely, if a cell continues to move at a normal rate in the presence of a compound, it is a reasonable indication that the cells are not suffering severe functional impairment.

Proteins in the fluid phase will bind rapidly to surfaces such as clean glass or plastic, and this should be borne in mind when handling such media. The rapid adsorption of protein probably means that it is unnecessary to precoat substrata if the cells are added in a protein-containing medium.

Gradients in the fluid phase are relatively difficult to stabilize, and to overcome this problem a variety of special systems has been devised. Essentially, the solution to the problem lies in restricting convection (micropore filters, orientation chambers) or in generating a steep gradient *in situ* (*Candida* assay).

C. Cell–Cell Interactions

Although most studies of cell locomotion have been carried out with homotypic cell populations—or populations which are notionally of this type—this is really rather unrealistic. For all leukocytes the interactions between different cell types is likely to be of particular interest. Two particular kinds of interaction, contact inhibition of locomotion and effects on the geometry of the substratum, will be discussed. Other interactions have been filmed—for example, the contact lysis of cells by cytotoxic T lymphocytes (Sanderson and Glauert, 1977). Many indirect interactions have been investigated—the effects of cell-released factors on the behavior of a test population being very commonly studied—by almost all methods but especially by capillary-tube emigration, a notoriously uninterpretable test. We will not attempt to review this area.

1. Contact Inhibition of Locomotion

The collision of two fibroblasts leads to an alteration in their direction of movement so as to prevent their overlapping one another (Abercrombie, 1979; for a review, see Heaysman, 1978). This behavior upon collision leads, for example, to radial outgrowth of cells from a tissue explant, since cells deviating from an outward radial path will have a much higher probability of colliding with other cells and having to turn. It is now fairly generally recognized that two distinct phenomena may operate to bring about an inhibition of overlapping. The first involves contact inhibition type 1 in which collision leads to the formation of a cell–cell adhesion, subsequent paralysis of the locomotory machinery, and the eventual retraction of one or both cells, often accompanied by an abrupt break-down of the adhesion. The second, contact inhibition type 2, depends on the balance between cell–substratum and cell–cell adhesion; if cells adhere more strongly to the substratum than to the dorsal surfaces of other cells, it will be difficult for them to relinquish their hold upon the substratum in order to overlap another cell. Thus the extent of overlapping will be affected by cell–substratum adhesion.

Studies of cell–cell contact behavior, particularly in the case of cells known to cooperate in immunological interactions, are overdue.

2. Effects on the Geometry of the Substratum

One intriguing aspect of leukocyte behavior is the leukocyte's willingness to penetrate cell monolayers and move around underneath. This behavior has been observed for leukocytes of almost all classes with monolayers of endothelium, fibroblasts, or reticular cells (Armstrong and Lackie, 1975; Beesley *et al.*, 1979; Haston, 1979). It seems that movement beneath cells is in fact easier in that rates of movement are significantly higher, and for lymphocytes this is one of the few situations in which rapid movement can be observed directly. The relatively restricted nature of the "channel" in which the cell is moving may force the underlapping leukocyte onto the substratum, improving the coupling of the locomotory machinery to the substratum. Obtaining traction upon both ceiling and floor may also contribute to the more rapid movement observed.

This sort of behavior has been observed simply by setting up filming chambers with cell monolayers on one coverslip, adding cell suspensions and watching—via time-lapse film—the interactions. As with much of this sort of work, the observations are relatively easy to make—but very much harder to interpret and analyze.

REFERENCES

Abercrombie, M. (1979) Contact inhibition and malignancy, *Nature (London)* **281**:259.

Albrecht-Buehler, G. (1977) Phagokinetic tracks of 3T3 cells: Parallels between the orientation of track segments and of cellular structures which contain actin or tubulin, *Cell* **12**:333.

Allan, R. B., and Wilkinson, P. C. (1978) A visual analysis of chemotactic and chemokinetic locomotion of human neutrophil leucocytes. Use of a new chemotaxis assay with *Candida albicans* as gradient source, *Exp. Cell. Res.* **111**:191.

Armstrong, P. B., and Lackie, J. M. (1975) Studies on intercellular invasion *in vitro* using rabbit peritoneal neutrophil granulocytes (PMNS). 1. Role of contact inhibition of locomotion, *J. Cell. Biol.* **65**:439.

Beesley, J. E., Pearson, J. D., Carleton, J. S., Hutchings, A., and Gordon, J. L. (1979) Granulocyte migration through endothelium in culture, *J. Cell. Sci.* **38**:237.

Bessis, M., and Burté, B. (1965) Positive and negative chemotaxis as observed after the destruction of a cell by U.V. or laser microbeams, *Tex. Rep. Biol. Med.* **23**:204.

Boyden, S. V. (1962) The chemotactic effect of mixtures of antibody and antigen on polymorphonuclear leucocytes, *J. Exp. Med.* **115**:453.

Bradbury, S. (1977) Quantitative image analysis, in: *Analytical and Quantitative Methods in Microscopy,* Society of Experimental Biology Seminar Series, Volume 3 (G. A. Meek and H. Y. Elder, eds.), Cambridge University Press, pp. 91–116.

Bryant, R. E., de Prez, R. M., Van Way, M. H., and Rogers, D. E. (1966) Studies on leukocyte motility. I. Effects of alterations of pH, electrolyte concentration and phagocytosis on leukocyte migration, adhesiveness and aggregation, *J. Exp. Med.* **124**:483.

Bryant, R. E., Sutcliffe, M. C., and McGee, Z. A. (1972) Effect of osmolalities comparable to those of the renal medulla on function of human polymorphonuclear leukocytes, *J. Infect. Dis.* **126**:1.

Carpenter, R. R., Barsales, P. B., and Gauchan, R. P. (1968) Antigen-induced inhibition of cell migration in agar gel, plasma clot and liquid media, *J. Reticuloendothel. Soc.* **5**:472.

Carter, S. B. (1967) Haptotaxis and the mechanism of cell motility, *Nature (London)* **213**:256.

Chenoweth, D. E., Rowe, J. G., and Hugli, T. E. (1979) A modified method for chemotaxis under agarose, *J. Immunol. Meth.* **25**:337.

Comandon, J. (1917) Phagocytose *in vitro* des hématozoaires du calfat (enregistrement cinematographique), *Compt. Rend. Soc. Biol.* **80**:314.

Comandon, J. (1919) Tactisme produit par l'amidon sur les leucocytes; enrobement du charbon, *Compt. Rend. Soc. Biol.* **83**:1171.

Cutler, J. E. (1974) A simple *in vitro* method for studies on chemotaxis, *Proc. Soc. Exp. Biol. Med.* **147**:471.

Dahlgren, C. (1979) Modulation of polymorphonuclear leukocyte locomotion *in vitro*, Linköping University Medical Dissertations, No. 66.

David, J. R., and Remold, H. G. (1976) Macrophage activation by lymphocyte mediators and studies on the interaction of macrophage inhibitory factor (MIF) with its target cell, in: *Immunobiology of the Macrophage* (D. S. Nelson, ed.), Academic Press, New York, pp. 401–426.

Dixon, H. M., and McCutcheon, M. (1935) Absence of chemotropism in lymphocytes, *Arch. Pathol.* **19**:679.

Dunn, G. A. (1981) Chemotaxis as a form of directed cell behaviour: Some theoretical considerations, in: *Biology of the Chemotactic Response* (J. M. Lackie and P. C. Wilkinson, eds.), Cambridge University Press. Cambridge, pp. 1–26.

Dunn, G. A., and Ebendal, T. (1978) Contact guidance on oriented collagen gels, *Exp. Cell. Res.* **111**:475.

Dunn, G. A., and Heath, J. P. (1976) A new hypothesis of contact guidance in tissue cells, *Exp. Cell. Res.* **101**:1.

Falk, W., Goodwin, R. H., and Leonard, E. J. (1980) A 48-well micro chemotaxis assembly for rapid and accurate measurement of leukocyte migration, *J. Immunol. Meth.* **33**:239.

Folger, R., Weiss, L., Glaves, D., Subject, J. R., and Harlos, J. P. (1978) Translational movements of macrophages through media of different viscosities, *J. Cell. Sci.* **31**:245.

Gail, M. H., and Boone, C. W. (1972) Cell–substrate adhesivity: A determinant of cell motility, *Exp. Cell. Res.* **70**:33.

Gallin, J. I., Clark, R. A., and Kimball, H. R. (1973) Granulocyte chemotaxis: An improved *in vitro* assay employing ^{51}Cr-labelled granulocytes, *J. Immunol.* **110**:233.

George, M., and Vaughn, J. H. (1962) *In vitro* cell migration as a model for delayed hypersensitivity, *Proc. Soc. Exp. Biol. Med.* **111**:514.

Harrington, J. T., and Stastny, P. (1973) Macrophage migration from an agarose droplet: Development of a micromethod for assay of delayed hypersensitivity, *J. Immunol.* **110**:752.

Harris, A. (1973) Cell surface movements related to cell locomotion, in: *CIBA Foundation Symposium,* (R. Parter and D. W. Fitzsimons, eds.), Volume 14, Elsevier/North-Holland, Amsterdam, pp. 3–26.

Harris, H. (1954) Role of chemotaxis in inflammation, *Physiol. Rev.* **34**:529.

Haston, W. S. (1979) A study of lymphocyte behavior in cultures of fibroblast-like reticular cells, *Cell. Immunol.* **45**:74.

Heaysman, J. E. (1978) Contact inhibition of locomotion: A reappraisal, *Int. Rev. Cytol.* **55**:49.

Horwitz, D. A., and Garrett, M. A. (1971) Use of leukocyte chemotaxis *in vitro* to assay mediators generated by immune reactions. I. Quantitation of mononuclear and polymorphonuclear leukocyte chemotaxis with polycarbonate (Nucleopore) filters, *J. Immunol.* **106**:649.

Keller, H. U., and Sorkin, E. (1967) Studies on chemotaxis. IX. Migration of rabbit leucocytes through filter membranes, *Proc. Soc. Exp. Biol. Med.* **126**:677.

Keller, H. U., Borel, J. F., Wilkinson, P. C., Hess, M., and Cottier, H. (1972) Reassessment of Boyden's technique for measuring chemotaxis, *J. Immunol. Meth.* **1**:165.

Keller, H. U., Gerber, H., Hess, M. W., and Cottier, H. (1975) Studies on the regulation of the neutrophil chemotactic response using a rapid and reliable method for measuring random migration and chemotaxis of neutrophil granulocytes, in: *Future Trends in Inflammation,* Volume II (J. P. Giroud, D. A. Willoughby, and G. P. Velo, eds.), Birkhäuser Verlag, Basel, pp. 326–339.

Keller, H. U., Wissler, J. H., Damerau, B., Hess, M. W., and Cottier, H. (1980) The filter technique for measuring leucocyte locomotion *in vitro*. Comparison of three modifications, *J. Immunol. Meth.* **36**:41.

Lackie, J. M. (1982) Aspects of the behaviour of neutrophil leucocytes, in: *Cell Behaviour* (R. Bellairs, A. S. G. Curtis, and G. A. Dunn, eds.), Cambridge University Press, Cambridge, pp. 319–348.

Lackie, J. M., and Smith, R. P. C. (1980) Interactions of leukocytes and endothelium, in: *Cell Adhesion and Motility* (A. S. G. Curtis and J. D. Pitts, eds.), Cambridge University Press, Cambridge, pp. 235–272.

Lauffenburger, D. A., and Zigmond, S. H. (1981) Chemotactic factor concentration gradients in chemotaxis assay systems, *J. Immunol. Meth.* **40**:45–60.

Lewis, W. H. (1931) Locomotion of lymphocytes, *Bull. Johns Hopkins Hosp.* **49**:29.

Lewis, W. H. (1934) On the locomotion of the polymorphonuclear neutrophils of the rat in autoplasma cultures, *Bull. Johns Hopkins Hosp.* **55**:273.

McCutcheon, M. (1946) Chemotaxis in leucocytes, *Physiol. Rev.* **26**:319.

Moss, V. A., Simpson, H. K. L., and Roberts, J. A. (1979) A semi-automatic method for measuring leucocyte movement, *J. Immunol. Meth.* **27**:293.

Nelson, R. D., Quie, P. G., and Simmons, R. L. (1975) Chemotaxis under agarose: A new and simple method for measuring chemotaxis and spontaneous migration of human polymorphonuclear leukocytes and monocytes, *J. Immunol.* **115**:1650.

Nind, A. F. (1981) Neutrophil chemotaxis: Technical problems with nitrocellulose filters in Boyden-type chambers, *J. Immunol. Meth.* **49**:39.

Orr, W., and Ward, P. A. (1978) Quantitation of leukotaxis in agarose by three different methods, *J. Immunol. Meth.* **20**:95.

Palmer, R. M. J., Stepney, R. J., Higgs, G. A., and Eakins, K. E. (1980) Chemokinetic activity of arachidonic acid lipoxygenase products on leucocytes of different species, *Prostaglandins* **20**:411.

Peterson, S. C., and Noble, P. B. (1972) A two-dimensional random-walk analysis of human granulocyte movement, *Biophys. J.* **12**:1048.

Rabinovitch, M., De Stefano, M. J., and Dziezanowski, M. A. (1980) Neutrophil migration under agarose: Stimulation by lowered medium pH and osmolality, *J. Reticuloendothel. Soc.* **27**:189.

Ramsey, W. S. (1972) Analysis of individual leucocyte behaviour during chemotaxis, *Exp. Cell. Res.* **70**:129.

Ramsey, W. S., and Harris, A. (1973) Leucocyte locomotion and its inhibition by antimitotic drugs, *Exp. Cell. Res.* **82**:262.

Riddle, P. N. (1979) *Time Lapse Cinemicroscopy,* Academic Press, London.

Sanderson, C. J., and Glauert, A. M. (1977) The mechanism of T cell mediated cytotoxicity. II. Morphological studies of cell death by time-lapse microcinematography, *Proc. Roy. Soc. Lond. B* **192**:241.

Schreiner, A. (1978) *In vitro* locomotion of human leukocytes. Photographically determined distribution in millipore filters at different incubation times, *Acta Pathol. Micro. Scand. C* **86**:117.

Schreiner, A., and Vaula, D. (1978) Kinetics of locomotion of human granulocyte populations, *Acta Pathol. Micro. Scand. C* **86**:205.

Smith, M. J. H., and Walker, J. R. (1980) The effects of some antirheumatic drugs on an *in vitro* model of human polymorphonuclear leucocyte chemokinesis, *Br. J. Pharmacol.* **69**:473.

Swanson, M. J., and Becker, E. L. (1976) Measurement of chemotaxis of human polymorphonuclear leukocytes in filters by counting the number of cells in a single plane and comparison with leading front method, *J. Immunol. Meth.* **13**:191.

Todd, G., and Dowdle, E. B. (1978) Neutrophil chemotaxis: The kinetics of cellular locomotion *in vitro, Int. Arch. Allergy Appl. Immunol.* **57**:165.

Turner, S. R. (1979) Acdas: An automated chemotaxis data acquisition system, *J. Immunol. Meth.* **28**:355.

Valerius, N. H. (1977) Neutrophil granulocyte chemotaxis *in vitro, Acata Pathol. Micro. Scand. C* **85**:289.

van Dyke, T. E., Reilly, A. A., Horoszewicz, H., Gagliardi, N., and Genco, R. J. (1979) A rapid semi-automated procedure for the evaluation of leukocyte locomotion in the micropore filter assay, *J. Immunol. Meth.* **31**:271.

Weese, J. L., McCoy, J. L., Dean, J. H., Ortaldo, J. R., Busk, K. R., and Herberman, R. B. (1978) Technical modifications of the human agarose microdroplet leukocyte migration inhibition assay, *J. Immunol. Meth.* **24**:363.

Wilkinson, P. C. (1982) *Chemotaxis and Inflammation,* 2nd ed., Churchill Livingstone, Edinburgh (in press).

Wilkinson, P. C., and Allan, R. B. (1978) Assay systems for measuring leukocyte locomotion, in: *Leukocyte Chemotaxis* (J. I. Gallin and P. G. Quie, eds.), Raven Press, New York, pp. 1–24.

Zigmond, S. H. (1974) A modified filter method for assaying polymorphonuclear leukocyte locomotion and chemotaxis, *Antibiot. Chemother.* **19**:126.

Zigmond, S. H. (1977) Ability of polymorphonuclear leukocytes to orient in gradients of chemotactic factors, *J. Cell. Biol.* **75**:606.

Zigmond, S. H., and Hirsch, J. G. (1973) Leukocyte locomotion and chemotaxis, *J. Exp. Med.* **137**:387.

7

Sizing of Cells by the Electrical Resistance Pulse Technique

Methodology and Application in Cytometric Systems*

VOLKER KACHEL

I. INTRODUCTION

A. Reasons for Measuring Size of Biological Cells

Cells are the constructive elements of all complex biological organisms. It is evident that the size and number of the basic elements define the size of the complete organism or the size of the organs of the organism. Growth of an organism is caused by the growth of the cells, by an increase of the number of cells, or by both. By counting and sizing cells, such fundamental events can be investigated.

Pathological changes of cell populations in many cases are connected with increase or decrease of their mean cellular volume or their volume distribution. Particularly with diseases of red or white blood cells, the volume distribution curves of those cells are changed.

Samples of cells may contain a mixture of different cell types differing in volume. Treatment with drugs or reagents may stimulate or depress the growth of such cell types in many different ways. Size distribution measurements may manifest such differences.

Cell aggregation can be recognized from cell volume distributions. An

*Dedicated to Prof. Dr. G. Ruhenstroth-Bauer.

VOLKER KACHEL • Max-Planck-Institut für Biochemie, D-8033 Martinsried, West Germany.

aggregate of cells—depending on the number n of aggregated cells—is measured with about n times the volume of the single cell.

The examples mentioned characterize cell sizing alone as a valuable tool in cytology. Cell sizing, however, also plays an important role as basic parameter in multiparameter analyzing systems in which simultaneously different cell features are determined, e.g., DNA content, immunological surface features of cells, or cell volume. In many cases, the optically measured parameter is directly influenced by the volume or the surface of the cell considered, e.g., if we are interested in determining the immunological surface density of an antigen which is marked by a fluorochrome. In this case, it is not sufficient to measure the fluorescence of the antigen alone because its amounts are directly proportional to the cellular surface. Small and large cells of identical surface densities are measured with completely different fluorescence values. If all fluorescence measurements are related to the cell surface of each cell, which in most cases can be derived from cell volume, correct density measurements are possible.

The points mentioned indicate that sizing of biological cells is of great interest in biology and medicine as an independent single parameter or as a basic parameter in multiparameter analyzing systems, and electrical resistance pulse sizing is a suitable method in this field.

Electrical sizing is a method of flow cytometry. Flow methods are particularly marked by a throughput of a few thousand cells per second. Therefore, a large number of cells can be measured in a short time with high statistical relevance.

B. A Short Review on the History of Electrical Sizing

Historically, there are two main phases in the use of the electrical sizing method, which was introduced by W. H. Coulter (1953). Each cell passing through the sizing transducer generates an electrical pulse, which contains in its height the information about the volume of the cell. Initially, most investigators accepted direct proportionality between pulse height and cell volume and the shape of the distribution curves; especially, the skewness of blood cell distributions was explained as biological (Doljanski et al., 1966; Grant et al., 1960; Lewis and Goldman, 1965; Weed and Bowdler, 1967; Winter and Sheard, 1965). One of the first to question the direct relation of pulse height to volume was Mattern et al. (1957), who presumed that the area under the pulse rather than its height should correlate with the volume of the particle. Coincidence problems were the only possible sources of error considered during this period (Wales and Wilson, 1961, 1962; Coulter, 1966; Princen and Kwolek, 1965; Princen, 1966; Strackee, 1966). In the second phase, artifacts of the Coulter method itself received increased consideration.

The first theoretical analysis of the pulse height particle volume problem was performed by Kubitschek (1960). He concluded that sizing of small particles is independent of particle shape. But, he points out, a more precise derivation could show that the pulse amplitude is not strictly independent of particle shape. van Dilla *et al.* (1967) reduced the skewness of particle size distribution curves by the use of longer orifices and amplifiers whose frequency response was reduced. Gregg and Steidley (1965) introduced the shape factor 3/2 for spheres and made the first model measurements in an enlarged electrolytic tank; but the field patterns they determined were not correct, and their treatment of nonspherical particles is questionable. Gutmann (1966) explained the skewness of red cell volume distribution curves by the different orientation of the cells; only two discrete orientations of red cells were considered in his theory. Harvey and Marr (1966) and Harvey (1968) used an electronic integrating and differentiating device to influence the pulse shapes, and the first experimental hydrodynamic focusing device was described by Spielman and Goren (1968), but without evident theory. Based on experiments with a "capillary-directed flow system," the skewness and bimodality of erythrocyte volume distributions were explained by Shank *et al.* (1969) as resulting from different deformation of the cells at different radii in the orifice. A comprehensive survey of the existing physical literature applicable to resistance pulse sizing is given by Grover *et al.* (1969a). Independently, Thom (Thom, 1968, 1972a,b; Thom and Kachel, 1971; Thom *et al.*, 1969) and Kachel (Kachel, 1970, 1972, 1973; Kachel *et al.*, 1970, 1974) introduced the focusing technique in Europe. Thom's (Thom *et al.*, 1969) model experiments made it evident that edge effects in the electric field of the sizing orifice strongly disturb Coulter volume distribution curves. Our group demonstrated these effects in unique orifices (Kachel *et al.*, 1970; Kachel, 1972); the relationship between pulse signal and particle volume was explained. Particle orientation, rotation, and deformation were demonstrated by a direct photographic technique (Kachel *et al.*, 1970; Kachel, 1972, 1974a, 1976, 1979). Electronic methods to reduce edge artifacts (Kachel, 1972, 1973) have been developed and tested, and a family of new single and multiparameter flow cytometric instruments with incorporated electrical sizing capabilities and electrical calibration have been designed (Kachel, 1974b, 1975, 1976; Kachel and Glossner, 1976, 1977; Kachel *et al.*, 1977, 1980a,b). Another edge artifact reducing transducer with a flow collar was developed by Karuhn *et al.* (1975).

Potential sensing transducers were described by Salzman *et al.* (1973), Leif and Thomas (1973), and Thomas *et al.* (1974). In this technique current electrodes and potential sensing electrodes are separated from each other.

Zimmermann *et al.* (1974, 1975) developed a method which uses the electrical sizing process for studying breakdown effects of cell membranes. This technique recently was extended by Groves (1980) to breakdown measurements of single cells.

Electrical sizing also has been incorporated within flow cytometric sorters (Fulwyler, 1965; Menke *et al.*, 1977; Steinkamp *et al.*, 1973).

In the past, the electrical resistance pulse method was used for counting and sizing of red blood cells (Grant *et al.*, 1960; Brecher *et al.*, 1962; Lusbaugh *et al.*, 1962b; Nevius, 1963; Winter and Sheard, 1965; Doljanski *et al.*, 1966; Adams *et al.*, 1967; Weed and Bowdler, 1967; Bull, 1968; Buckhold *et al.*, 1969; Wilkins *et al.*, 1970; Breitmeyer *et al.*, 1971; Miller *et al.*, 1972; Valet *et al.*, 1972a,b, 1974, 1975a; Hanser *et al.*, 1974; Boss *et al.*, 1973, 1975; Valet and Opferkuch, 1975), white blood cells (van Dilla *et al.*, 1967; Ben Sasson *et al.*, 1974; Ross, 1978; Steen and Nielsen, 1979), platelets (Schulz and Thom, 1973; Paulus, 1975; Haynes, 1980), bacteria (Kubitschek, 1958; Harvey and Marr, 1966; Smither, 1975; Zimmermann *et al.*, 1975), spermatozoa (Brotherton, 1975; Weber and Mueller, 1978), vaginal cells (Cassidy *et al.*, 1975), virus (De Blois *et al.*, 1974), liver cell nuclei (Valet *et al.*, 1975b), heart cells (Nash *et al.*, 1979), nerve cells (Chaussy *et al.*, 1981), plant cells (Kubek and Shuler, 1978), mitochondria (Gebicki and Hunter, 1964), particles in milk (Newbould, 1974), and industrial particles (Spielman and Goren, 1968).

II. BASIC EVENTS AND PROBLEMS IN ELECTRICALLY SIZING TRANSDUCERS

A. The Basic Coulter Effect

Coulter's principle of electrical counting and sizing of particles in the microscopic range is shown in Fig. 1, which shows a typical transducer of a so-called Coulter counter.

The transducer, which counts the cells and transforms the size information into an electrical signal, consists of a glass tube closed at the bottom which is immersed into a container in which the cells to be counted or sized are suspended in an electrically conducting fluid.

The only connection between the interior of the tube where a suction is applied and the outer container holding the particle suspension is a small orifice. In general, it is of cylindrical shape with a diameter between 20 and 200 μm. As a result of the suction, the electrolyte flows from outside the tube to its interior, transporting the cells through the orifice. Simultaneously, a constant electric current flows through the orifice between electrodes that are placed at both sides of the orifice.

Each cell, which must differ in its electrical resistivity from that of the suspending electrolyte, changes the electrical resistance of the orifice if it is passing through it. The resistance change is transformed by the current into a pulse-shaped electrical signal. The height of this pulse generated by each cell is a measure of its volume.

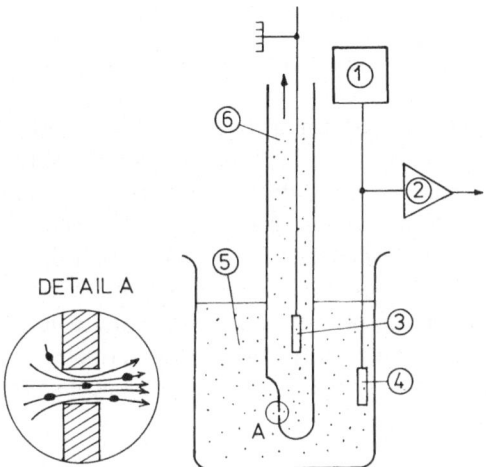

FIGURE 1. The basic Coulter cell counting and sizing device. (1) Current source; (2) amplifier; (3) grounded electrode; (4) active electrode; (5) highly diluted cell suspension; (6) tube where suction is applied. Detail A shows the orifice itself with random particle paths.

The height of the pulse, however, which in most electrically sizing instruments primarily is a voltage pulse, depends not only on the volume of the cells but also on other features of the cells and of the sizing system. The insufficient knowledge of the basic events during the sizing process had led to misinterpretations of electrically measured size distribution curves.

The rest of this chapter attempts to elucidate how electrical sizing works, what the advantages and limitations of this technique are, and how it can be applied and combined with other—particularly with optically working—cytometric methods.

B. Fundamental Relations of Particle Volume, Electrical Resistivity, Resistance Change, and Pulse Height

1. The Mathematical Relation of Pulse Height and Particle Volume

The determination of the particle volume is based on the fact that the electrical resistance of the orifice is changed during the time a particle is passing through it. We first assume that the particles to be sized are of spherical shape and the electrical field in the orifice is uniform, i.e., the electrical current flowing through the orifice is uniformly distributed over the entire cross section of the orifice. In this case, a mathematical relation derived by Maxwell (1883) can be applied for calculation of the pulse height–particle volume relation:

$$\rho = \frac{2\rho_1 + \rho_2 + p(\rho_1 - \rho_2)}{2\rho_1 + \rho_2 - p(\rho_1 - \rho_2)} \rho_2 \tag{1}$$

This relation was derived primarily for calculation of the total resistivity of a medium of the resistivity ρ_2 which has spheres of the resistivity ρ_1 embedded into it. The volume fraction of the embedded spheres related to the volume of the embedding medium is $p = \Delta V/V$. The distance of the spheres from each other is such that the influence of one sphere on others can be ignored.

In order to apply the equation to our problem, we consider only one particle in a cylindrical orifice of a cross section q and a length l, whereby the diameter of the orifice is large in relation to the diameter of the particle.

For the volume of the orifice, where a homogeneous field exists, we introduce the equation

$$V = ql \tag{2}$$

The resistance of the orifice without particles is R_2. However, its resistance during the time a particle is passing through is

$$R = R_2 + \Delta R_2 \tag{3}$$

or, expressed in terms of resistivity,

$$\rho = \rho_2 + \Delta R_2 (q/l) \tag{4}$$

The voltage change detectable at the electrodes is

$$\Delta U = i \Delta R_2 \tag{5}$$

By transforming equation (1) and introducing equations (2) through (5), we obtain for the height of the voltage pulse at the electrodes

$$\Delta U = \frac{\Delta V (\rho_2 i) \left(\dfrac{3[1 - (\rho_2/\rho_1)]}{(\rho_2/\rho_1) + 2} \right)}{q^2} \tag{6}$$

$$\Delta U = \frac{\Delta V(\rho_2 i f_s)}{q^2} \tag{7}$$

Equation (6) identifies the factors that influence the pulse height in a sizing transducer if the field is homogeneous and the particles are spheres: The pulse height is proportional to the current i and the particle volume V and inversely proportional to the square of the cross section of the orifice. The pulse height depends on the resistivites of the particles ρ_1 and the electrolytes ρ_2.

The expression

$$f_s = \frac{3[1 - (\rho_2/\rho_1)]}{(\rho_2/\rho_1) + 2} \tag{8}$$

is the particle shape and conductivity factor. For nonconducting spheres it takes the value 3/2.

As will be discussed below, many types of biological cells—particularly nucleated cells—can be considered spheres for sizing.

The theoretical background for treatment of ellipsoidal particles can be taken from the work of Fricke (1924, 1953) and of Velick and Gorin (1940).

By introducing equations (2) through (5) into the basic equation of Velick and Gorin, one obtains the following for prolate ellipsoids of revolution (which are of particular interest, as will be shown below):

$$\Delta U = \frac{\Delta V \rho_2 i \left(\dfrac{2[1 - (\rho_2/\rho_1)]}{2 + ab^2 \, La \, [(\rho_2/\rho_1) - 1]} \right)}{q^2} \tag{9}$$

This equation is identical to equation (6) with the exception that the shape and conductivity factor of ellipsoids of revolution is

$$f_e = \frac{2[1 - (\rho_2/\rho_1)]}{2 + ab^2 \, La \, [(\rho_2/\rho_1) - 1]} \tag{10}$$

whereby

$$ab^2 \, La = 2 - \frac{2}{1 - (a/b)^2} + \frac{(a/b)\left(\ln \dfrac{\dfrac{a}{b} - [(a/b)^2 - 1]^{1/2}}{\dfrac{a}{b} + [(a/b)^2 - 1]^{1/2}} \right)}{[(a/b)^2 - 1]^{3/2}} \tag{11}$$

where a is the long axis and b is the short axis. Graphical representations of equation (11) are shown in Fig. 2a and b.

Still more complicated is the evaluation of the factor f_e of ellipsoids with three different axes. In this case, elliptical integrals must be evaluated. Examples of form factors of nonconducting ellipsoids with three different axes and variable orientations in a homogeneous field are given by Velick and Gorin (1940).

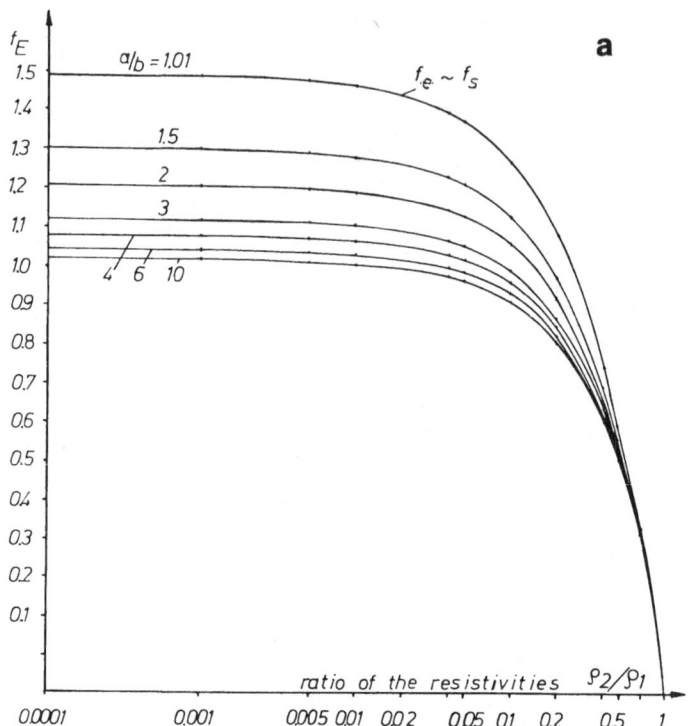

FIGURE 2. (a) The shape and conductivity factor f_e of prolate ellipsoids or revolution as a function of the ratio of the electrolyte resistivity 2 to particle resistivity 1. The ratio of the long axis to the small axis of the ellipsoids is a/b. A sphere is characterized by $a/b = 1$. (b) f_e as a function of the axial ratio of prolate ellipsoids of revolution. For a nonconducting sphere ($a/b = 1$), $f_e = f_s = 1.5$. For considerably conducting particles, the influence of the shape is more and more reduced.

The most practicable way to determine the shape factor of nonspherical or nonellipsoidal cells is to measure identically shaped model particles in an enlarged electrolytic tank model system, as will be shown below.

Equations (6) through (9) describe the interconnection of pulse height, particle volume, diameter of the orifice, resistivity of electrolyte and cells, etc. These equations can be used for establishing a calibrating system, as will be shown below.

If the original assumption that the particle is small with respect to the diameter of the orifice is not valid, the derived formulas are no longer accurate. From calculations by Smythe (1961, 1964), the overestimated volume of large nonconducting particles can be corrected.

Figure 3 shows the formal increase of the shape factors f_s and f_e with increasing ratio of particle diameter to orifice diameter. Smythe has shown that the increase depends on the shape of the particles.

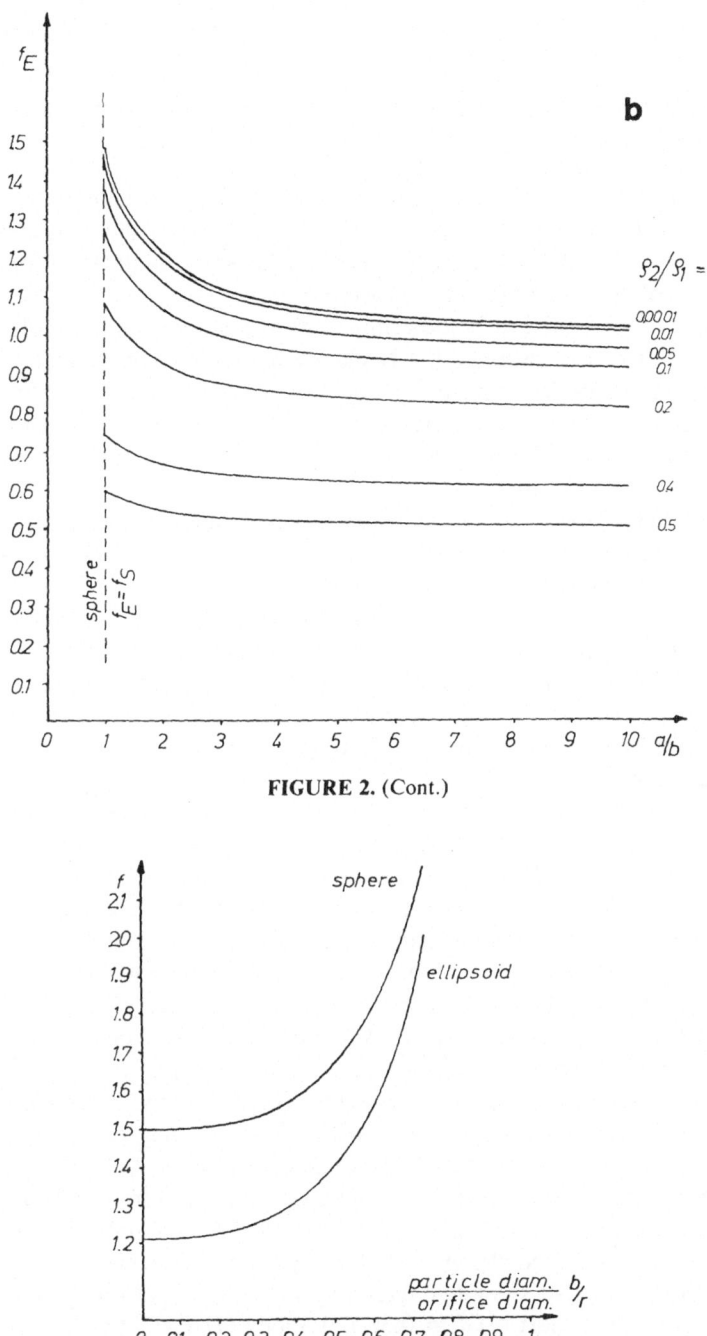

FIGURE 2. (Cont.)

FIGURE 3. Overestimation of large particles, which formally can be treated as an increase of the form factor. If, for example, in a 100-μm orifice nonconducting spheres of 50 μm diameter are measured, a shape factor of 1.65 must be applied instead of 1.5 for correct interpretation of the measured pulse height.

2. Pulse Height and Electrical Field Measurements in Enlarged Model Orifices

a. The Electrolytic Tank Instrumentation. The mathematical relations derived above are true only if the electric field inside the sizing orifice is homogeneous, i.e., if the electric current that flows through the orifice is uniformly distributed over the entire cross section.

The electric field can be determined by measurements in an electrolytic tank model or by evaluation of the Laplace equation by a numerical relaxation method (Wendt, 1958). In a similar way, the height of electrical volume pulses can be studied in an electrolytic tank model system. Such experiments are described in detail by Kachel (1972). The most important results are summarized below.

The model experiments have been performed in enlarged orifices of 11 cm diameter and six different lengths between 2.75 cm and 22 cm corresponding with length to diameter ratios of 0.25 to 2.

By reasons of symmetry, only the lower halves of the orifices had been modeled. In such a way it was possible to measure the axially symmetrical electric field distribution with small needles from the surface of the electrolyte and the resistance change by immersing model particles into the electrolyte. Figures 4 and 5 show schematic graphs of the field and resistance change measuring devices.

Resistance change and fields strength have been measured either in the natural electric field of the orifices or in the field that was generated by the same current but made homogeneous by putting parallel plates of stainless steel at both ends of the orifices. In such a way, resistance change as well as field strength in the natural and homogenized fields could be compared directly. The basic idea was to have always a relationship to the homogeneous field in which the calculations described above are true. Figure 6 shows the model particles and the field sensor. A model orifice with homogenizing plates and the particle-immersing mechanism are shown in Fig. 7.

b. Results of the Model Experiments. Figure 8 shows a typical plot of a resistance change model experiment with a sphere having a volume of 1 cm^3 in an orifice having a diameter of 11 cm and a length of 7.75 cm. The sphere was moved centimeter by centimeter along the axial path from outside into the orifice. At each position the sphere was immersed several times in the electrolyte. Thus the resistance change was directly evident at each position.

Figures 9 and 10 show two examples of field strength and pulse height measurements in a short ($l/d = 0.5$) and a long ($l/d = 2$) orifice. In the (c) portions of both figures, the equipotential lines of the natural field are plotted in steps of 2% of the potential between the electrodes. The distance of the lines is a measure of the field strength. The field is homogeneous if the lines are parallel.

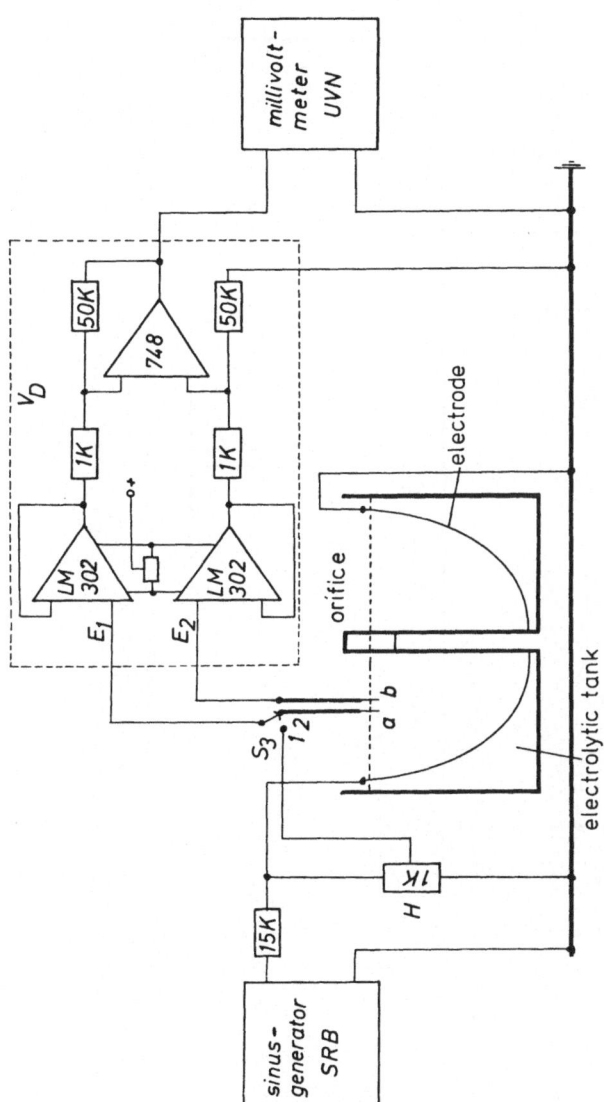

FIGURE 4. The electrolytic tank circuitry for measuring the electric field in and around model orifices. For measuring equipotential areas, S_3 was switched to position 1 and the potential of the equipotential line was determined by the potentiometer H. Probe b was used to sense the potential of the surface of the electrolyte. With S_3 at 2, both probes mounted in a distance of 0.45 mm sensed the field strength directly. The a.c. supply voltage frequency was in the 100–1000 Hz range. By reasons of symmetry the measured values are also valid for the three-dimensional space in and around the orifice.

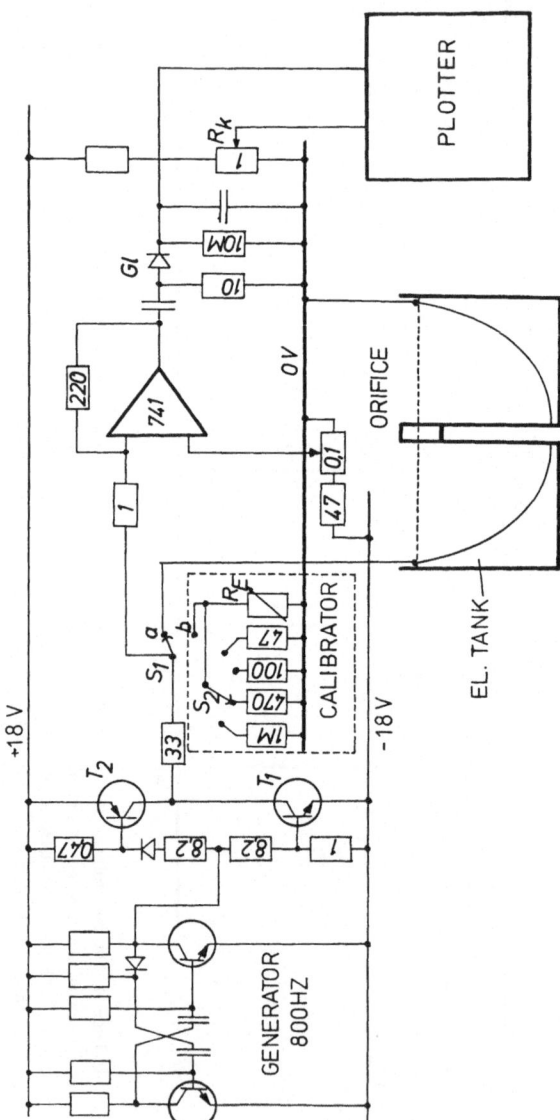

FIGURE 5. Circuitry for the resistance change measurement. Model particles were immersed into the model orifice from top. The resistance change was sensed by the amplifier 741, rectified and plotted at the plotter S. The circuit was absolutely calibrated by simulating the orifice by R_E (50-Ω range) and switching different parallel resistors.

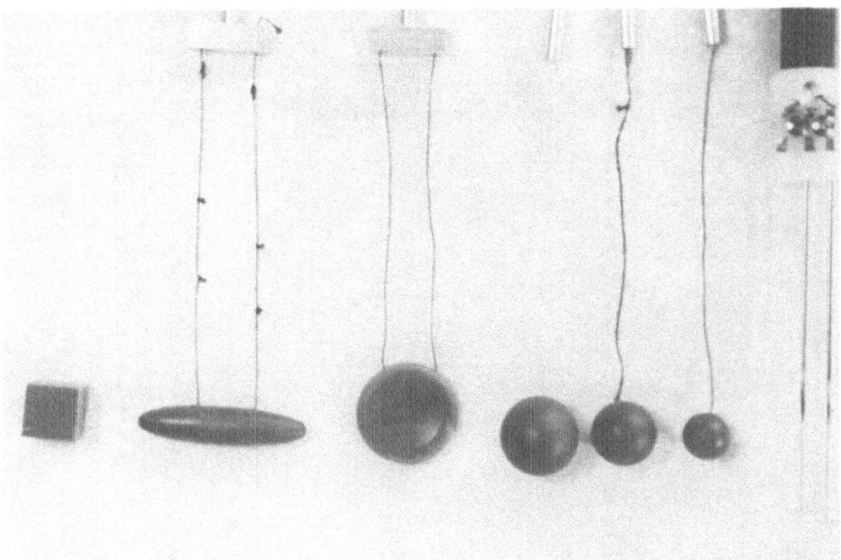

FIGURE 6. The model particles for the resistance change experiments, made from nonconducting PVC. The cube at right is the reference body of very precisely 1 cm³. The other bodies were referenced to the cube by weight. The axial ratio of the ellipsoid of revolution is 4:1. The field sensor is shown at right.

FIGURE 7. The particle-immersion mechanism over a model orifice with parallel plates put on its inlet and outlet. By putting on and taking off the stainless steel plates, a natural or homogenized field was produced in the orifice. By reasons of symmetry, only the lower part of the orifice system was modeled. The electrodes, placed about five orifice diameters outside the orifice, were made of half-spherical stainless steel mesh. The particle automatically was immersed several times at each position.

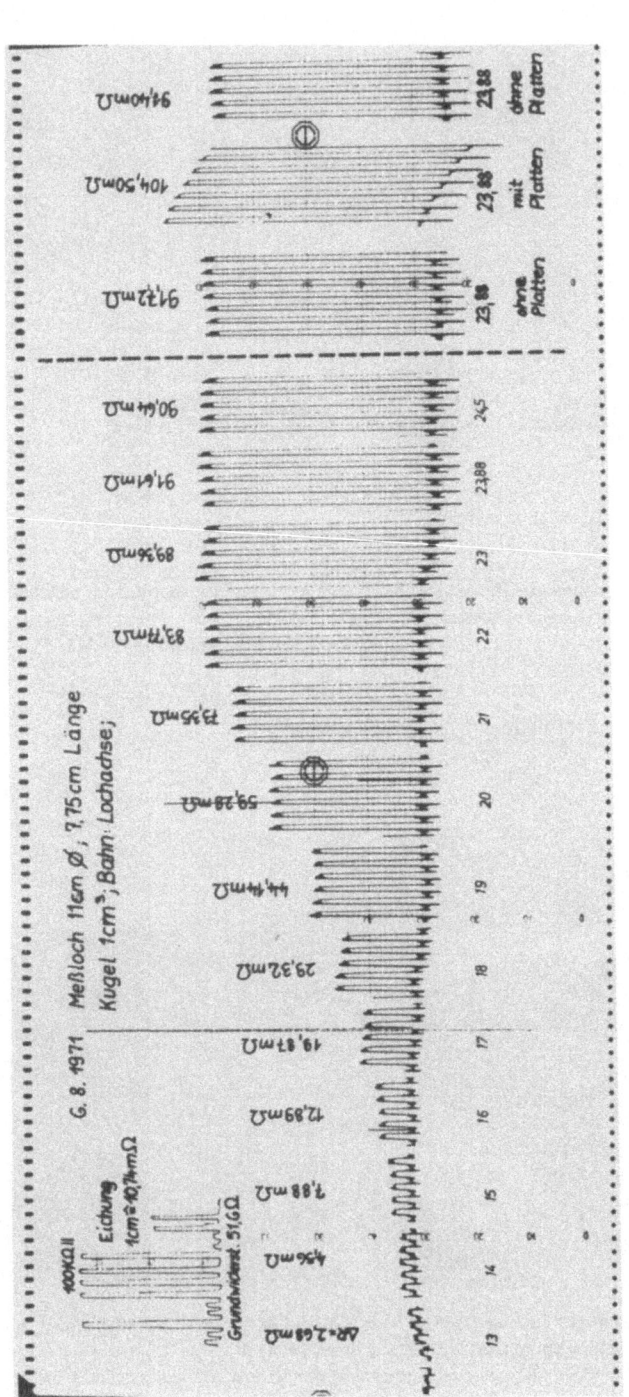

FIGURE 8. A typical plot of a resistance change model measurement along the axial path of an orifice with a length-to-diameter ratio of 0.705. At left, the particle is outside the orifice. By moving to the right, it is wandering into the orifice. By reasons of symmetry, only one-half of the orifice from outside to the middle area was measured. At each position the resistance change was reproduced several times. The numbers at bottom mark the distance scale in centimeters. The orifice inlet is arbitrarily set to 20. At the right-hand side of the broken line, the comparison of pulse heights in the natural field (91.7; 91.4 mΩ) and the field homogenized by the parallel plates (104.5 mΩ) are shown. At the top left is the calibration: a parallel switch of 100 kΩ to 51.6 Ω results in a resistance change scale of 10.74 mΩ/cm in the original protocol.

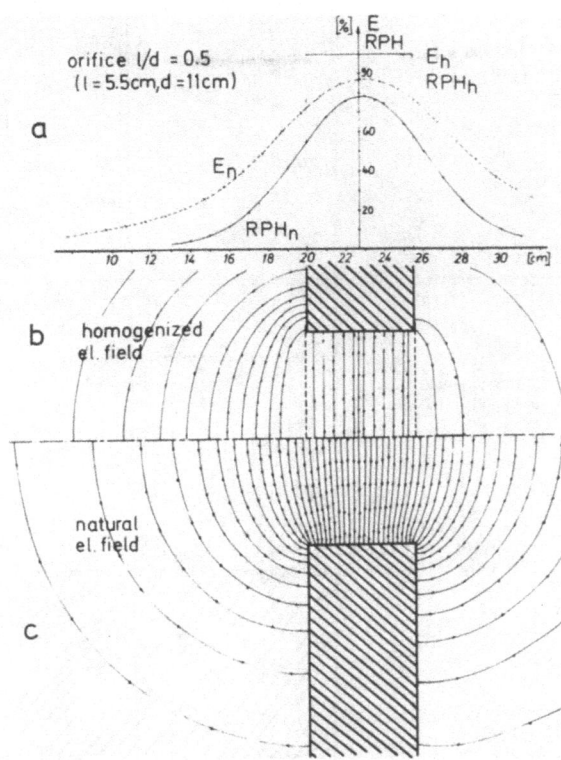

FIGURE 9. Model measurements in a short orifice with a length-to-diameter ratio of 0.5. (a) Field strength E_n and resistance change RPH_n measured along the axial path. Both curves are normalized to the corresponding values in the homogenized electrical field, i.e., the y scale shows the field strength and resistance change in percent of the values in the homogenized field. The center of the orifice is found at 22.75 cm of the x scale and the inlet and outlet at 20 cm and 25.5 cm. Field strength and resistance change curves differ in shape and height. The E^2 curve, however, equals the RPH curve. (b) Pattern of equipotential lines if the field inside the orifice is made homogeneous by putting parallel plates of stainless steel at both inlet and outlet. (c) Pattern of the equipotential lines of the natural electric field. In the central area of the orifice ($x = 22.75$ cm), the 50% potential line is found. The potential difference of the lines is 2% and the local distance of the lines is a measure of the field strength.

At no point in the natural field of the short orifice does a homogeneous field exist. Inside the homogenizing plates the field is homogeneous as expected and as shown in the (b) portions of the figures. But also in the inner portion of the long orifice the natural field is homogeneous, as is indicated by the parallel lines. The (a) portions show field strength E and resistance change pulse height (RPH) along the axial path. The 100% values on the scales mark the homogeneous field strength, which is measured with the plates at both ends of the orifices.

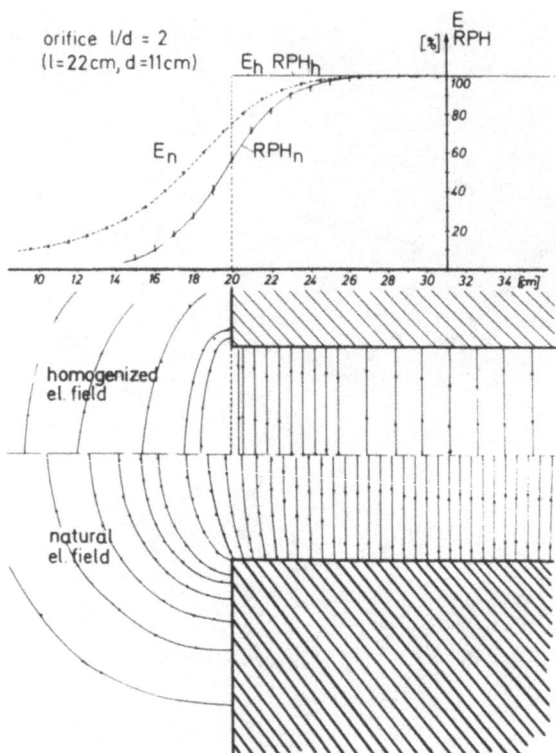

FIGURE 10. Model measurements in a long orifice with a length-to-diameter ratio of 2. Inside the orifice, even in the natural field a homogeneous electrical field is found. Pulse height and field strength reach 100% in the central area of the orifice.

The natural field strength at the axial path of the short orifice reaches only 87% of the homogeneous field strength. In the long orifice, both the natural and the homogenized field strength match 100%.

The pulse height curves RPH are measured in the axial path of the orifices. We notice two striking facts: even outside the orifice a resistance change is produced by the particles, and the field strength and pulse height curves are of different height and shape. The pulse height curve follows the E square curve, as is shown in the (a) portions of Figs. 9 and 10.

This conspicous fact is explained by the following consideration. According to equations (7) or (9),

$$\Delta U = \frac{\Delta V(\rho_2 i f)}{q^2} \tag{9'}$$

$$E = \frac{\rho_2 i}{q} \tag{12}$$

By introducing equation (12) into equation (9′), we obtain

$$\Delta U = \frac{E^2 \Delta V(f)}{\rho_2 i} \tag{13}$$

Equations (9′) and (13) indicate that the field strength E is inversely proportional to the cross section which is passed by the current i. The pulse height,
however, is inversely proportional to the square of the cross section q. The cross
section q in this case is not the cross section of the orifice (this is true in a
homogeneous field) but the area over the entire path of the current i, where an
identical current density with respect to field strength exists. In our orifice system, the areas of identical field strength within a considerable distance from
the orifice inlet are surfaces of halfspheres. For a very small particle, a homogeneous field can be assumed to be everywhere over such an area of constant
field strength.

Equation (13) indicates that the pulse height is proportional to the square
of the field strength if the resistivity ρ_2 and the current i are kept constant and
the particle is moving through the orifice. The constancy of i and ρ_2 is important insofar as E itself depends on ρ_2 and i, according to equation (12). This
square relation makes it possible to construct a pulse height topograpy over the
entire orifice from the field strength topograpy.

Figure 11a summarizes the pulse height measurements on the axial path
in the model orifices of different l/d ratios. Above l/d ratios of 1, the field
distribution in the central area of the orifices is such that in the natural field,
field strength and pulse height reach the values of the homogeneous field.

Figure 11b shows the pulse heights depending on the radial position in the
central planes of the orifices. In long orifices in which a homogeneous field
exists, the pulse height in the central plane is independent of the radial position
of the particle path. If the l/d ratio is reduced more and more with the consequence of a growing inhomogeneity of the electric field, the pulse height
depends more and more on the radial position of the particle path.

Figure 12 illustrates how the model system can be used to determine form
factors of particles of any shape or orientation. The model particles are identical in volume but differ in shape and orientation. The scale shows the measured resistance changes related to the true volume of 100%.

3. The Calculated Electrical Field in the Sensitive Zone

Another way to determine electrical fields is to solve the Laplace equation
by a numerical relaxation method (Wendt, 1958). With this method, a net of
field points is calculated by iteration. With a computer program (Koller, 1970),
three dimensional fields of axial symmetry can also be determined. The net

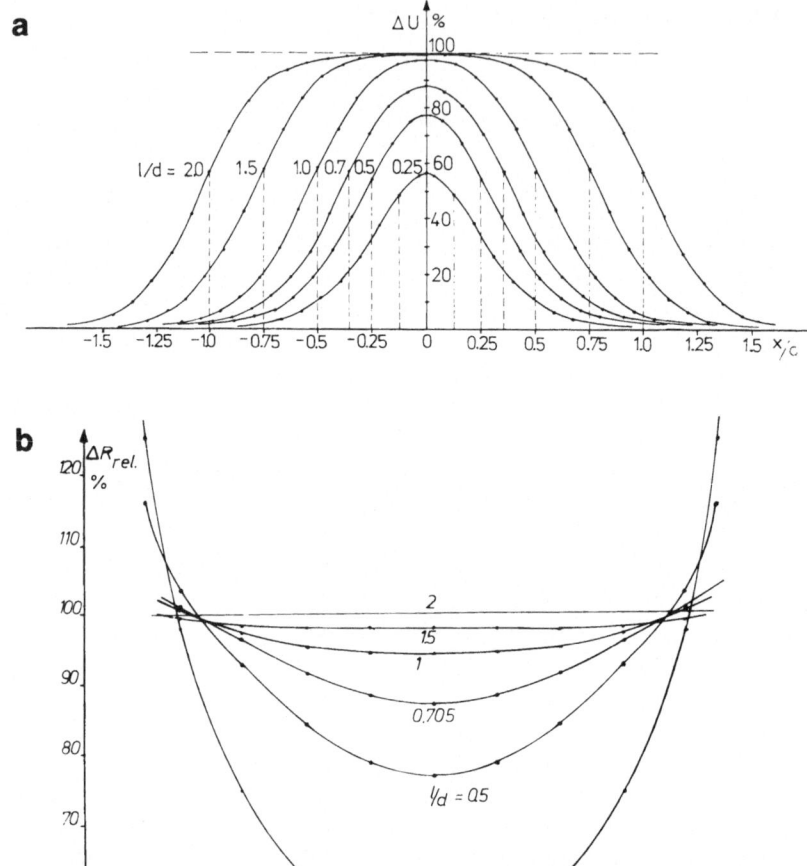

FIGURE 11. (a) The result of the pulse height measurements with spheres at the axial paths of the model orifices. The origin of the x axis is in the central area of the orifices; the x distances are normalized to the orifice diameter d. The percent scale of the y axis indicates the pulse height in relation to that of the homogenized field. The broken lines mark the orifice inlet and outlet for each pulse. The pulse maxima of the short pulses are far below the amplitudes expected in the homogeneous field. (b) The pulse heights in the model orifices as a function of the radial position of the particles in the central area of orifice of different l/d ratios. The edge distortions are not considered in this graph. The shorter the orifice is, the more the pulse height depends on the radial position of the particle path.

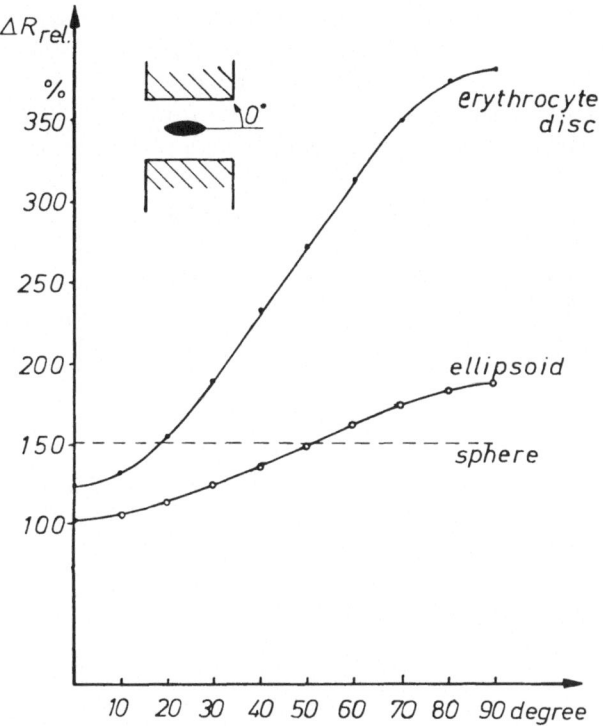

FIGURE 12. The change of the shape factors of nonspherical particles if rotated in the homogeneous field of a long orifice ($l/d = 2$). The particles are shown in Fig. 6.

points of the electrical fields of cylindrical orifices of different l/d ratios are tabulated in Appendix A. In these tables, E/Em is the relative field strength (in percent) related to the field strength in the homogeneous field, and U/Um is the pulse height (in percent) related to the pulse height in the homogeneous field. The distances in axial direction x and radial direction r are related to the orifice diameter d. The x scale is defined such that the orifice inlet is found at $x/d = 0$ and the negative values mark the region outside the orifice. With the radial scale, the orifice axis is found at $r/d = 0$ and the orifice wall at $r/d = 0.5$. The normalization to the diameter of the orifice makes the tables applicable to orifices of any diameter.

For reasons of symmetry, only a quarter of the entire field of the sensing zone of the orifices is calculated.

4. The Calculated Topography of Pulse Amplitudes in Cylindrical Orifices

The pulse amplitudes are proportional to the square of the field strength at each point in the orifice as indicated by equation (13). As in the model experiments, the field strength and pulse height values should be normalized to the adequate homogeneous field, i.e., to the mean field strength and to the pulse height in the field of the mean field strength.

The mean field strength E_m is calculated from the field strength distribution in the following way. The current through the orifice is:

$$i = j_m q = E_m q (1/\rho_2) \tag{14}$$

where j_m is current density. This current is also equal to the sum of the partial currents that flow through the ring-shaped areas of the orifice with the field strength E_j. The current in each ring area equals

$$i_j = E_j (2\pi r_j c)(1/\rho_2) \tag{15}$$

where $1/\rho_2$ is resistivity of the electrolyte, E_j is the partial field strength taken from the field calculations, and $2 r_j c$ is the ring-shaped area of the thickness c where E_j is valid. We obtain

$$q E_m = c\pi (E_1 \, 2 r_1 + E_2 \, 2 r_2 + \ldots\ldots\ldots)$$

and

$$E_m = \frac{2 \, c \, \pi}{q} \sum_1^n E_i r_i \tag{16}$$

Figures 13–15 are plotted from a net of pulse height points whose relative pulse height is calculated according to

$$\Delta U_x[\%] = \frac{\Delta U_x}{\Delta U_m} = \frac{E_x^2}{E_m^2} \cdot 100 \tag{17}$$

The U_x/U_m values (in percent) for orifices of different length-to-diameter ratios are tabulated in the Appendix. By connecting all points of identical height, contour lines such as those on a map are generated. From such a pulse height distribution for each particle path, the adjoined pulse shape and height can be constructed. The pulse shape, however, is also influenced by the flow velocity at the path considered. The comparison of the calculations with the model experiments show that both determinations are in good agreement (Fig. 11).

The pulse curves of Fig. 14b are constructed from the particles' paths PI–PIV of Fig. 14a. The scale of the abscissa reflects distances from the center

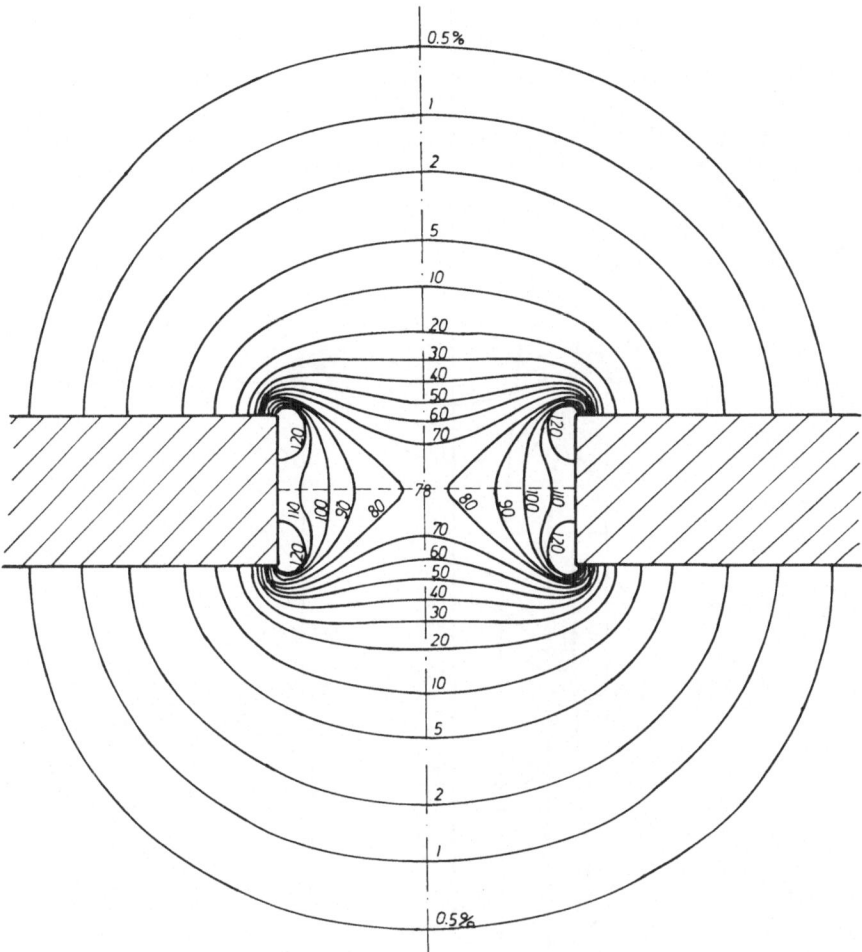

FIGURE 13. Pulse height topography in and around an orifice with an l/d ratio of 0.5. This and the patterns of Figs. 14 and 15 are derived by the E square relation from the field strength values tabulated in the Appendix. Due to the symmetry of revolution of the electric field, this topography is valid in the three-dimensional space around the orifice axis. The lines mark areas of identical pulse height in percent of the pulse height that would be produced by a particle in a homogeneous field of the same orifice. A homogeneous field exists if the current flowing through the orifice is uniformly distributed over the entire cross section of the orifice. In this short orifice the field and pulse height distribution is far from being uniform, resulting in a very different valuation of the particles at the orifice axis and at its wall.

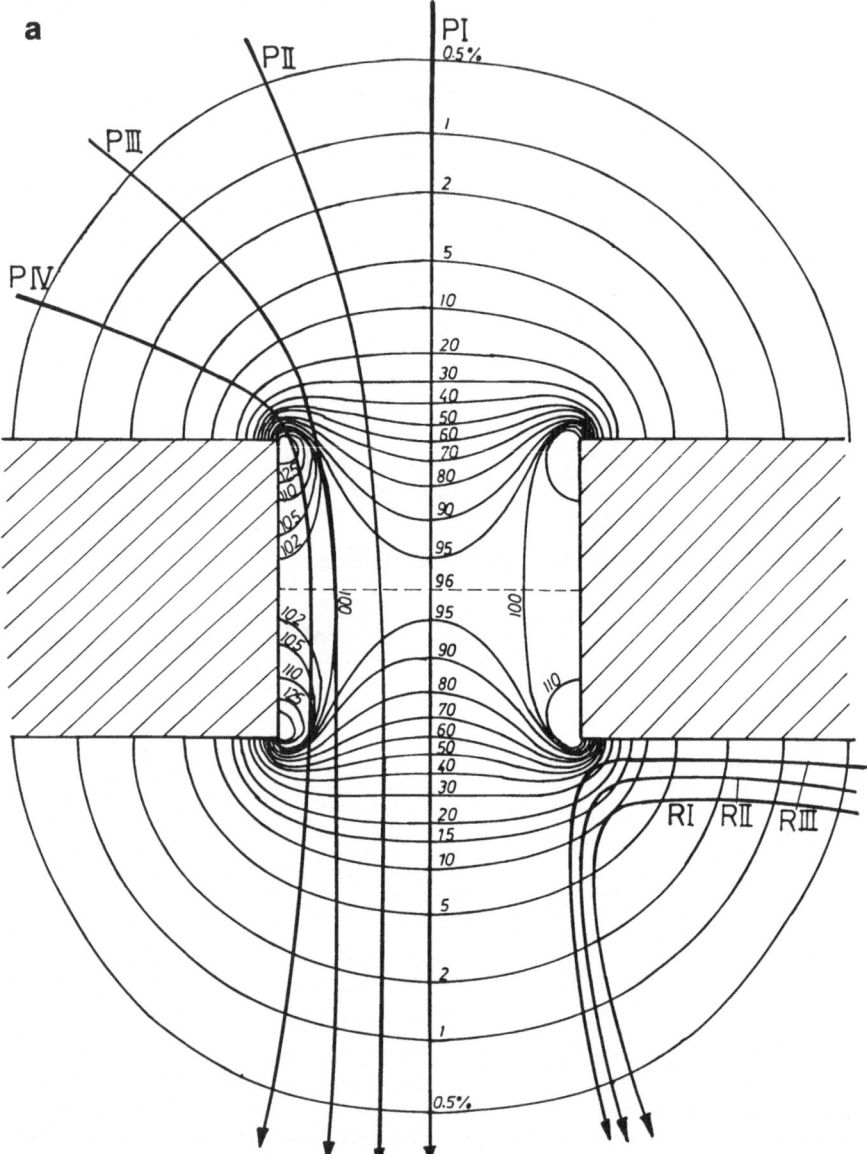

FIGURE 14. (a) Pulse height topography of an orifice with an l/d ratio of 1. Along the central area of the orifice marked by the broken line, the field approximates homogeneity. PI–PIV are presumable particle paths. Pulses constructed from these paths are shown in (b). RI–RIII are paths of recirculating cells, i.e., of cells which are just measured before and reenter the sensitive zone from the back. Pulses from such paths are shown in Fig. 33. (b) Pulses constructed from

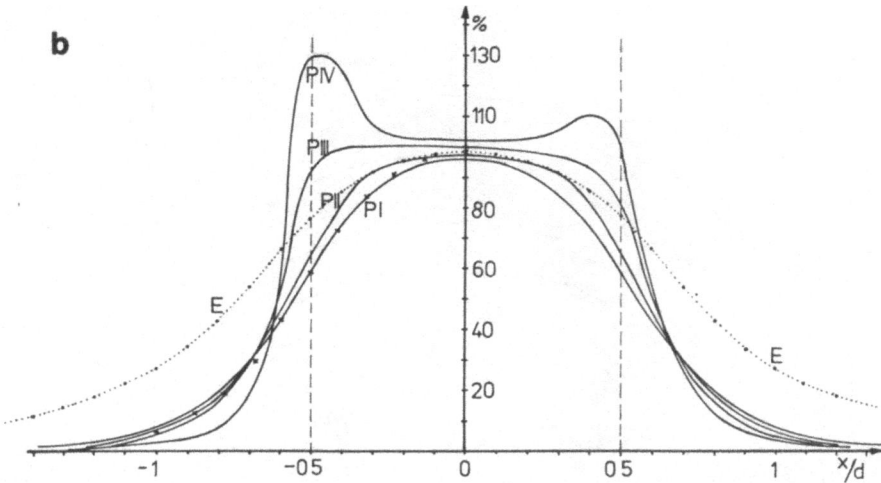

the particle paths PI–PIV of (a). The pointed curve shows the field strength along the axial path. The origin of the x axis is in the center of the orifice. Orifice inlet and outlet are marked by the broken lines.

of the orifice normalized to the orifice diameter d. For comparison with the oscillographic pulse patterns of Fig. 21, in which the abscissa is the time, the flow velocity profile (see Fig. 19) must be taken into consideration. Pulses from paths near the orifice wall are delayed by the friction.

C. Flow Conditions and Flow Line Coordination in the Sensitive Zone

1. General Remarks

An electrically conducting fluid transports the cells suspended in it through the sensing zone of the transducer. Therefore, the flow conditions play an important role in cytometric, and particularly in electrically sizing, transducers.

In general, two types of flow are distinquished: laminar flow and turbulent flow. In a laminar flow path, flow lines and flow layers do not mix. They remain in coordination even if the flow path changes its shape. In a turbulent flow, the flow lines are mixed and distorted in a very complicated way.

Flow in a sizing transducer must be laminar. A dimensionless number found by Reynolds defines whether flow can be considered as laminar. The Reynolds number is calculated as

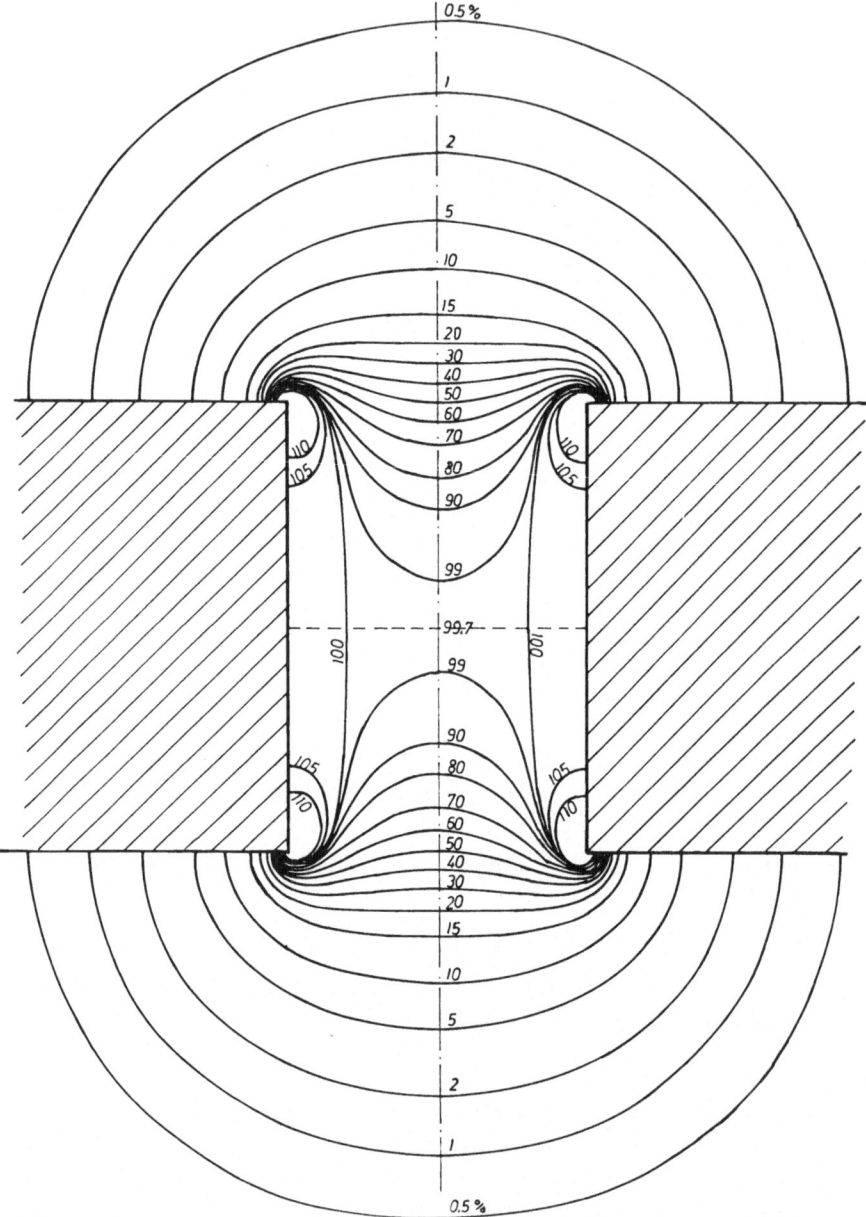

FIGURE 15. The pulse height topography of a long orifice with an l/d ratio of 2. The electric field in the central area is nearly completely homogeneous, effecting there pulse amplitudes which are independent of the radial position of the particle path. Particles at paths near the orifice wall also produce distorted two-peak pulses.

$$R = \frac{d \cdot v_m}{v} \tag{18}$$

where d is diameter of the flow path, v_m is mean flow velocity, and v is dynamic viscosity of the fluid ($= 0.01$ cm/sec at $20°$C for water).

The mean flow velocities in electrical sizing instruments are in the range of 1–5 m/sec. The range depends mainly on the pressure difference that is applied to the transducer. The mean flow velocity is determined by collecting the flow output V_t of the transducer during the time t. With the cross section q of the orifice,

$$v_m = \frac{V_t}{q\,t} \tag{19}$$

Mean flow velocities of usual orifices, as a function of the pressure applied, are shown in Fig. 20. The so-called critical Reynolds number, $Re_{kr} = 2300$, describes the transition between laminar and turbulent flow. Below Re_{kr}, the flow is laminar. In general, Reynolds numbers in small sizing orifices are far below the critical number.

2. The Sizing Flow Path in The Orifice

In a typical sizing transducer, the fluid, coming from a wide space, is concentrated into a thin hole. Such a sink trap flow is characterized by straight flow lines directed toward the orifice. These flow lines describing the direction of flow are coordinated with areas perpendicular to them which connect all points of identical flow velocity.

In a field of potential flow, where the friction of the walls is neglected, the areas of identical velocity outside a distance of $s > 2r$ from the inlet of the orifice are of half-spherical shape (r = radius of the orifice). By approaching and entering the orifice, the flow is transformed into a parallel flow with plane circular-shaped areas of equal velocity. Between $s = 2r$ and the inlet of the orifice, these areas can be approximated by half-ellipsoids or segments of spheres.

Figure 16 explains the flow line coordination and the velocity relations. From the law of mass conservation in laminar flow follows the equation

$$v_A\,q_A = v_m q \tag{20}$$

The flow velocity in a distance s from the inlet of the orifice is calculated as

$$v_A = \frac{q}{q_A}\,v_m \tag{21}$$

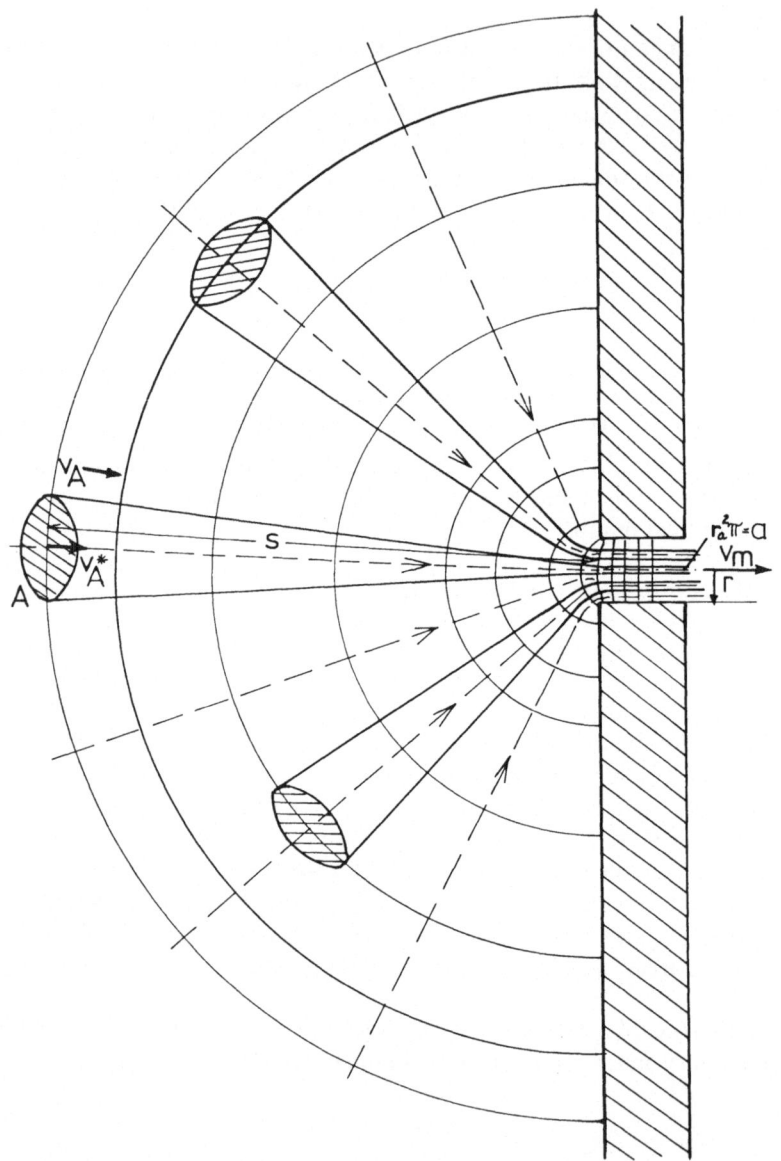

FIGURE 16. The flow line coordination that explains the hydrodynamic focusing principle.

for $s > 2r$, with $q = r^2\pi$ and $q_A = 2\pi s^2$.

$$\frac{v_A}{v_m} = \frac{r^2}{2s^2} \tag{22}$$

For $s > 2r$, q_A can be approximated by surfaces of half-spheroids. Figure 17 shows the flow velocity and flow acceleration in the axial region of the inlet portion of an orifice. The main acceleration takes place at $x/d = -1$, i.e., about one orifice radius in front of the orifice inlet.

From Fig. 16, the principle of flow constriction can be derived. The laminar flow lines forming the boundary of the cross section A on the velocity area v_A are constricted in such a way that they delimit the cross section a inside the orifice. By the principle of not mixing flow lines, the flow constriction is calculated:

$$\frac{A}{a} = \frac{2s^2}{r_a^2} \tag{23}$$

$$a = A\frac{r_a^2}{2s^2} \tag{24}$$

For a given orifice and a given input area A, the constriction depends only on the distance s if the fluid flowing through A is also moving with $v_A^* = v_A$ (Fig.

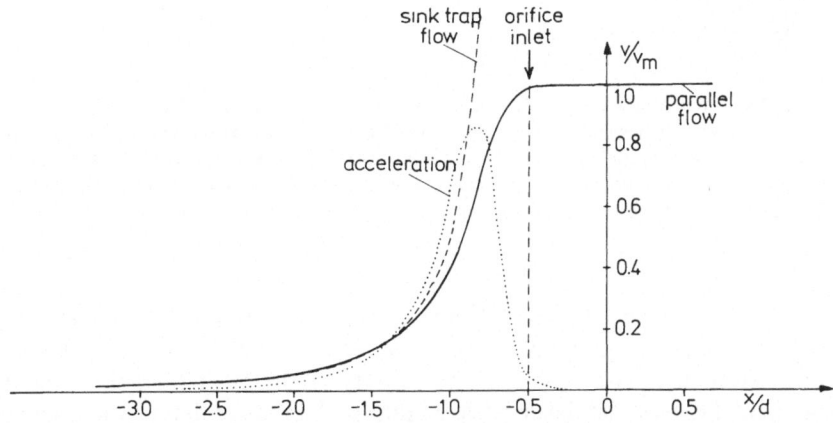

FIGURE 17. The flow velocity along the axial path in the sensitive zone. Outside of $x/d = -1.5$, i.e., a distance of one orifice diameter outside the orifice inlet, a sink-trap flow can be assumed that follows the $(1/s)^2$ law. Inside the orifice the flow is parallel. In the transition zone between both defined flow types, the maximum flow acceleration is shown by the pointed curve.

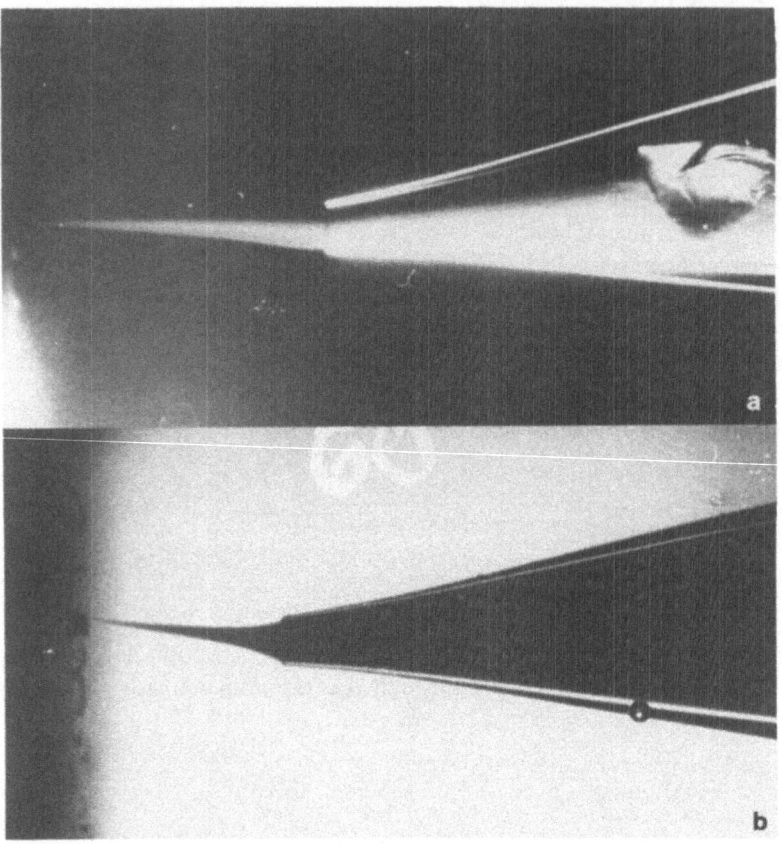

FIGURE 18. Practical illustration of the hydrodynamic focusing principle. (a) The glass tip at right ejects a highly concentrated suspension of fixed erythrocytes that is steadily focused by the flow forces on its way to the orifice at left (not seen). (b) The glass tip ejects black ink, whereby the flow velocity at the outlet of the glass tip is lower than the flow velocity in the focusing fluid, causing an additional constriction of the ink stream immediately after immersion from the tip.

18a). If $v_A^* < v_A$ the introduced flow path immediately is constricted (Fig. 18b), and if $v_A^* > v_A$ it is bulged out (Fig. 18c) until the velocity of the fluid flowing through A is adapted to the velocity of the other fluid. Moving the position of A moves also the constricted flow path in the orifice, or using several input areas A simultaneously—several completely separated—constricted flow paths in one orifice may be generated (Fig. 18d).

This principle of constricting streams of fluids, which can also bear particles, is called "hydrodynamic focusing" (Spielmann and Goren, 1968). It is

(c) In contrast to (b), the flow velocity in the tip outlet is higher than the velocity of the focusing flow. The ink stream is bulged out. (d) A double-focusing device. Both ink streams are separately focused without interaction. Even in the area behind the orifice, the ink threads remain undistorted. This device can be used for comparative investigations on different samples.

extensively used in flow cytometric instruments to direct cells to the points of measurement.

3. The Velocity Profile in the Orifice

In a tube with laminar flow, the characteristic parabolic flow develops over the so-called inlet length $x = 0.06\ d\mathrm{Re}$ (Langhaar, 1942). The flat core of the flow profile over this length is transformed more and more to the typical parabolic flow profile of a tube (Fig. 19). In the core of the flow path, the fluid is moving entirely with identical velocity. No shear forces exist between flow lines, which could affect particles moving in the fluid. Near the orifice wall the flow velocity decreases to zero. In this region, adjacent flow lines have different flow velocities, which can rotate particles flowing there (see below). Figure 19, derived from Tatsumi (1952), explains how the flow profile develops depending on the distance from the inlet length, which is normalized to (dRe).

Figure 20 shows how the mean flow velocity in different sizing orifices depends on the suction applied.

D. Behavior of Biological Particles in the Sizing Transducer

In the previous sections, the theoretical situation in sizing transducers was considered. This section describes how cells behave in practice in sizing transducers.

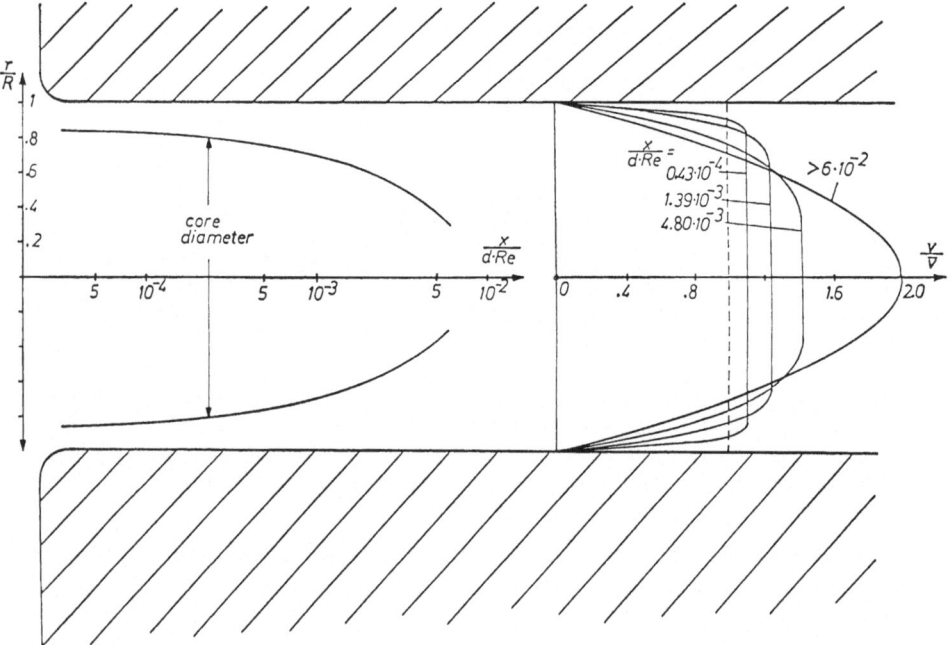

FIGURE 19. The flow velocity profile in the inlet length of cylindrical tubes according to Tatsumi (1952). In the axial region of the orifice, a flat profile is found with constant flow velocities. Near the orifice wall the flow velocity decreases to zero, an effect which is caused by the friction of the wall. On the way along the tube, the profile is transformed into a fully developed tube flow with a parabolic profile. In the limited length of sizing orifices, a flat profile exists around the orifice axis.

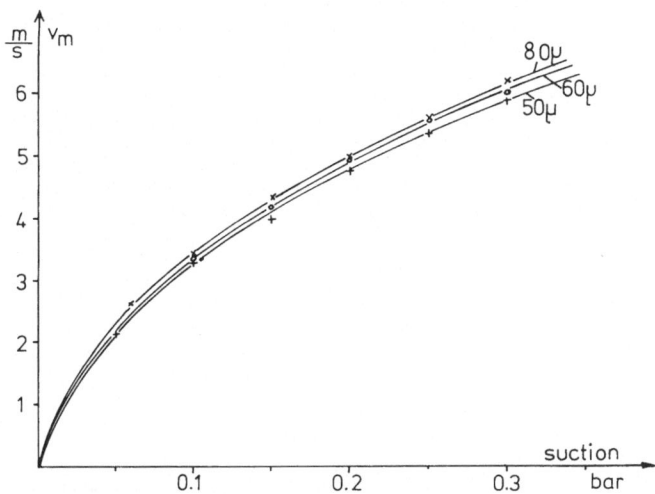

FIGURE 20. The mean flow velocities in orifices of different diameters as a function of the pressure applied.

1. The Equipment Used for the Particle Experiments

Figure 21 shows schematic diagrams of two types of flow chambers in which the sizing orifice can be observed by a high-resolution microscope. In the vertical chamber, the flow path is observed in axial direction, i.e., the cells passing through the orifice are seen from the back, and in the horizontal chamber the cells observed from radial direction are seen from the side.

By using the principle of hydrodynamic focusing, distinct particle paths are generated. Both chambers are covered by a coverglass, which is bearing the injection tip of the cell suspension. By moving the coverglass, the injection point of the cells and, as explained above, the particle path inside the orifice can be moved to different radial positions. The volume pulses picked up from electrodes in the chambers have been amplified, made visible at the screen of an oscilloscope, and stored in a pulse height analyzer.

For either to make visible or to photograph the fast-moving cells, a very short illumination time of 40 nsec is required. Our photographing device was built from a Zeiss standard microscope, equipped with a 24 × 36 mm camera and a Nanolite 18N18 flash tube with an Argon pressure chamber that was activated by a Nanolite driver (Impulsphysik, Hamburg) and a particle triggered circuit. The standard sizing instrumentation is described below.

2. Pulse Shape and Height at Different Paths in a Sizing Orifice

From the inhomogeneities of the electric field inside short sizing orifices as shown in Figs. 9–15, differences in pulse height are expected if particles are

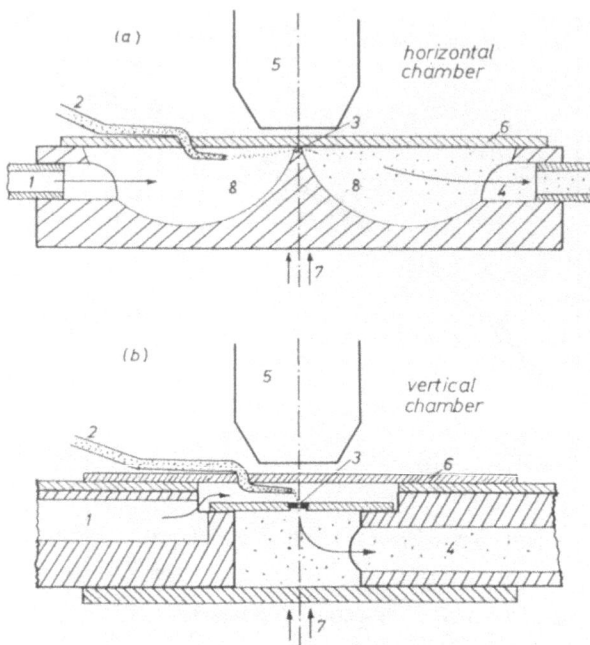

FIGURE 21. Two types of experimental transducer chambers which are specially suited for optical investigations of flow events and particle behavior in the orifice. (1) Supply of particle-free solution and current path form the electrodes not seen; (2) tube conducting the particle suspension; (3a) orifice with rectangular cross section (100 μm × 100 μm), (3b) orifice with circular cross section (100 μm diameter); (4) tube connected to the pump and the grounded electrode; (5) objective of the microscope; (6) movable cover glass; (7) illumination; (8) semispherical excavations connected by the orifice. The electrodes are located outside tubes (1) and (4).

passing at different paths through such an orifice. An experiment performed with fixed human erythrocytes in the vertical chamber is shown in Figs. 22 and 23. The stream of particles, which normally cannot be recognized if illuminated by permanent light, is made visible by mixing the suspension with black ink.

The pulse shape changes from that of a bell at the axial path to that of a trapezoid and then to two peaked m shapes from cells at paths near the wall of the orifice. The friction-caused retardation of flow speed elongates the pulse duration at paths near the orifice wall.

The orifice used in this experiment has a l/d ratio of about 0.8. The inhomogeneous field makes the pulse height dependent on the radial position of the particle path. The moderate increase of the mean values from 2.44 to 2.67 between K6 and K17 is caused by the field in the central area. Here the pulses are bell-shaped or trapezoidal.

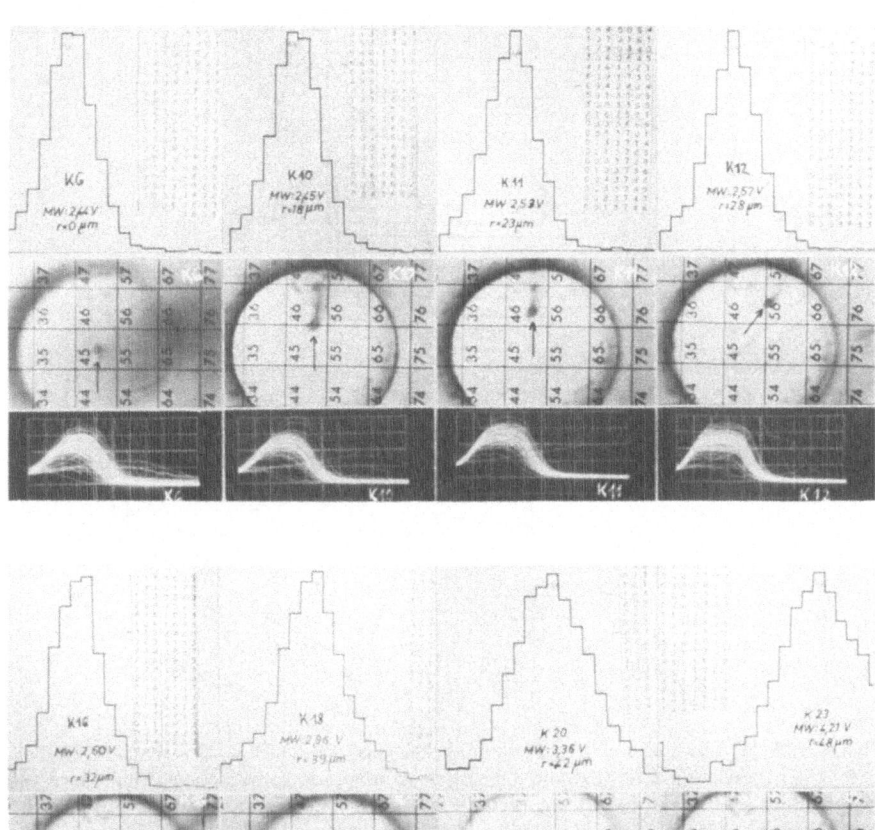

FIGURE 22. An experiment in the vertical chamber that demonstrates how pulse height and shape depend on the position of the particle path in the orifice. The lower pictures present axial views through the 100-μm-diameter orifice taken just when the cells were sized. The suspension of the fixed erythrocytes is made visible by black ink. MW, Mean value of the pulse heights in V; r, distance of the particle stream from the orifice axis. Current through the orifice 0.46 mA; suction is 0.08 bar; time axis is 10 μsec/division.

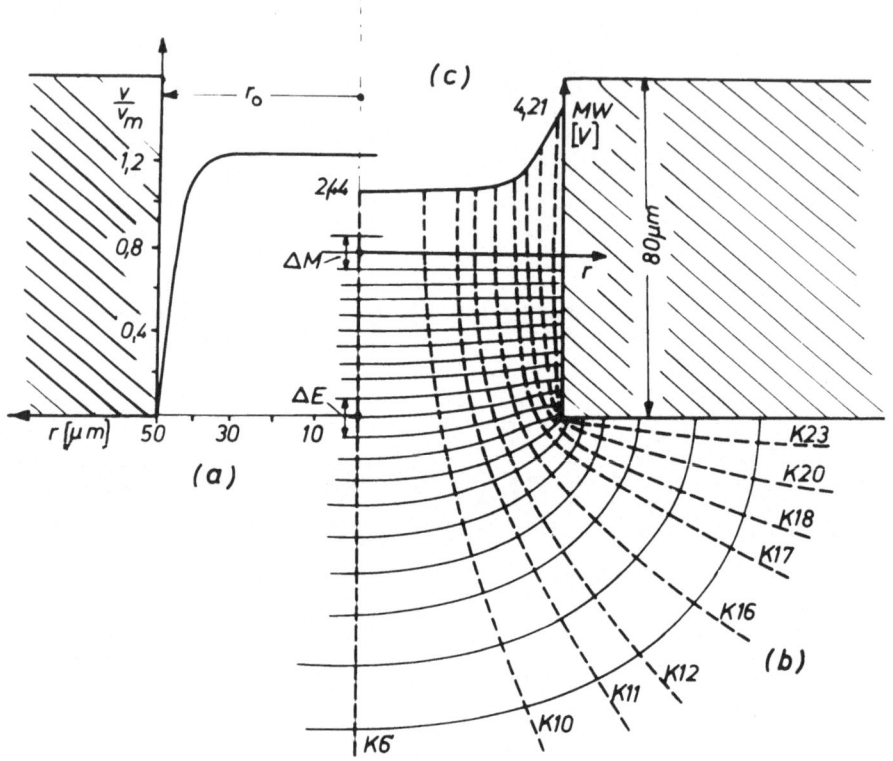

FIGURE 23. (a) The flow profile in the orifice of the experiment of Fig. 22 (calculated for x/d = 0.8). (b) Presumable side view onto the particle paths of Fig. 22 with equipotential lines. (c) The mean values of the distribution curves of Fig. 22 as a function of the radial position of the particles in the orifice.

The steep increase at the outer paths is caused by the edge distortions of the field, recognizable by the two-peaked shape of the pulses.

The results of this experiment are in good agreement with the predictions that can be made for different paths from the pulse height topography curves of Fig. 14a,b.

The pulse height curves generated at distinct paths can be summed to a composite curve, whereby the flow velocity profile in the orifice must be taken into consideration. Such a composite curve is in good agreement with a general distribution measured without use of the focusing device (Kachel *et al.,* 1970) (Fig. 24).

This experiment clearly demonstrates, that the hydrodynamic focusing technique must be applied if correct pulse height distributions are to be measured.

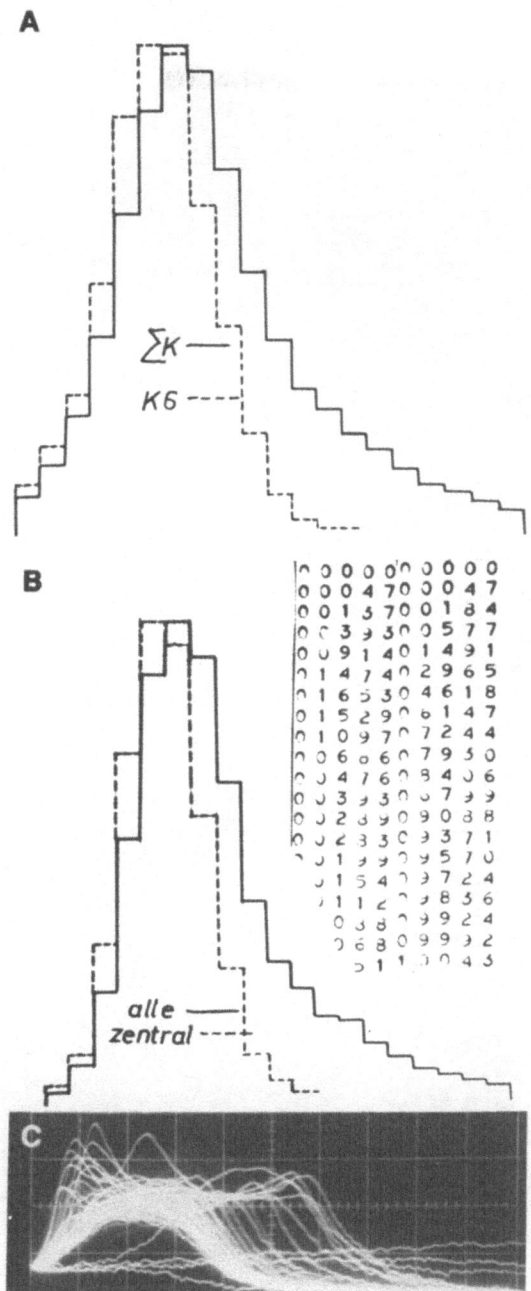

FIGURE 24. (A) Comparison of a curve composed from the partial distributions of Fig. 22. ΣK with the curve measured at the axial path (K6). The composed curve agrees well with the curve shown in (B) measured without focusing (Kachel, 1970). (C) Pulse pattern from a measurement without hydrodynamic focusing. The different types of pulses purely shown in Fig. 22 are mixed together.

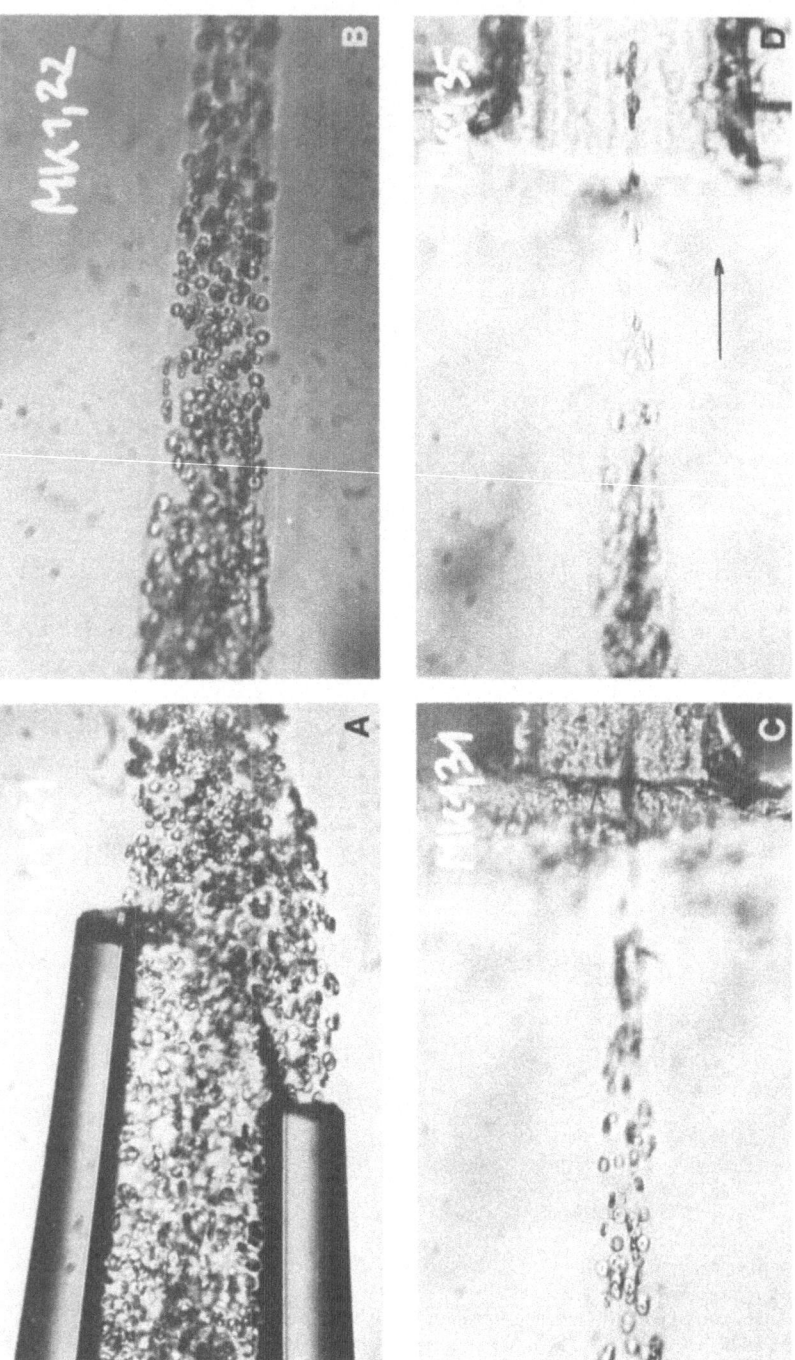

FIGURE 25. Demonstration of how a crowding stream of native erythrocytes injected into a flow in front of an orifice is constricted to a thin thread of lengthwise-oriented cells. Deformation of the cells takes place directly in front of the orifice (horizontal chamber, suction 0.2 bar).

3. Orientation and Deformation of Cells

a. Direct Experimental Determination. The model experiments shown above have explained that shape and orientation of particles during sizing in the orifice affect the pulse height signal that is picked up at the electrodes of the transducer.

This theoretical knowledge is of limited value if living biological cells, which in some cases are flexible in shape, are sized. It is to be expected that their shape is transformed by the accelerating and constricting flow forces acting along the flow path.

With the optically accessible transducers and the ultrashort flash equipment, the rapidly moving cells can be studied directly during the sizing process. Figure 25 demonstrates how a stream of cells injected into the flow in front of an orifice by a glass tip is constricted to a narrow path in the orifice and how a crowding of cells injected at a point of moderate flow velocity is thinned by the substantial increase of flow velocity. The cell number per unit of time is identical at each cross section of the stream. Figures 26–30 show different types of biological cells in the core stream and near the orifice wall. From such photographs we learn about the orientation and deformation behavior of cells: Nonspherical cells are uniformly oriented in the inner region of the orifice

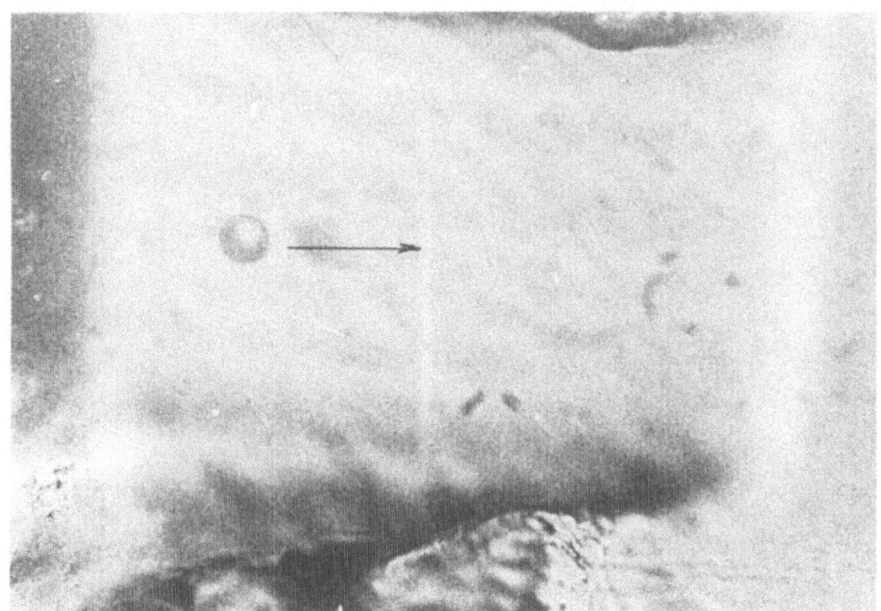

FIGURE 26. A latex particle passing the orifice of the horizontal chamber.

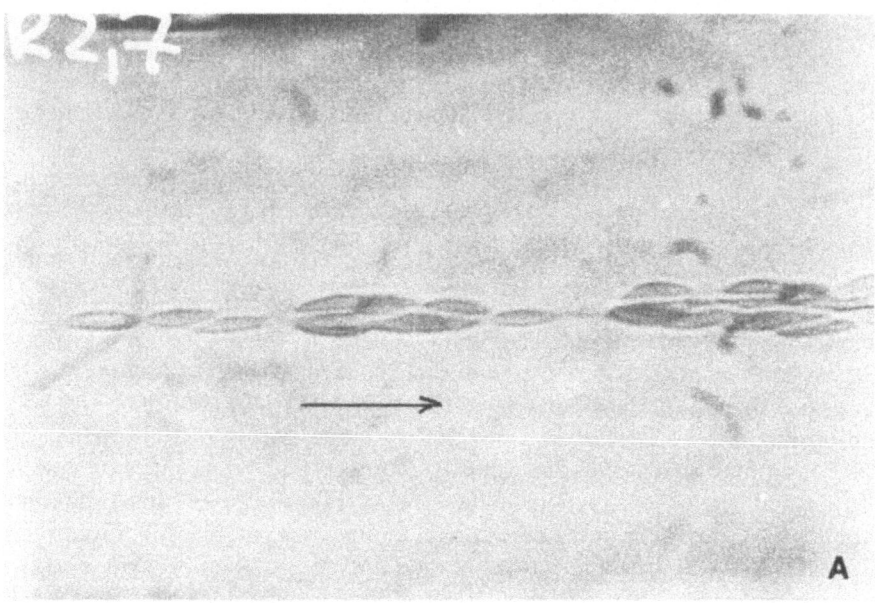

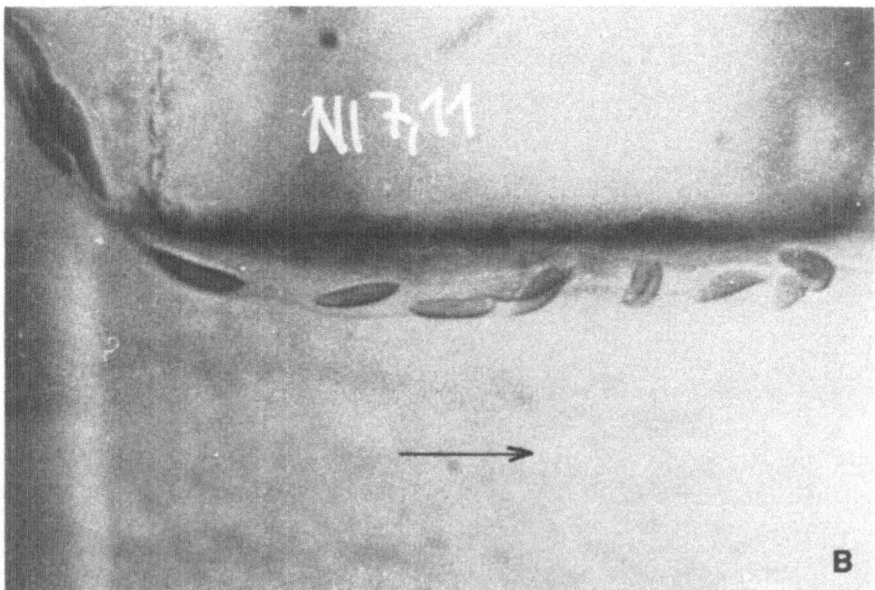

FIGURE 27. (A) A stream of native rat erythrocytes that are lengthwise oriented and deformed to ellipsoidlike bodies. (B) Native human erythrocytes passing the horizontal orifice near the wall. The cells rotate and are deformed very individually.

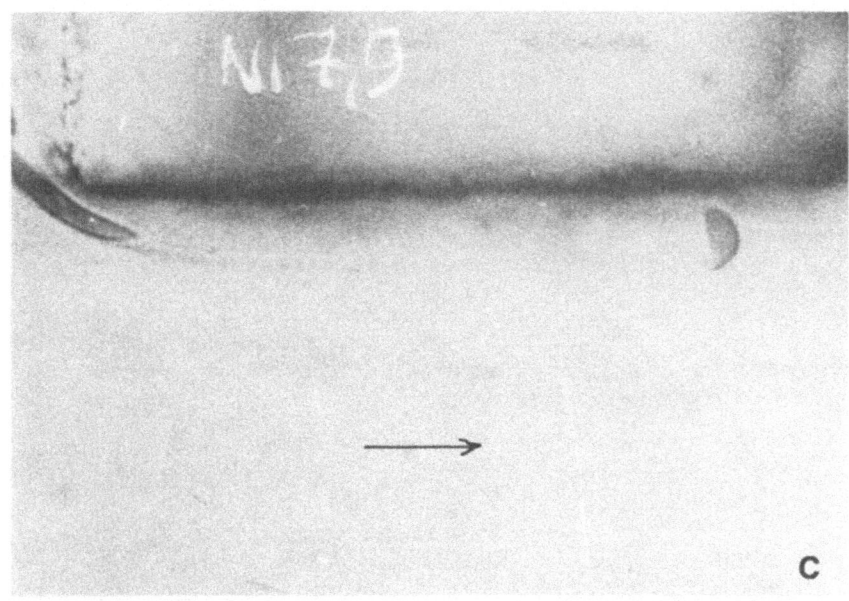

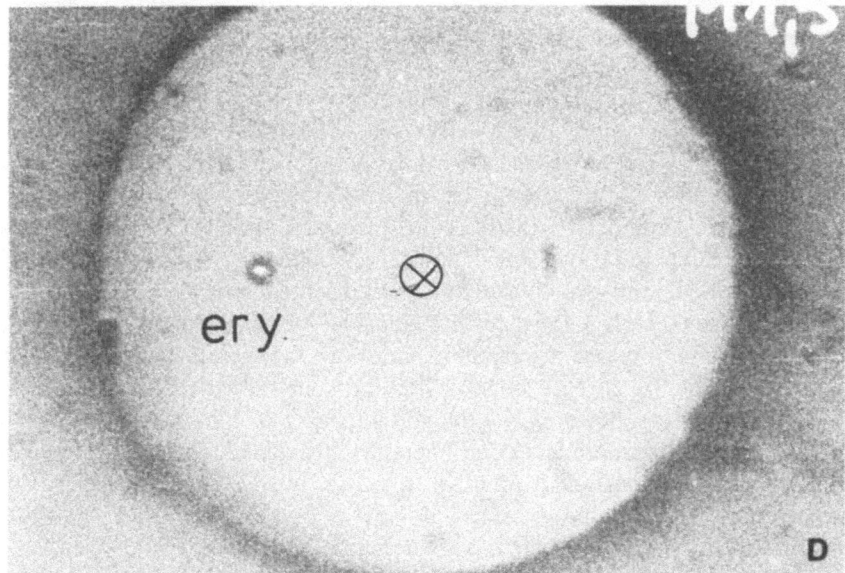

(C) A single native human erythrocyte rotating near the orifice wall. (D) Axial view into the orifice of the vertical chamber with a native erythrocyte deformed to a body of circular cross section.

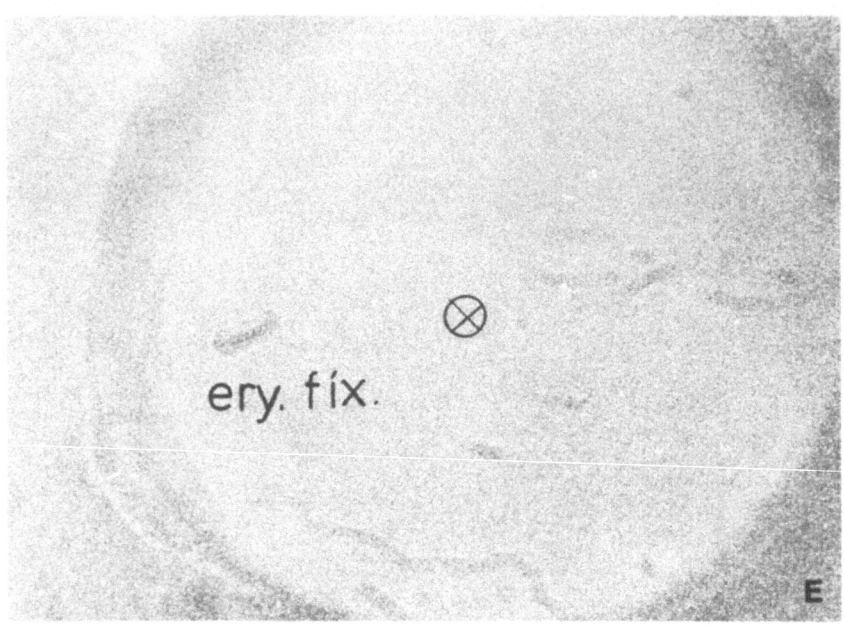

FIGURE 27. (Cont.) (E) A fixed human erythrocyte in the vertical orifice showing the biconcave cross section.

where the shear-free core stream exists. Near the orifice wall, the uniformity of orientation is disturbed by the shear forces, which are generated by the friction-retarded flow. In this region rotating cells are observed.

Electrical sizing derives double benefit from focusing the cells to the axial zone of the orifice. First, the edge distortion and inhomogeneities of the electric field are avoided, and second, the uniformly oriented cells are evaluated with identical shape factors. A third advantage of the focusing technique is that the cells are in contact for only a very short time with the sheath electrolyte, which conducts the electric current to the orifice and could affect the cells.

The flow forces, however, not only orient the cells. Flexible cells are elongated to prolate ellipsoidlike bodies. Therefore, the shape factor determination of such bodies is of particular interest. In the shear zone near the orifice wall, the cells are very individually deformed—one more reason to avoid this region if quantitative volume determination is desired.

The axially oriented alignment of the cells takes place in the acceleration zone in front of the orifice (Figs. 17, 25) and not inside, where a flat, shearless flow profile exists over the core region of the cross section.

The orienting process is explained by the velocity and pressure conditions

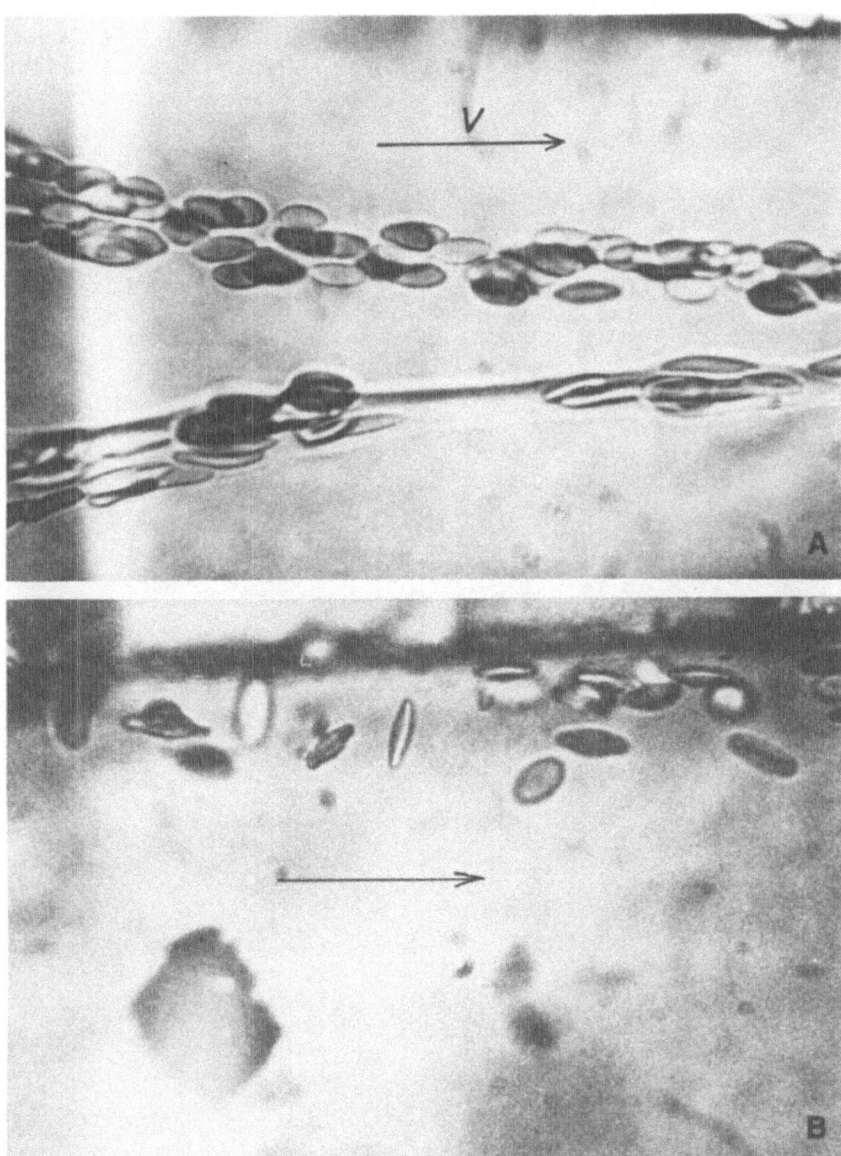

FIGURE 28. (A) Comparative study of differently treated human erythrocytes using the double-stream device. Lower stream: The cells suspended in their own plasma are distorted to ellipsoid-like bodies. The stria indicates the delamination of the plasma. Upper stream: The same cells suspended in 0.45% NaCl solution. The cells are swollen and less deformable than they are in the native state. (B) Fixed chicken erythrocytes rotating in the shear flow near the orifice wall.

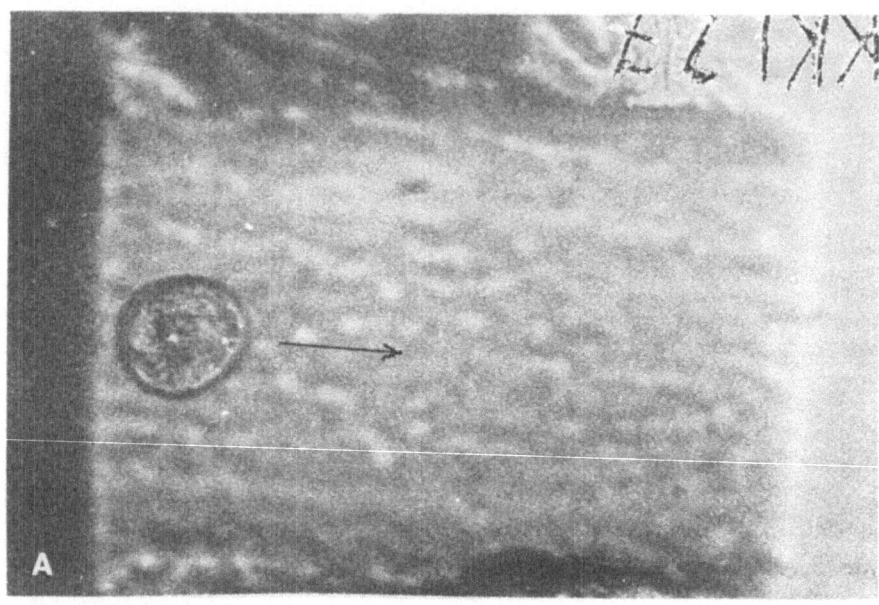

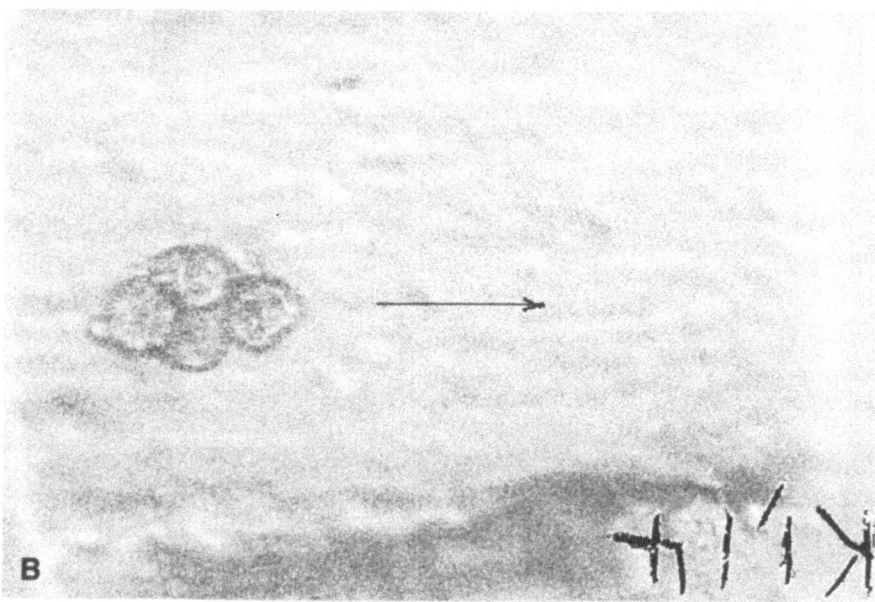

FIGURE 29. (A) An isolated liver cell passing the orifice of the horizontal chamber. The cell is not deformed. (B) A quadruplet of liver cells in the horizontal chamber. Cell preparation according to Howard and Pesch (1968).

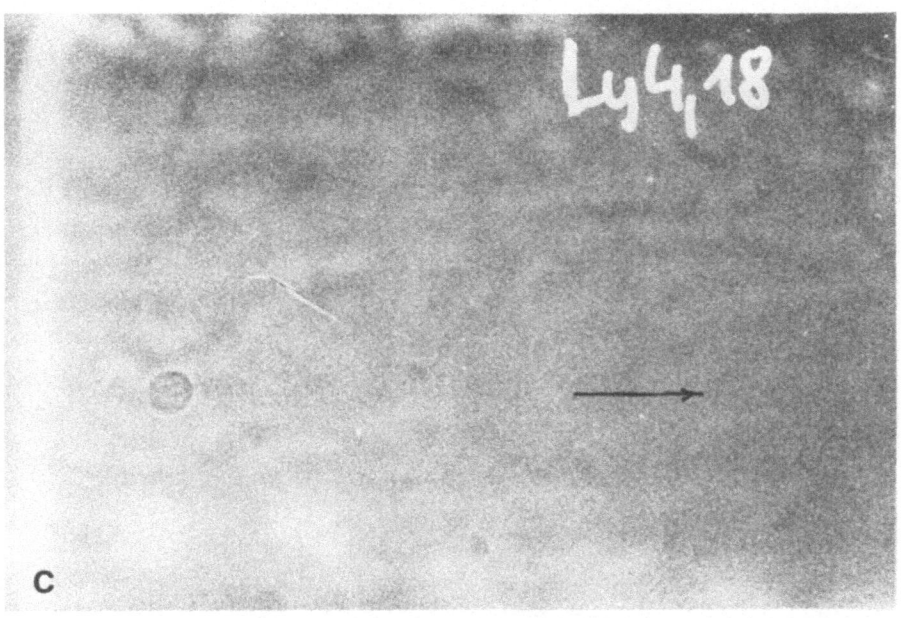

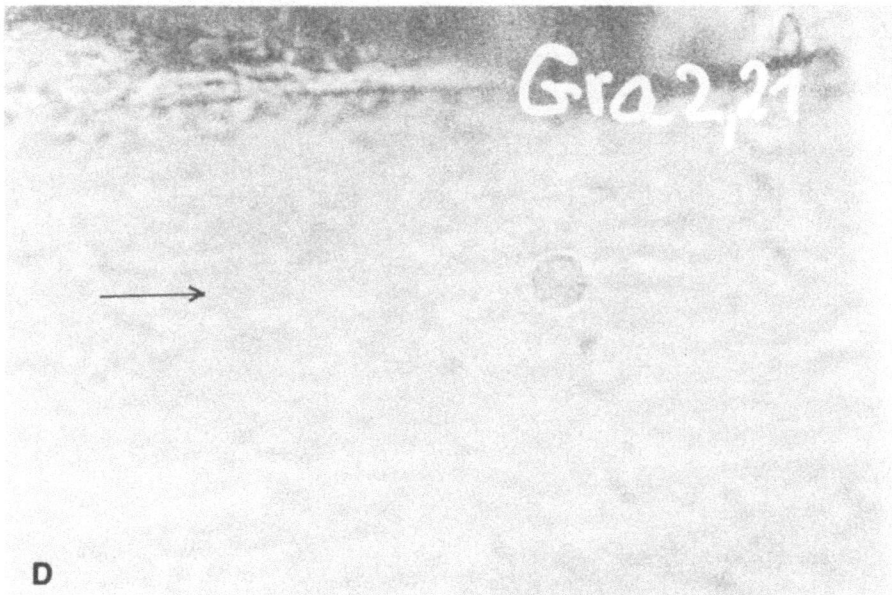

(C) Native human lymphocyte passing the horizontal orifice (v_m = 3.5 m/sec). Cell preparation according to Otto and Schmid (1970). (D) Native human granulocyte passing the horizontal orifice showing no deformation (v_m = 3.5 m/sec). Cell preparation according to Otto (1970).

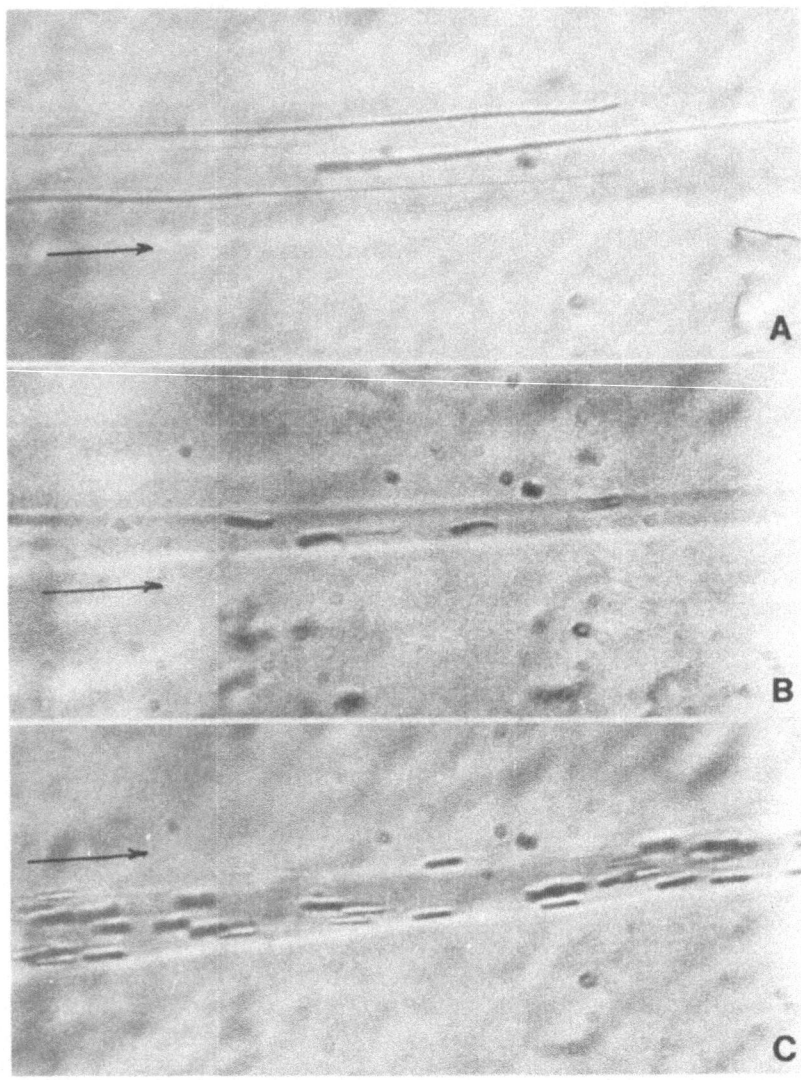

FIGURE 30. (A) Bull sperm cells with tails in the horizontal orifice. (B) Bull sperm cells having lost the tails in the horizontal orifice. (C) Abalone sperm cells in the horizontal orifice.

in the accelerating and constricting flow, which mathematically is described by the Bernoulli equation

$$P_A + \frac{\rho_d V_A^2}{2} = P_B + \frac{\rho_d V_B^2}{2} \tag{25}$$

where ρ_d is the density of the fluid with flow velocities at area A and B.

According to Fig. 31, the points A and B of a particle moving in the accelerating flow have different velocities, and at their momentary positions different pressures exist.

$\alpha = 0$ characterizes an unstable equilibrium. It is very likely that random events torque the particle out of the equilibrium. One of the points, here point B, enters a zone of increased velocity and decreased pressure in relation to point A. These differences, which can be calculated from equation (25), rotate

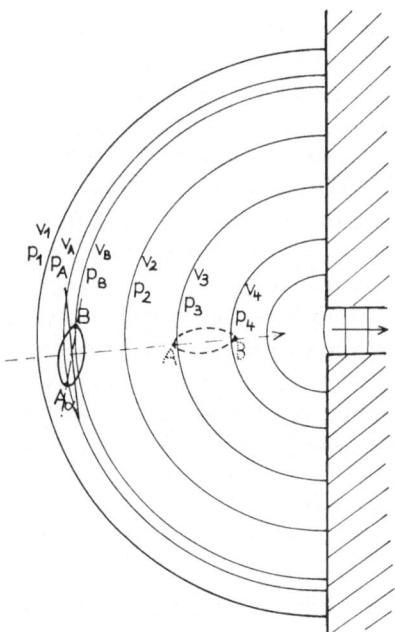

FIGURE 31. Explanation of the orienting forces on nonspherical particles. With $\alpha = 0$ the particle is in an unstable equilibrium. If random events rotate the particle out of the equilibrium as shown here, point B is in a zone of increased flow velocity and decreased pressure in relation to point A. The forces generated by this difference rotate the particle until the stable position marked by the broken line is reached. In this position the forces over the particle are maximal and stretch the deformable cells.

the cell until the stable position shown by the broken lines in Fig. 31 is reached wherever the pressure difference and velocity difference over the particle is maximal. Flexible particles that have reached this stable orientation are stretched over their length by the pressure.

b. *Indirect Methods to Determine Particle Shape Factors.* Several authors attempted to study cell orientation and deformation by indirect methods without direct observation. Grover *et al.* (1972) proposed that the minimum shape factor of cells and thus their axial ratios may be determined by measuring the difference between the maximal and minimal pulse height produced by the rotation of cells in the orifice. Apart from the difficulty for cells to pass the orifice with their long axis perpendicular to the orifice axis (it is possible only near the orifice wall, and there the pulse heights are distorted by the inhomogeneities of the electric field and the proximity of the wall), this method would give true shape factors only if the axial and transaxial shape of the cells is the same. Figure 27 demonstrates that an identical shape cannot be assumed for deformable cells moving in the core stream and others rotating near the wall. Therefore, shape factors of deformable cells cannot be determined by this method, and the minimal shape factor of nondeformable cells, which is necessary for volume calculations with focusing systems, can be determined more easily with a normal microscopic examination.

The two-orientation method of shape determination was also used by Golibersuch (1973) and Thomas *et al.* (1974), who interpreted the highest and lowest pulse amplitudes of presumably rotating cells. However, these investigators ignored the fact that native red blood cells lose their biconcave shape.

Thom (1972b) proposed a method for determining shape factors by using two orifices in sequence with different geometric properties and comparing the pulse heights produced in these orifices by the same cell. Thom did not define the exact pulse-height-to-shape-factor relationship in short orifices which must be known for such quantitative measurements, and electrical breakthrough effects of cell membranes (see Section II-D-6) may have also influenced his results.

The method of Waterman *et al.* (1975), which compares the microhematocrit mean volume of red cells with their resistance-pulse-measured mean volume, is correct if both measurements are performed in the same electrolyte and if the flexibility of the cells investigated is such that they are optimally packed for the hematocrit determination.

Deformability studies were carried out by Mel and Yee (1975) with a mixture of fixed and native cells at high and low flow rates. They drew conclusions about the deformability of the native cells from the peak difference between such measurements but did not take into consideration possible orifice edge and cell rotation effects in their unfocused device.

4. Coincidence of Particles

One of the basic conditions for a useful application of electrical flow sizing is that the particle concentration in the suspension is such that with a high probability only one particle is in the sensitive zone of the transducer at any time.

With a mean flow velocity v_m in the orifice, a cross section of the particle path of q' and a mean particle rate of N particles/sec, we obtain for the particle concentration in the suspension:

$$D = \frac{N}{v_m q'} \tag{26a}$$

In a nonfocusing system, where q' is the entire cross section of the orifice, D must be smaller than in a focusing system in which q' is the cross section of the focused path (see Section II-C-2).

Coincidence is due to the presence of two or more nonadherent particles in the orifice in such a manner that they are not resolved as single events. Electronic coincidence is caused by analyzing devices with a too-long processing time, e.g., analog to digital conversion time for one cell. If only the volume distribution of a cell population is considered, the electronic coincidence can be ignored if always the first pulse of coincident pulses is evaluated, regardless of whether it is the smaller or larger one. In this case the statistical quality of the distribution is not affected because only a random number of cells is omitted. This type of coincidence can be reduced by use of fast electronic analyzers.

Particle coincidence is caused by a too-small distance between the particles themselves. Wales and Wilson (1961, 1962) and Princen and Kwolek (1965) have described two types of coincidence: the so-called horizontal interaction, where two particles enter the orifice in such a way that two pulses are generated but only the larger one is evaluated by the peak detector, and the so-called vertical interaction, where both particles are in such proximity that they form one pulse with a double pulse height (Fig. 32a).

Coincidence theories concerning the count loss of Coulter counters have been published by Wales and Wilson (1961, 1962), Princen and Kwolek (1965), Princen (1966), Mercer (1966), and Helleman (1972). If only particles are counted, there are methods of pulse processing, e.g., with electronic differentiating circuits, that can resolve coincidence caused by horizontally interacting particles.

With horizontal interaction, the pulse heights lie between the true pulse heights of the single particles and the sum of both. Depending on the triggering circuit, either the first of the two or the greater pulse is evaluated. The true

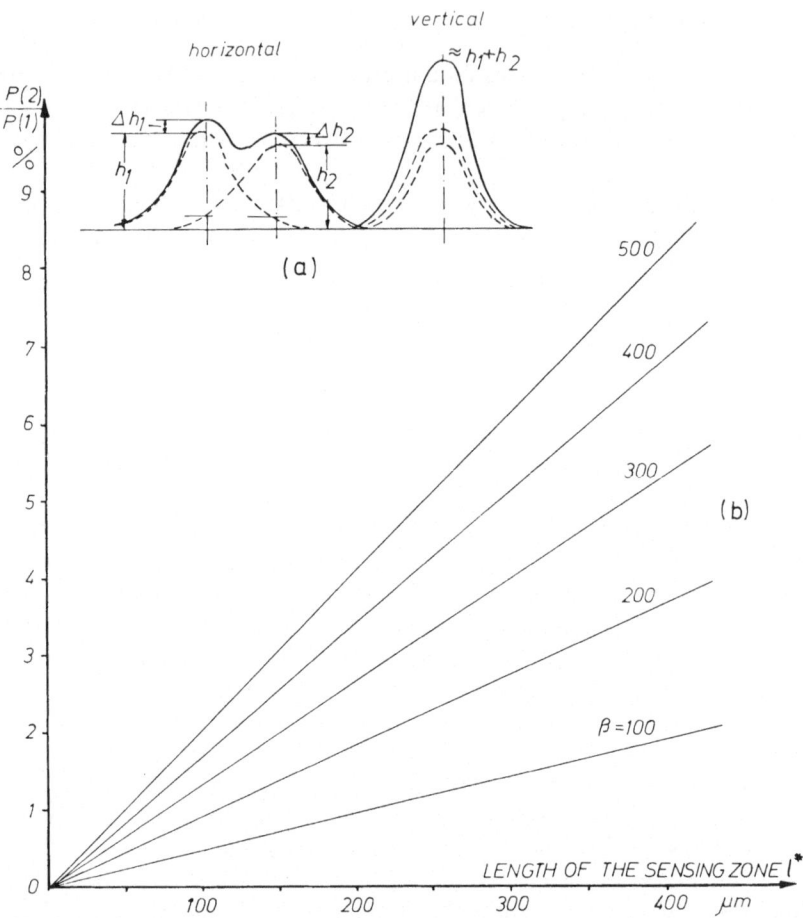

FIGURE 32. (a) Explanation of the horizontal and vertical coincidence. (b) The coincidence probability as a function of the length of the sensing zone of the orifice l*, which consists of the length of the orifice itself and the length of the region outside the orifice where the slopes of the pulses are generated. The particle density at the focused path, measured in particles per meter, is used as a parameter.

pulse height of both particles in all cases is distorted by superposition. With vertical interaction the sum of both pulses is always measured.

In long orifices there is increased probability of horizontal interaction. If we assume (1) a pulse rate of N particles per second which pass the orifice in a random distribution per unit of time on a focused path with a flow velocity of v_m, and (2) the sensitive length of the orifice is x (the physical length of the orifice and distances outside the orifice where a considerable field strength exists as defined in Section II-B), then $\gamma = x(N/v_m)$ particles are present in the orifice on the average (Fig. 32b).

The probability of the presence of n particles in the sensitive zone of the orifice is expressed by Poisson's law (Wales and Wilson, 1961):

$$p(n) = \gamma^n e^{-\gamma}/n! \tag{26b}$$

This coincidence probability depends on the flow velocity, the average number of particles passing through the orifice (which can be determined by counting the pulses per unit of time, and the length of the sensitive zone). The particle density related to the focused particle path is defined as $\beta = N/v_m$ particles/ unit of length.

In Figure 31b, $p(2)/p(1)$, i.e., the probability of the presence of two particles related to the probability of one particle in the orifice, is plotted as a function of the length of the sensitive zone and different particle densities β. This ratio indicates how many doublets are expected if the probability of singlets is 100%. For example, with $v_m = 5$ m/sec, $N = 1000$ particles/sec, and with a sensitive length of the orifice of 100 μm, the coincidence probability is near 1%.

Most of the pulses of coincident cells are characterized by increased pulse length. Pulse length sensing circuits may decrease the coincidence effect by ignoring long pulses. This method fails with widely varying volume distributions because with an absolute trigger level increased pulse height is connected with an increased pulse length (see Fig. 44).

5. Recirculation of Cells at the Orifice Outlet

From Figs. 13–15 it is evident that the sensitive zone of a sizing orifice is not restricted to its interior but also covers a considerable field outside of it which surrounds the inlet and outlet portion.

The high-velocity jetlike flow leaving the orifice has a water jet pumplike effect on the surrounding fluid, which contains particles that have just been measured. A recirculation flow that transports such cells back into the expiring active field of the orifice is kept going. Figure 14 shows three such recirculation paths (RI–RIII), and pulses presumably generated at such paths are shown in Fig. 33a.

The frequency of the recirculation pulses depends on the cell concentration in the waste fluid of the transducer. The recirculation problems become important when both large and small cells are present in one suspension, particularly if there are relatively few small cells. In this case the secondary pulses of large cells will interfere with the primary pulses of the small cells and con-

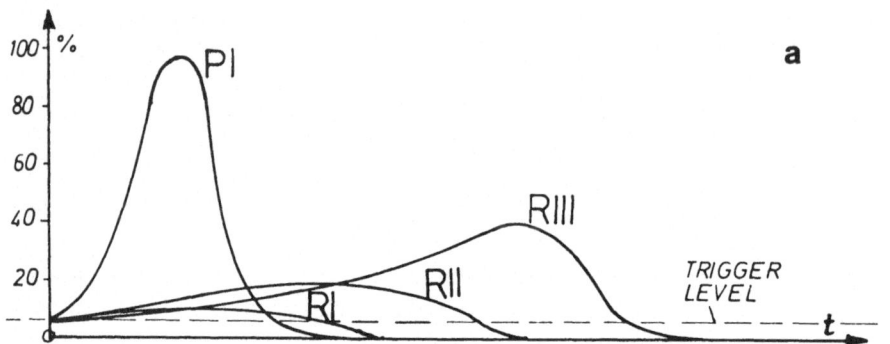

FIGURE 33. (a) RI–RIII are constructed pulses from recirculating cells which correspond with the particle path RI–RIII of Fig. 14. The length of the pulses recognized by the trigger device depends on the height of the pulses. Therefore, rejection of these unwanted pulses by length detection cannot be successfully applied. (b) Short bell-shaped primary pulses of latex particles mixed with recirculation pulses of erythrocytes in a Metricell transducer. (c) The pulse pattern of the Metricell analyzer after the antirecirculation flow is switched on.

siderably impair the distribution of the small cells. Due to the low flow velocity at the recirculation path, the secondary pulses are significantly longer as the wanted primary pulses.

Recirculation pulses may be removed in two ways: by electronically excluding pulses exceeding a limit in time duration or by hydromechanically preventing the recirculation of cells into the sensitive region of the low-pressure side of the orifice.

The electronic method is problematic insofar that the length of a pulse known to the evaluation unit is determined by the time distance between the two trigger points at the rising and falling slope of the pulses. Pulses originating from particles at path RI are detected in the length range of the primary pulses despite the fact that they are recirculation pulses.

With hydromechanical prevention, a second sheath flow is used to direct the cells leaving the measuring orifice to a second, so-called catcher orifice or tube (Haynes, 1980). Another approach realized in our Metricell instruments is to wash the sensitive outlet region of the orifice with particle-free electrolyte, thus keeping the recirculating cells away (see Section III-A-3).

Figure 33b shows primary pulses of 3.34-μm latex particles in a 30-μm diameter orifice overlapped by secondary pulses of fixed erythrocytes that recirculate in the waste fluid of a Metricell transducer. With the antirecirculation flow washing the orifice outlet, the secondary pulses are eliminated (Fig. 33c).

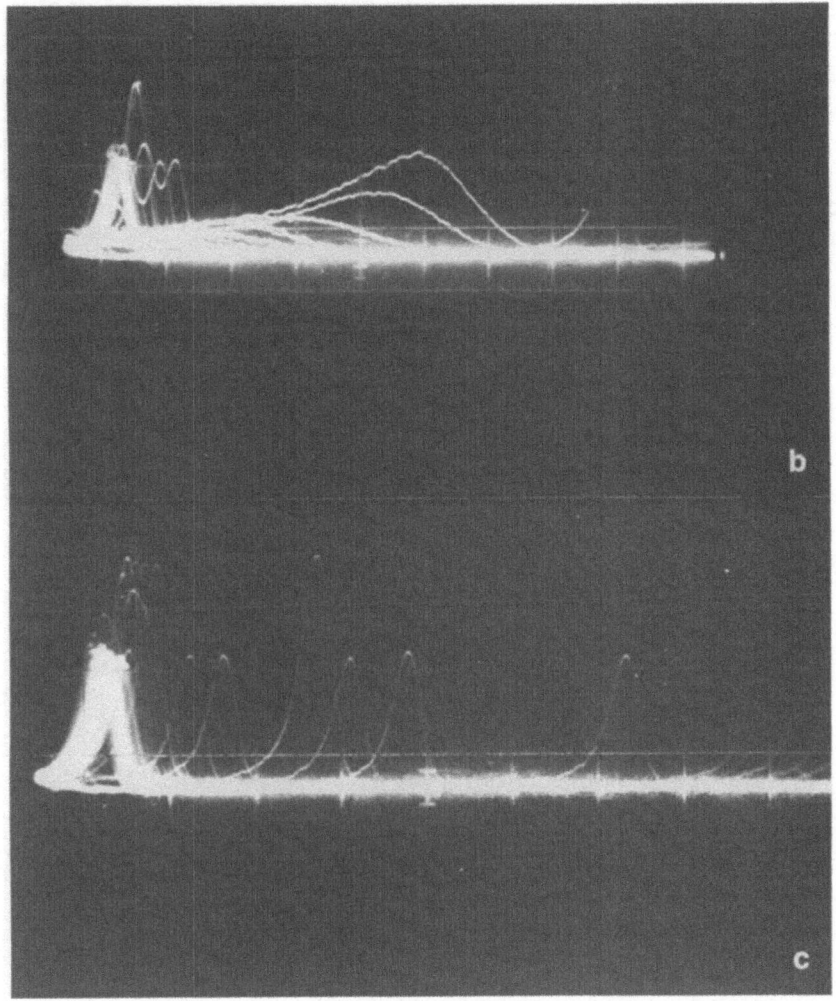

FIGURE 33. (Cont.)

6. Electrical Breakthrough Effects of Cells

In a sizing transducer, the cells are coming from a region of low field strength into the high field strength of the orifice. Normally, the interior of a cell is isolated by the cell membrane from the surrounding electrolyte. Above a distinct field strength, however, the cell membrane electrically breaks through, and the electric current can also flow through the interior of the cell. This breakthrough effect, reported by Schulz and Nitsche (1972), was investigated in detail by Zimmermann *et al.* (1974, 1975) and Jeltsch and Zimmermann (1979).

Such breakthrough effects, if existent, strongly influence the electrical cell volume determinations in a nonlinear way. With a field strength above the critical point, a smaller pulse height is determined as with a subcritical field strength.

According to equation (12), the field strength in the sizing orifice is determined as

$$E = \frac{\rho_2 \cdot i}{q}$$

if an orifice of $l/d > 1$ is used.

According to Zimmermann *et al.* (1974), the cell membrane breaks

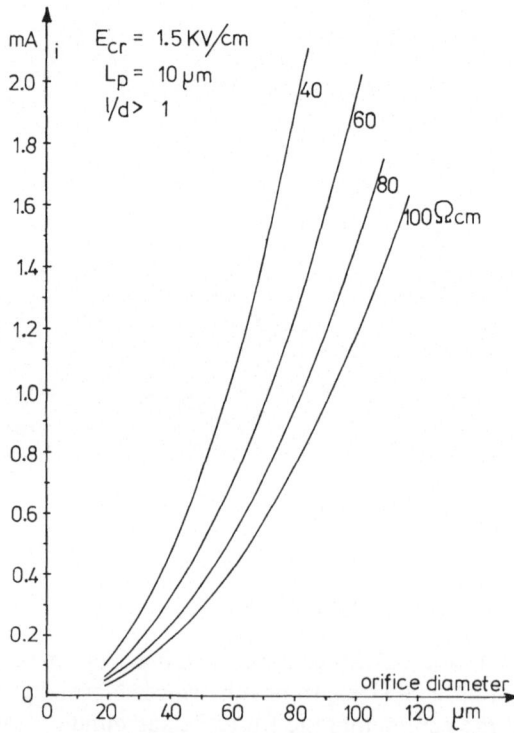

FIGURE 34. The breakthrough point of cells 10 μm in length as a function of the orifice current, orifice diameter, and electrolyte resistivity ρ_2 if a critical field strength of 1.5 kV/cm is assumed.

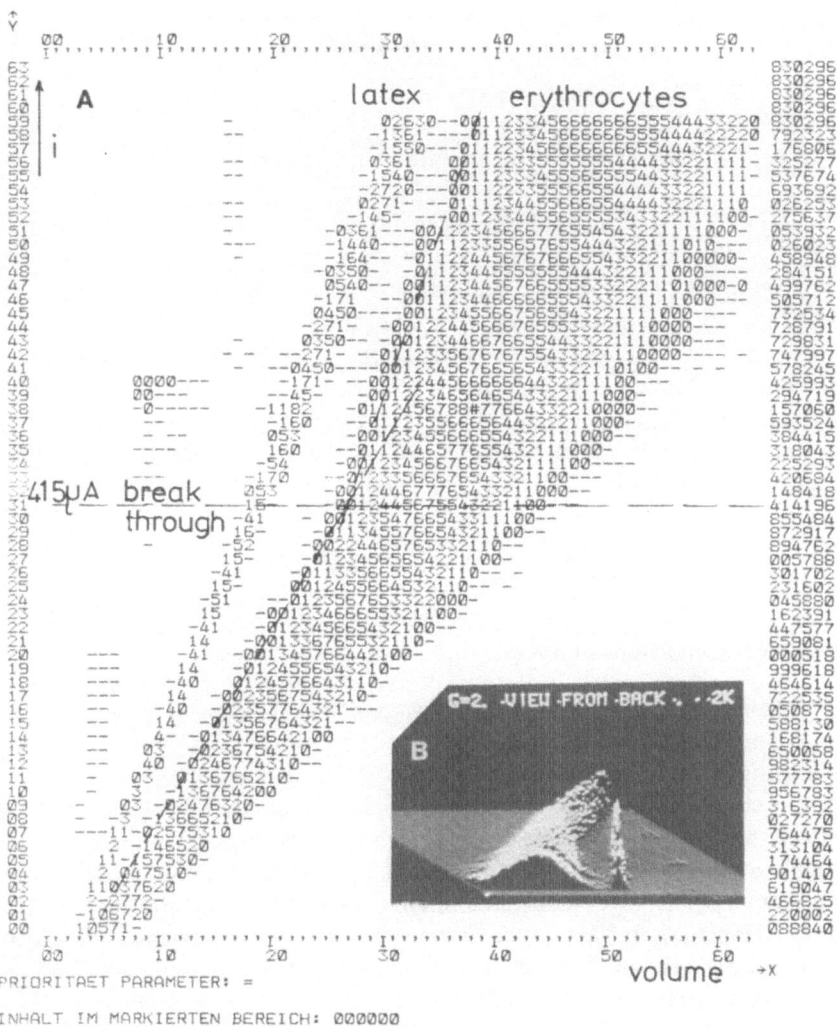

FIGURE 35. (A) Print of a breakthrough experiment with native human erythrocytes, shown by a pseudo two-parameter sequence of 60 one-parameter curves. The analysis was performed with the new CYTOMIC analyzer (Kachel *et al.*, 1980b). The analysis starts with the first distribution in line 0, measured with a current of 100 μA. The current for the distributions in the following lines was increased successively by 15 μA. The bend in the course of the erythrocyte sequence at 415 μA (corresponding with a breakthrough field strength of 1.26 kV) indicates the beginning breakthrough of the cells. Above the ciritical current, the cell volume measured is too small. The distribution of latex particles mixed with the erythrocytes and measured simultaneously does not show a breakthrough bend. (B) Isometric view onto the sequence of curves of the breakthrough experiment.

through if a critical voltage ($U_\sigma > 1.6$ V) is applied over the cell. From this voltage the critical field strength $E_\sigma = 1.6/L_p$ can be determined. L_p is the length of the cell in the direction of the field.

An estimation of the critical range can be taken from Fig. 34. The curves are calculated for a critical field strength of 1.5 kV/cm and a cell length of 10 μm.

In Fig. 35a, sequences of size measurements are shown in which the field strength in the orifice is increased by increasing the orifice current. A mixture of nonconducting latex spheres and native human erythrocytes was measured simultaneously with increasing current. The straight course of the narrow latex curves indicates that with these particles the pulse height increases linearly without breakthrough. The bend in the course of the native erythrocytes, however, indicates the point at which the breakthrough of these cells begins. Figure 35b shows an isometric view onto the sequence of the 60 volume distributions that are taken with the CYTOMIC analyzer (see Section VI-A).

III. INSTRUMENTATION

A. General Conditions and Review

A resistance pulse flow system consists of four main portions.

1. The transducer, or sensor, picks up the biological information "cell size" and transforms it into an electrical signal.

2. The weak voltage or current pulses are amplified in the analog portion of the system and made visible at the screen of an oscilloscope, which is a very valuable tool in supervising the function of the transducer; most irregularities in the sensor system are manifested in changes of the pulse pattern.

3. The pulse height analyzer portion converts the amplified analog pulses into digital signals, which are classified and stored in a digital memory and are made visible in the form of histograms on a screen.

4. Processing of the stored data—e.g., calculation of statistical values, comparison with other theoretical or measured histograms, and conservation of the data on long-time storing media—is performed in the computer portion of the system.

The best-known resistance pulse device, mostly used for counting particles, is the Coulter counter. It is of limited accuracy in sizing cells. Many investigators have modified the commercial Coulter counter or constructed instruments of their own. Doljanski *et al.* (1966), Lusbaugh *et al.* (1962a), van Dilla *et al.* (1967), and Paulus (1975) equipped Coulter counters with multichannel analyzers. Bull (1968) and Steen and Nielsen (1979) equipped a Coulter counter with a delay circuit; a delayed gating device was also used by

Grover *et al.* (1969b) in an instrument of their own design. Other specially designed particle analyzers without focusing were used by Gutmann (1966), Adams *et al.* (1967), and Harvey (1968). Focusing devices are described by Spielmann and Goren (1968), Thom *et al.* (1969), Shank *et al.* (1969), Kachel *et al.* (1970), Kachel (1970, 1972, 1976), Merrill *et al.* (1971), von Behrens and Edmondson (1976), and by Haynes and Shoor (1978), who use fluid resistors in their fluid system. Shuler *et al.* (1972) added a focusing device to an instrument with a one-threshold analyzer, a problematic combination because the suspension flow is not constant with time. A system with a so-called "flow-directing collar" was developed by Karuhn *et al.* (1975) and Davies *et al.* (1975). Electronic signal analysis techniques are described by our group (Kachel, 1972, 1973), by Coulter Electronics patents (1971, 1972), and by Waterman *et al.* (1975). A technique similar to the four-electrode impedance bridge devices (Ferris, 1963) used to measure impedances of electrolytes and tissues has been introduced into Coulter sizing by Salzman *et al.* (1973), Leif and Thomas (1973), and Thomas *et al.* (1974). Superior accuracy and resolution of these so-called "potential sensing transducers" have not yet been demonstrated as have perfected two-electrode devices. An electrical sizing system that combines size determination and breakdown measurements of cells is described by Groves (1980).

Multiparameter analytical and sorting instruments that size electrically are described by Fulwyler (1965), Steinkamp *et al.* (1973), and our group (Kachel and Glossner, 1976; Kachel *et al.*, 1977; Menke *et al.*, 1977), which also has built a system for fast imaging in flow of cells selected by electrically measured volume (Kachel *et al.*, 1979, 1980a) (see below).

In the following, the requirements of the instrumental portions of a sizing system will be discussed. The system elements of our Metricell devices will be used to point out a practical volumetry system for biological cells.

B. Transducer

The transducer is the most critical part of a sizing system. The thin orifice necessary in this technique tends to be blocked if uncleaned fluids are used. The complete current path between the electrodes is sensitive to electrical interferences or vibrations. Therefore, the transducer must carefully be shielded and gas bubbles must be taken away from the main current path.

The first requirement of a high-accuracy transducer is hydrodynamic focusing equipment. Figure 36 shows how a usual Coulter counter capillary can be modified to a hydrodynamically focusing device. The capillary is now immersed into a beaker filled with particle-free electrolyte (sheath fluid) rather than the suspension itself. The suspension, whose electrolyte must be electrically identical with the sheath fluid, is supplied from a small vessel attached to

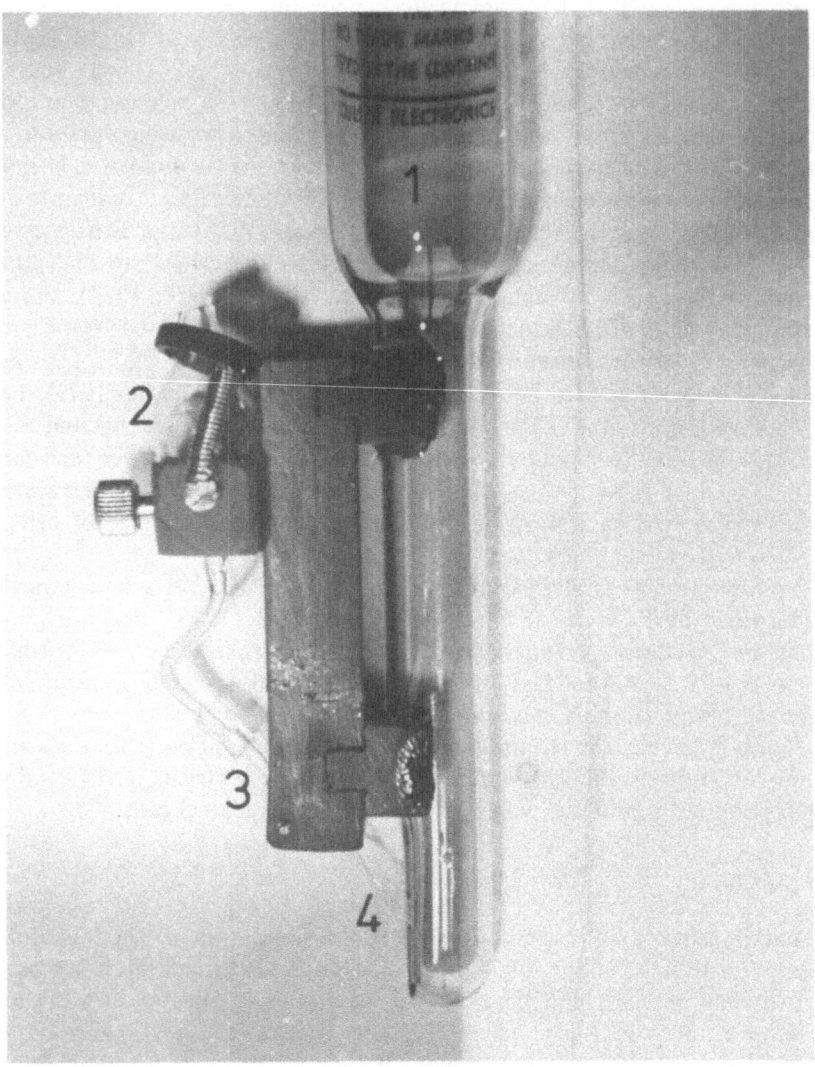

FIGURE 36. The easiest way of building a hydrodynamically focusing sizing transducer (Kachel *et al.*, 1970). (1) Original Coulter capillary; (2) suspension vessel; (3) holder for the focusing device; (4) injection tip for the suspension.

the capillary via the glass tube that ends in a small tip in front of the orifice. The device will work well even if the orifice axis and the injection tube axis include an angle as shown in Fig. 36. The particle path in the orifice develops according to the principle explained in Section II-C-3. The amount of suspension, injected through the tip, is determined by the level pressure difference of sheath fluid and particle suspension level.

The more advanced Metricell flow system is shown in Fig. 37. Here the

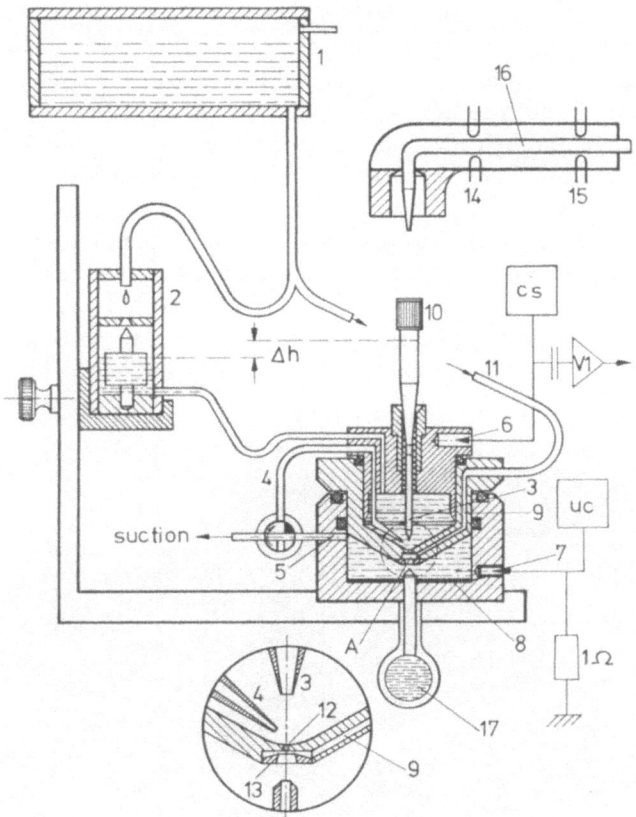

FIGURE 37. The standard Metricell transducer system. (1) Electrolyte tank; (2) regulation chamber, which electrically isolates the electrolyte tank from the system; (3) particle injection tube; (4) cleaning channel; (5) suck connection for waste fluid exit; (6) active electrode block; (7) ground electrode connection; (8) ground electrode; (9) channel for the antirecirculation fluid; (10) pipette tip containing the cell suspension; (11) tube conducting the antirecirculation fluid; (12) orifice; (13) antirecirculation device; (14) stop-light barrier; (15) start-light barrier; (16) particle tube of the counting head; (17) cleaning rubber ball. (cs) Current source; (V1) preamplifier; (uc) calibration pulse generator.

particle-free sheath fluid is supplied from supply tank 1 via the regulation
chamber 2. This has a twofold function: Its dropping chamber isolates the storage tank from the electrically active part of the sensor system, and the ball
valve in its lower part holds a fluid level which is constant in relation to the
regulation chamber. This can be raised and lowered for controlling the suspension output from tube 3 into the focusing sheath fluid. The proper measuring
chamber consists of an electrically active upper part which is connected with
the current source and the amplifier via the electrode block 6 and a lower part
connected with ground via electrode 8 and the 1-Ω calibrator resistor.

The cell suspension prepared in a pipette tip (10) is injected into the focusing sheath fluid via tube 3. For cleaning the orifice, the suction normally
applied to 5 is switched to tube 4. By simultaneously pressing the rubber ball
(17), a backflow through the orifice removes the dirt, which is immediately
sucked out via tube 4. The particle tube (3) is cleaned by sucking back particle-
free electrolyte.

The detail A, together with channel 9, explains the built-in antirecirculation device which prevents distortions caused by recirculating cells (see Section II-D-5).

Optionally, a counting head, shown in Fig. 38, can be attached to the
chamber. Instead of a pipette tip, a calibrated glass tube (16) with two light

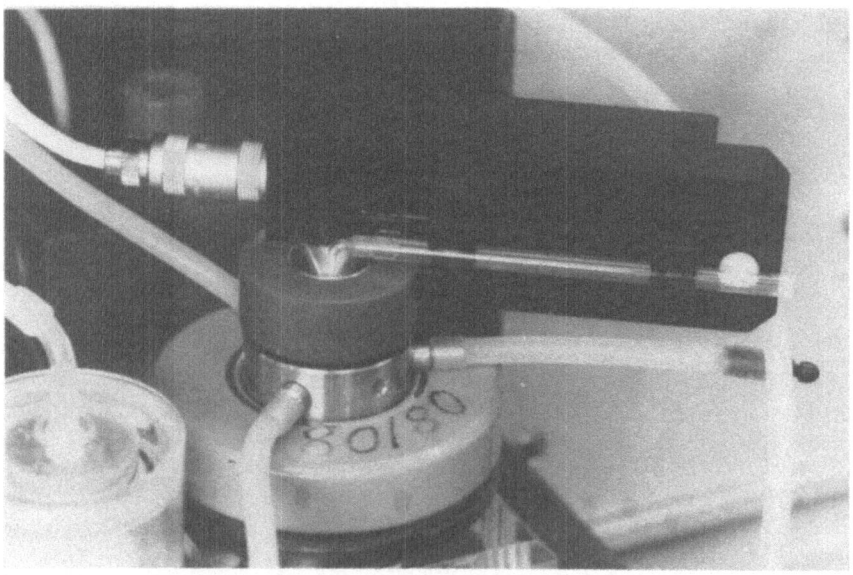

FIGURE 38. The counting head with its two light barriers plugged to the transducer.

barriers (14 and 15) defining a distinct volume of suspension is connected to the suspension tube (3). The meniscus of the suspension starts the analysis if it is passing the first light barrier (15) and stops it if it is passing the second (15). The histogram stored relates to the defined volume of suspension.

This type of sizing chamber can be dismounted easily and different orifice diameters can be implemented.

A recently developed sizing flow chamber particularly suited for rapid exchange of samples and kinetic measurements is shown in Figs. 39 and 40. For such measurements, a direct access to the orifice without long tubes must be available. In the new chamber, the function of the pipette tip and the particle tube fixed in the chamber is combined in a reaction vessel, which is plugged into the chamber. It directly outputs the cell suspension into the focusing sheath flow. By adding reagents to the suspension during the size analysis, the reaction of the cells, e.g., lysis of blood cells, can be studied immediately. There is no long tube in which the different states of the reaction can mix.

With this transducer, the problem of cleaning the particle tube no longer exists, since each sample can be prepared and measured in its own vessel.

This transducer can also be used for simultaneously counting cells if the particle vessel is filled with an exactly measured amount of suspension and measured until empty.

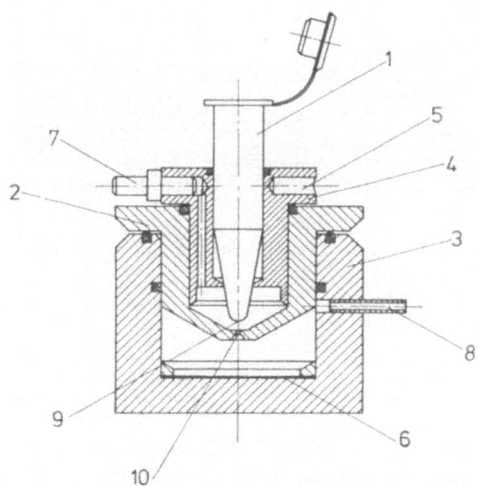

FIGURE 39. Newly developed transducer for rapid sample exchange and kinetic measurements. (1) Reaction vessel; (2) orifice holder; (3) holder; (4) active electrode block; (5) connector hole; (6) ground electrode; (7) supply channel for particle-free electrolyte; (8) suck connection; (9) injection hole; (10) orifice.

FIGURE 40. The chamber of Fig. 39 mounted into a Metricell instrument.

C. Electronic Instrumentation

1. Current Source

The current source generates the constant current i which flows through the orifice and forms the electric field. The current generates a voltage U_o between the electrodes over the basis resistance of the orifice system R_o. This basis voltage U_o, which also covers the polarization potentials at the electrodes, is increased by ΔU if a particle increases the resistance of the orifice by ΔR. The pulse voltage ΔU, which solely contains the cell volume information, is separated from the basic voltage U_o and coupled by the capacitor C_c to the amplifier.

This a.c. coupling results in the fact that slow changes and high-frequency noise of the current influence the sizing system differently. According to equation (7), the pulse height generated in the orifice is proportional to the direct constant current i. From this point of view, a 1% constancy of the current is sufficient if a 1% accuracy of the pulse height ΔU is desired. This is true as long as the frequency of the current fluctuations is below the lower cut-off frequency of the amplifier system. If, however, the noise frequency is within the bandwidth of the amplifier, the noise current is directly transformed to a noise voltage at the basis resistance R_o and coupled to the amplifier.

Let us assume a basis resistance of 20 kΩ, a current of 0.5 mA, and a resistance change (ΔR) of 2 Ω if a particle is passing through the orifice. The pulse height signal received from the amplifier is $\Delta U = i\Delta R = (0.5 \text{ mA})(2 \Omega) = 1$ mV.

If the current slowly changes by 1% to 0.505 mA, the amplifier receives $\Delta U = (0.505 \text{ mA})(2 \Omega) = 1.01$ mV. The pulse height signal increases, as expected, by 1%.

If, however, the current is superposed by a fast, noisy fluctuation of 1%, we calculate with $U_n = R_o\Delta i$: $U_n = (20 \text{ k}\Omega)(0.005 \text{ mA}) = 100$ mV. In this case the distortion is 10^5% of the wanted signal. The example explains that only very-low-noise current sources are suited for sizing transducers.

The lowest possible noise limit is defined by the noise generated by the resistance of the orifice itself ($U_n = 4kTR_of$; where f is the bandwidth considered, T is temperature, and k is the Boltzmann constant). It can be assumed that this noise is increased by the noise of the current transition from the electrodes into the electrolyte and by the noise of the amplifier itself. The following example may point out practical noise conditions.

We assume an orifice of 60 μm diameter and 60 μm length and an electrolyte resistivity of 60 Ω-cm which forms a basic resistance of about 20 kΩ. With $i = 1$ mA, we wish to have a noise limit at a particle volume of 1 μm^3 (sphere).

According to equation (7), we obtain a noise pulse height equivalent to 1 μm^3 of $\Delta U = 0.11$ mV. The current fluctuations equivalent to this pulse height are $\Delta i = \Delta U/R_o = 5.5 \cdot 10^{-9}$ A.

A simple constant current source is built by connecting a stabilized source of a relatively high voltage over a resistor Rh with the orifice (Fig. 41). The current is determined by:

$$i = \frac{U_h - U_o}{R_h} = i_o - \frac{U_o}{R_h} \tag{27}$$

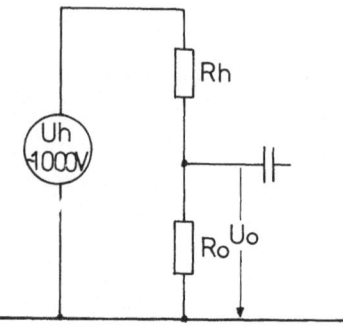

FIGURE 41. A simple current source for a transducer. $R_h > R_o$. R_o is the resistance of the orifice.

where i_o is the short circuit current if $R_o = 0$. With increasing R_h, the current
becomes more and more independent of U_o and R_o. In sizing circuits, the orifice resistance R_o is in the range of 5–50 kΩ. For a sufficient constancy, the
voltage U_h must be in the kV range.

A better current source for sizing is one which uses a low-noise field effect
transistor instead of U_h and R_h as shown. Such a source has an internal resistance far in the megaohms (Fig. 42). The stability of the current depends on
a carefully filtered and regulated supply voltage (U_B). By changing the resistance R_s, the current i can be adjusted in a wide range between zero and a few
milliamperes. The circuit works correctly if $U_o < U_g - U_g d$. Current sources
can also be built with operational amplifiers (Tietze and Schenk, 1971).

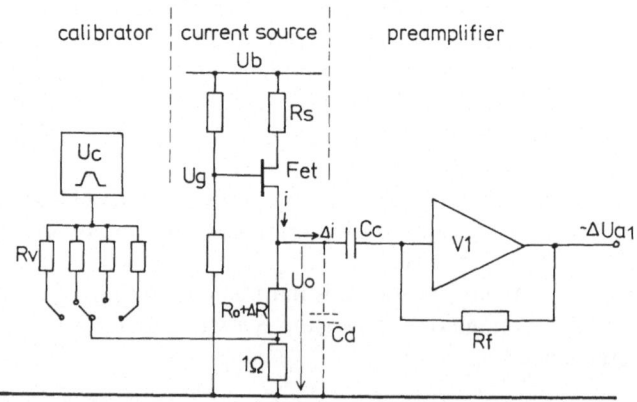

FIGURE 42. Schematic circuits of the calibrator, the field effect transistor current source, and
the preamplifier. U_g = gate voltage of the Fet.

2. Amplifiers

a. General Remarks. The amplifiers amplify the weak transducer signals (millivolt range and impedances of kiloohms) to a level suited for processing of the analog-to-digital converters (ADC). ADC input ranges are 0–2 V, 0–5 V, or 0–10 V, depending on the type used.

Amplifiers must work linearly without distorting the shape of the pulses. An amplifier is characterized by:

1. The amplification $v_a = U_a/U_e$, indicating the relation of the output amplitude U_a to the input amplitude U_e.
2. The bandwidth indicating the frequency range in which signals can be amplified. The lower and upper limits of the frequency range are determined by the 3-dB points of the characteristic amplification curve where the nominal amplitudes are reduced to about 70%. The higher the bandwidth is, the faster rising pulses can be amplified. At the other side, an amplifier with a bandwidth starting at $f = 0$ is also capable of amplifying d.c. signals.
3. The noise that is produced in it and added to the signal. Large bandwidth amplifiers add more noise. Therefore, the bandwidth of a sizing amplifier should be only as large as necessary for negligible distortion.

The pulse rise time conditions can be derived from the pulse height topography explained in Section II. In a distance of $s/d = -1$—that is, one orifice diameter outside the inlet of the orifice—the pulse amplitude is about 1% of the maximum height. From there to the point of the pulse maximum in the middle area of the orifice, the particle must proceed a distance of 1.5 orifice diameters if the orifice length-to-diameter ratio is 1.

The rise time of the pulse from 1% to 100% is defined by the time required to suck the fluid volume ($V_o = \frac{2}{3} 8\gamma^3$) which is contained in the half-sphere touching the 1% pulse height point, into the orifice, plus the time the particle needs to move from the orifice inlet to its middle area. The particle is moving with v_m and a fluid volume of $v_m r^2$ is sucked through the orifice per unit of time (r-radius of the orifice).

The pulse rise time is calculated as:

$$t = \frac{V_o}{v_m r^2 \sum} + \frac{l}{2v_m} \tag{28}$$

Small-diameter orifices produce shorter pulses if flow velocities are considered constant.

With a 30-μm diameter orifice if $l/d = 1$ and $v_m = 5$ m/sec, we obtain

$t = 19$ μsec. According to the usual rise-time definition (10–90% of the pulse slope), pulses below 10 μsec rise time are not expected. With an amplifier, rise time of one-tenth of the pulse rise time, i.e., about 1 μsec, the pulses are amplified with negligible distortion. This estimation, however, is restricted to particle paths near the orifice axis. Pulses from particles on edge paths (which are avoided by the focusing technique) rise much faster, as shown in Fig. 22.

b. Preamplifier. The preamplifier is directly coupled to the transducer and amplifies the weak resistance pulses from the 0.1–7 mV range to the 100-mV range. It must be designed with low noise elements, e.g., field effect transistors.

The state-of-the-art preamplifier in resistance pulse sizing is shown in Fig. 42. This type of amplifier is called "current-sensing" or "zero-input" amplifier. The inverted output voltage U_a is fed back to the input by the feedback resistor Rf. A particle passing through the orifice produces a resistance change whereby normally

$$U_o + \Delta U = (R_o + \Delta R)i \tag{29}$$

where i is constant. With this type of amplifier, U_o is kept constant, i.e., the voltage $U = i\Delta R$ is compensated by $U' = R_o\Delta i$ whereby the current i, a part of the constant current i, no longer flows through R_o but is branched off into the amplifier.

We obtain:

$$i(R_o + \Delta R) - \Delta i R_o = U_o \tag{30}$$

With $iR_o = U_o$,

$$i = \frac{i\Delta R}{R_o} \tag{31}$$

With a high open loop gain, the amplifier itself takes up negligible current and the output voltage $-U_a$ is calculated as

$$-U_a = \Delta i R_f = \frac{i\Delta R R_f}{R_o} \tag{32}$$

Since $i\Delta R = \Delta U$, we obtain for the amplification:

$$v_a = \frac{U_a}{\Delta U} = \frac{-R_f}{R_o} \tag{33}$$

The "zero-voltage-input" feature causes input capacitances, e.g., in the transducer itself or of shielded cables, not to affect the amplifier's rise-time response.

With $R_o = \rho_2 l+/q$ we obtain for the output voltage of the amplifier:

$$U_a = -\Delta U \frac{R_f q}{\rho_2 l+} \tag{34}$$

whereby $l+$ is an equivalent length which describes the real length of the orifice plus an additional fictive length which describes the resistance component of R_o outside the actual orifice. With ΔU from equation (7), we have

$$U_a = \frac{-\Delta V i f R_f}{q l+} \tag{35}$$

The resistivity ρ_2 is cancelled from equation (35), indicating that the output voltage U_a of a zero-input amplifier is independent of the resistivity of the electrolyte, i.e., the original pulse height increase is compensated by a decrease in amplification (see Figs. 50, 51, and 53).

c. Main Amplifier. The main amplifier amplifies the output of the preamplifier to the level of the evaluation unit. It mostly contains a gain control, a trigger, and a baseline restoration circuit (Fig. 43). When small pulse height ranges are of interest, linear amplifiers are best suited. With large ranges, logarithmic amplifiers are favorable.

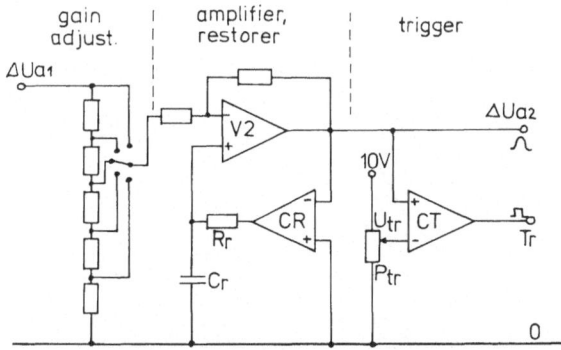

FIGURE 43. The main amplifier portion of the analog circuitry consisting of gain adjustment, amplifier and baseline restorer, and trigger device.

The amplification generally can be achieved by high-speed integrated operational amplifiers.

With the usual two-electrode technique, a coupling capacitor between the transducer and preamplifier separates the basis voltage U_o from the desired pulse. The differentiating effect of the capacitor produces an undershoot of each pulse, and with high pulse rates the baseline becomes totally negative. Baseline restoration circuits considerably reduce such effects. Restorers specially developed for nuclear pulse height measurements are described in the papers of Robinson (1961) and Chase and Poulo (1967). The restorer developed by Patzelt (1968) is well suited for use in sizing pulse amplifiers (Fig. 43). It consists of a comparator CR which compares the output of the operational amplifier U_a with zero potential. Its output is coupled over a low-pass filter consisting of R_r and C_r to the noninverting input of the amplifier V_2. This regulation circuit keeps the baseline of the output voltage U_a exactly at zero up to a pulse duty factor of 0.5.

Figure 43 also shows a simple trigger circuit. It consists of a comparator CT which compares the output voltage of the analog amplifier with the voltage level adjusted by the potentiometer P_{tr}.

By adjusting the trigger level, the operator can discriminate between wanted pulses and unwanted events, e.g., noise. Only pulses whose amplitude reaches the trigger level are considered existent for the evaluation unit.

The length of the trigger output signal can be considered a measure of the pulse length. With a fixed trigger level U_{tr}, however, the pulse length is modulated by the pulse height and thus by the volume of the particles, even if they pass along identical paths through the orifice. Pulse 2 in Figure 44 is identical in shape with pulse 1, but the length detected is considerably smaller. Pulses

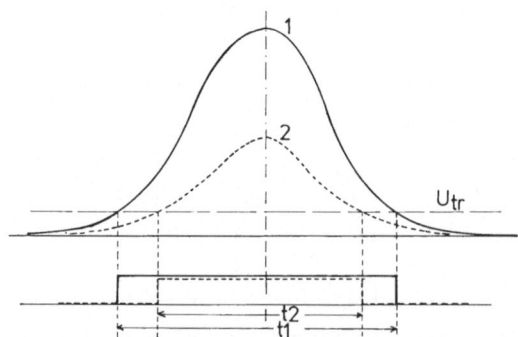

FIGURE 44. Explanation of how the pulse length detected depends on the pulse height. The trigger device reacts if the pulse voltage crosses the trigger voltage U_{tr}. High and small pulses at the same path are detected with different pulse length.

of different heights are triggered at different relative amplitudes. For exact pulse length determinations, a complicated trigger circuitry defining a trigger level that is related to the pulse maximum is necessary.

The main amplifier is followed by a peak-detect circuit, which senses and stores the pulse maximum and outputs it to the analog-to-digital converter.

According to the theory explained in Section II-B, the particle volume information is contained in the pulse amplitude and not in the pulse integral. Therefore, preferentially the pulse height and not the integral is measured in sizing instruments.

d. Digital Portion. A conventional digital circuitry of a sizing instrument consists of an analog-to-digital converter (ADC) and a multichannel pulse height analyzer (PHA).

The analog pulse height range is converted by the ADC into a range of successive addresses. Each address points to a location in an associated memory, and each time a pulse height address is determined by the ADC, the content of the associated memory cell is incremented by one. The stored histogram is displayed on a screen and can be plotted or printed or stored on a disk or tapes.

The histograms stored represent the frequency distribution over the volume scale of the cells that were analyzed.

The resolution of a PHA is limited by the number of channels available

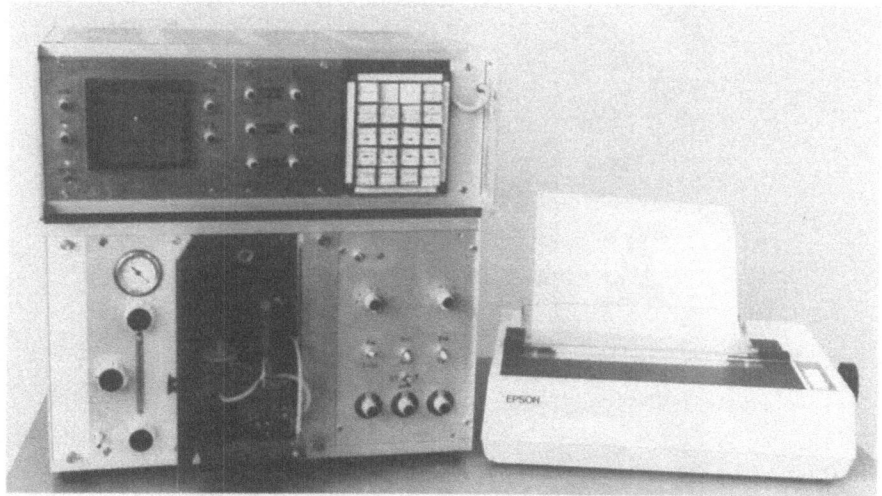

FIGURE 45. A Metricell cell volume analyzer. The lower box contains the fluid and transducer system and analog electronics; the upper box contains the CYTOMIC microprocessor evaluation unit. Histogram plots and digital output are made via the matrix printer at right.

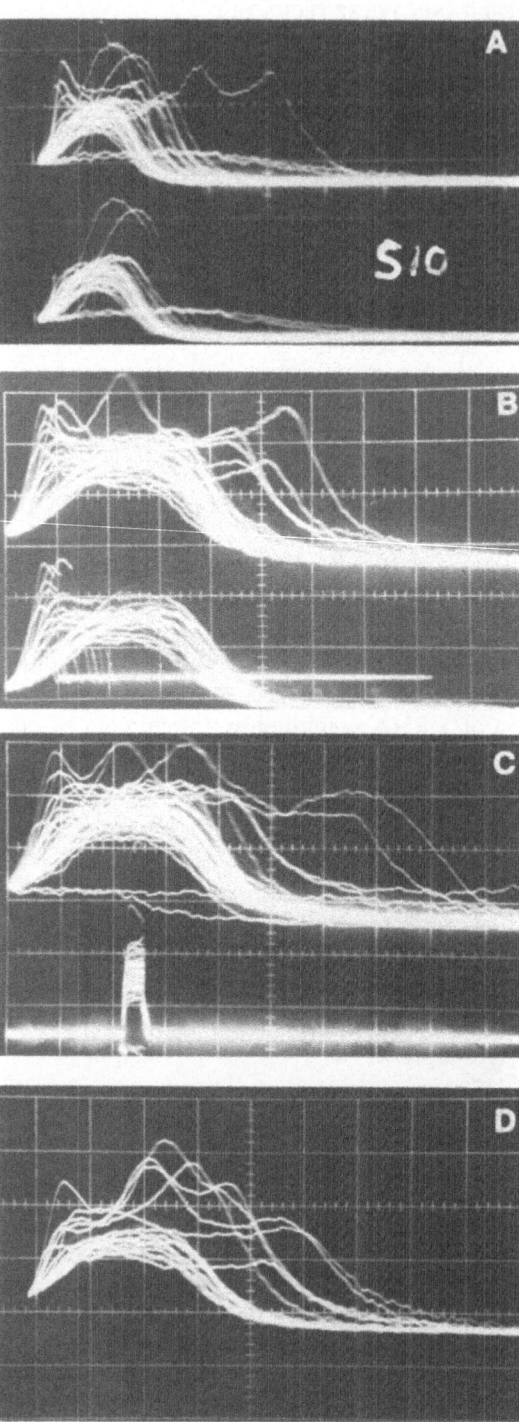

in each channel group. For resistance pulse height measurements of *biological cells*, 128 channels—or a maximum of 256 channels—are sufficient. More channels would provide more data to be handled and stored without a real increase of resolution since cell volume distributions normally are measured with coefficients of variation not below 10%.

The advanced-flow cytometric PHA "CYTOMIC", which includes data-processing facilities, is described in Section VI. Figure 45 shows a complete Metricell cell volume analyzer.

D. Electronic Pulse Handling Devices as Alternatives to Hydrodynamic Focusing

Apart from hydrodynamic focusing, attempts were made to improve the basic Coulter method by electronic pulse handling and elimination.

Pulses of particles on different paths in the orifice differ in shape. The shape differences may be employed to discriminate pulses that should be accepted and those that should be eliminated. Elimination of pulses implies that the resulting histograms no longer represent the total number of cells passed through the orifice. Use of such methods is problematic if the cells should be counted simultaneously.

The following criteria can be used:

1. Pulse duration. Pulses of particles at paths near the orifice wall are longer than particle pulses from the core region. Pulses above a limit in time duration are not evaluated (Coulter Electronics, 1971, 1972; Kachel, 1972; Waterman *et al.*, 1975).
2. Maximum trigger (MT). M-shaped pulses arrive considerably earlier at their pulse maximum than bell-shaped pulses from particles in the core region. Pulses with short maximum arrival time are rejected (Fig. 46b) (Kachel, 1972, 1973). A variation of this technique is the evaluation of the rise time differences of the different pulse types detected by pulse differentiation. Steep pulses are rejected (Fig. 46a).
3. Peak number evaluation. Only monopeaked pulses are accepted (Kachel, 1972).

←——————————————————————————————————————

FIGURE 46. (A) Evaluation of the pulse rise time differences for rejecting distorted pulses. The steepness of the pulses is sensed by a differentiating circuit. Steep pulses are rejected. Upper trace: the pulse pattern at the output of the amplifier. Lower trace: the pulse pattern at the output of the rejection device. (B) Maximum trigger (MT) method. Pulses with short maximum arrival times are rejected. (C) Delayed gating. The pulse amplitudes are evaluated when the particles are passing along the central area of the orifice. The edge distortions are avoided. Peaks of rotating cells, however, are evaluated. (D) Pulse pattern showing several three-peaked pulses caused by cell rotation near the orifice wall. Some of them have direct transition from the first peak to the second without a valley. Such pulses cannot be recognized by the MT method.

4. Gated pulse evaluation. With this technique—used by Bull (1968), Grover *et al.* (1969b), and Steen and Nielsen (1979)—the absolute pulse height is not evaluated, but after a fixed delay from the initial trigger point, a few-microseconds-long part of the pulse corresponding to the moment at which the particles are passing through the middle area of the orifice are gated out and evaluated (Fig. 46c).

The basic condition for a successful application of the electronic methods mentioned is the use of an orifice with a homogeneous field in its middle area ($l/d > 1$). Only with this condition fulfilled can the pulse height evaluation in the middle area be independent of the radial position of the particle path. Without this condition, the monopeaked pulses or gated pulse heights remaining after electronic processing show a particle-path-dependent spread in height.

The pulse duration and maximum trigger methods are sensitive to particle-flow velocities. For widespread pulse heights, the relative trigger point determination is critical, as shown above.

Maximum trigger and delayed gating are very critical in application with nonspherical cells. Such cells may rotate near the orifice wall and produce three peaked pulses which in part do not have a valley between the first and second peaks (Fig. 46d). The MT circuit cannot recognize such pulses, and the delayed gating device evaluates these unwanted peaks in particular. (Fig. 46c).

The rise time method, which defines the steepness of the pulses by differentiation, is especially sensitive to noisy fluctuations of the baseline.

IV. CALIBRATION PROBLEMS OF ELECTRICAL SIZING FLOW CYTOMETERS

A. Calibration in Long Orifices with Homogeneous Field

1. Mathematical Basis for Calibration

By calibration, a relative or absolute cell volume is coordinated with the pulse height measured in the transducer. With a relative calibration, the pulse height of cell populations with unknown volume are related to pulse heights of reference cells or particles whose absolute volume is also unknown. Suitable numbers for such relative calibrations are the mean values of the pulse height distributions considered. As a result, the unknown cell volume is expressed as a fraction number or a multiple of the volume of the calibrating particles. If the absolute volume of the calibrating particles is known, the calibration is absolute (see below).

From equation (7), a calibration procedure can be developed which is independent of calibration particles. Equation (7), resolved for the particle volume, reads:

$$\Delta V = \frac{q^2}{\rho_2 if} \Delta U \tag{36}$$

2. Determination of the Factors Influencing Calibration

a. *The Diameter of the Orifice.* To solve equation (36), q, i, and f must be determined in addition to ΔU. The resistivity ρ_2 of the electrolyte and the current i is easily measured by standard methods.

Problems arise from the determination of the cross section q of the orifice. The expression q^2 means that the diameter to the fourth is valid in the equation. An error of 1 μm in determination of the orifice diameter causes considerable errors in the volume calculated; e.g., if the diameter is measured as 31 μm instead of 30 μm, the error in volume is 10%. With larger orifices the relative error is less significant.

Equation (36) can be used to determine orifice diameters very sensitively if particles of known volume are available. We obtain with $l/d > 1$ for a homogeneous field in the orifice:

$$d = \frac{2}{\pi} \sqrt[4]{\frac{\Delta V(\rho_2 if)}{\Delta U}} \tag{37}$$

This method at present is of more theoretical value since the volume specifications given by the producers of particles are not of sufficient reliability.

Calibrating particles even with unknown absolute volume are valuable in relative determination of orifice diameters. Diameters of larger orifices are determined with smaller relative error. We measure identical particles in the known large and in the unknown small orifice (condition for both orifices is $l/d > 1$ and particle-to-orifice diameter < 0.2). Using equation (36), we can equate the volumes of both determinations:

$$\frac{q_x^2 \Delta U_x}{\rho_x i_x f} = \frac{q_1^2 \Delta U_1}{\rho_2 i_1 f} \tag{38}$$

With identical electrolytes ($\rho_2 = \rho_x$) in both measurements, results for the cross section of the small orifice are

$$q_x = q_1 \sqrt{\left(\frac{\Delta U_1}{\Delta U_x}\right)\left(\frac{i_x}{i_1}\right)} \tag{39}$$

Best suited for such determinations are uniform latex particles. The highest accuracy is achieved if ΔU_1 and ΔU_x are the means over the distributions.

b. The Shape and Electrical Resistivity of the Particles. The second problematic factor in equation (36) is the factor f describing the shape and resistivity of the particles. With rigid particles, the quiescent shape and the shape of the moving particle in the orifice can be assumed to be identical. The shape factor of lengthwise-oriented cells in a focused particle stream can be found by model experiments or by shape-factor calculations as described in Section II-B. Deformable cells must be studied under original flow conditions and their shape derived from the actual cell images.

A difficult problem is the determination of the electric resistivity of the particles. Fortunately, most biological particles are of very high electric resistivity. For practical use it is sufficient to ensure that the ratio of particle-to-electrolyte resistivity (ρ_1/ρ_2) has a negligible influence on the size calibration.

Two methods are suited for estimation of particle resistivity. With the first method (Metzger, 1970, personal communication) the particles of unknown resistivity are suspended and sized in a series of electrolytes with increasing resistivities. With nonrigid biological cells, the change in osmolarity has to be compensated by addition of nonionic substances. When the amplitudes of the pulses, which are observed with an oscilloscope, are reduced to zero, and the first pulses of inverse polarity appear, the resistivity of the electrolyte and of the particles is nearly equal.

This method can be applied without knowledge of the distribution curve of the particles. The field strength within the orifice ($E = i\rho_2/q$) increases with increasing electrolyte resistivity. Since the current i cannot be decreased proportionally to the increasing resistivity due to noise limitations, E might reach the point at which electrical breakthrough of the cells will occur (see Section II-D).

The other method is based on equation (7), which is provided to calculate particle volume if particle resistivity is known. If particle resistivity is unknown, we have two unknown variables in the equation—the particle volume ΔV and the resistivity ρ_1 of the particles.

The solution requires two independent equations that are obtained by sizing the cells in two electrolytes of different specific resistivities (ρ_2 and ρ_3) which are osmotically balanced by nonionic substances. The volumes of two characteristic points which can be identified in both volume distributions (e.g., the 50% of the rising and falling slopes of the distribution curves are equalized) and an equation with only the resistivity ρ_1 of the particles remain. For a long orifice with a homogeneous field, we obtain, according to equation (9), the measurement in electrolyte ρ_2:

$$\Delta V_1 = \frac{-\Delta U_1 q^2\{2 + ab^2 \mathrm{La}[(\rho_2/\rho_1) - 1]\}}{(\rho_2 i)\, 2[\rho_2/\rho_1) - 1]} \tag{40}$$

The measurement in electrolyte ρ_3:

$$\Delta V_2 = \frac{-\Delta U_2 q^2 \{2 + ab^2 \text{La}[(\rho_2/\rho_1) - 1]\}}{\rho_3 i \, 2 \, [(\rho_3/\rho_1) - 1]} \tag{41}$$

and the measurement whereby ΔV_1 equals ΔV_2:

$$\frac{\Delta U_1 \{2 + ab^2 \text{La}[(\rho_2/\rho_1) - 1]\}}{\rho_2 \{2[(\rho_2/\rho_1) - 1]\}} = \frac{\Delta U_2 \{2 + ab^2 \text{La}[(\rho_3/\rho_1) - 1]\}}{\rho_3 \{2[(\rho_3/\rho_1) - 1]\}} \tag{42}$$

with $s = (\Delta U_1/\rho_2)$ and $t = (\Delta U_2/\rho_3)$ and $A = ab^2 \text{La}$, we obtain:

$$\rho_1 = \frac{-b \pm (b^2 - 4ac)^{1/2}}{2a} \tag{43}$$

where

$$a = [(s/t) - 1] \, (2 - A) \tag{44}$$
$$b = \rho_3 \{A[(s/t) - 1] - 2s/t\} + \rho_2 \{A[(s/t) - 1] + 2\} \tag{45}$$
$$c = A\rho_2\rho_3[1 - (s/t)] \tag{46}$$

Positive real solutions give the particle resistivity.

The volume of the reference point of the distribution is calculated by introducing into equation (40) or (41):

$$V = -sq^2 \frac{2 + A[(\rho_2/\rho_1) - 1]}{i\{2[(\rho_2/\rho_1) - 1]\}} \tag{47}$$

A resistivity analysis of fixed erythrocytes was performed by computer program. The calculated resistivity was 3.4 kΩ/cm. Nonconducting latex particles with a resistivity of 156 kΩ/cm were determined (Kachel, 1979).

With this method, ρ_2 and ρ_3 can be chosen such that the increase in field strength is compensated for by a decrease in current, provided that cell breakthrough is avoided. With this method, it is theoretically possible to detect resistivity variations within the same particle population. Living cells are sensitive to changes of the electrolytes. Therefore, with unfixed cells, uncertainties regarding volume changes will remain even if the osmolarity of the electrolyte is balanced.

3. The Practical Particle-Independent Calibrator Device and Procedure

The pulse height analyzer primarily stores a pulse height distribution curve. The first requirement is to mark absolute voltage levels ΔU in the histogram, which can be related to the pulse amplitudes picked up at the electrodes.

By applying equations (7) or (9) for each ΔU value in the histogram, the corresponding particle volume ΔV can be calculated. Since ΔU and ΔV are linearly related, a volume scale can be developed by calculating a straight line equation in two calibrated points of the histogram. The calibrator points can consist of one calibrator point of defined amplitude and zero, or—independent of the instrument's zero—two defined pulse amplitudes.

A carefully stabilized pulse generator, producing trapezoidal voltage pulses of well-defined amplitudes in the millivolt range, is coupled in series to the orifice (Kachel, 1974b). This kind of coupling allows sizing and calibrating simultaneously and particle and calibrator pulses can be compared directly on the oscilloscope screen and in the histogram (Fig. 47).

Figure 42 explains how the calibrator pulses U_c are coupled to the transducer via the 1-Ω resistor.

The calibrator pulses are visible in the histogram as narrow peaks. The

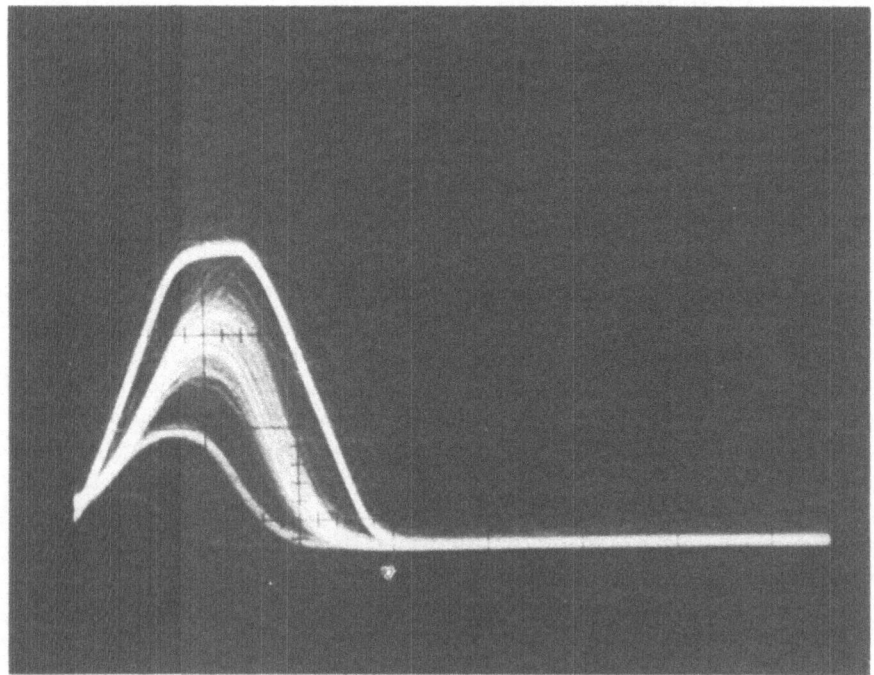

FIGURE 47. Pulse pattern taken during sizing with simultaneous calibration. The bell-shaped pulses are produced by the particles, the trapezoidal pulses by the calibrator.

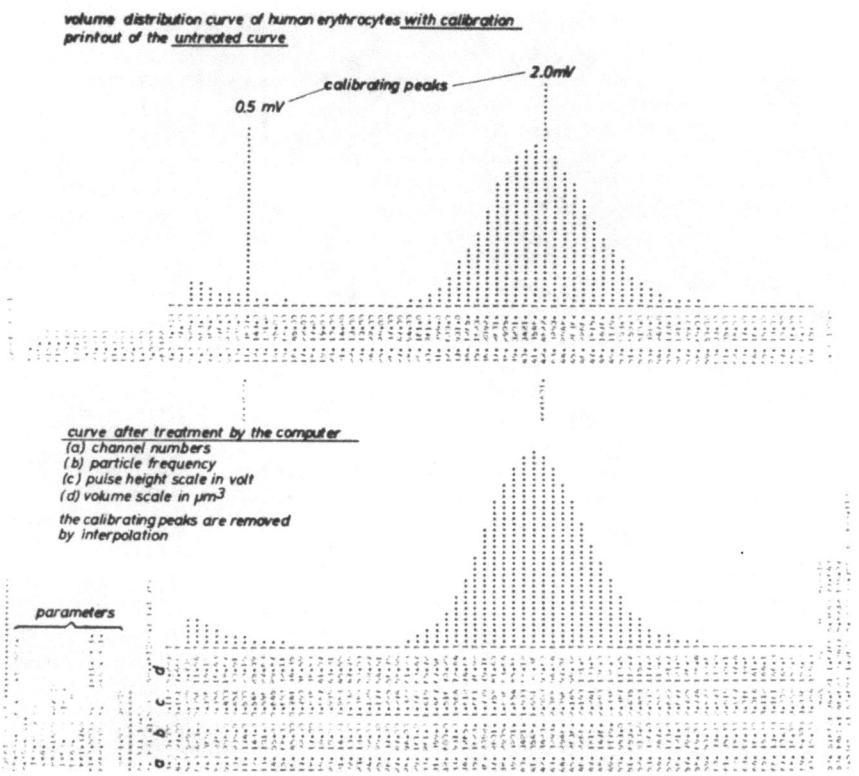

FIGURE 48. Volume distribution curve of native human erythrocytes demonstrating electrical calibration. Upper: printout of the untreated distribution containing the calibrator peaks of 0.5 mV and 2.0 mV. Lower: the same curve with pulse height scale (c) and volume scale (d) derived from the calibrator peaks by the computer. The calibrator peaks are removed by interpolation, and the number of calibrator pulses evaluated is indicated at the upper side of the printout.

pulse-height-to-cell-volume conversion is optimally done by a simple computer program which calculates the volume scale and subsequently removes the calibrator peaks from the histogram by interpolation (Fig. 48).

4. Use of the Calibrator for General Control Functions

a. Detection of Noise in the Sizing System. When the calibrator pulses are generated, they are completely uniform in height. By feeding in these pulses during the actual sizing, they are exposed to the same distortions as the particle pulses. Therefore, the distribution curve of the calibrator pulses rep-

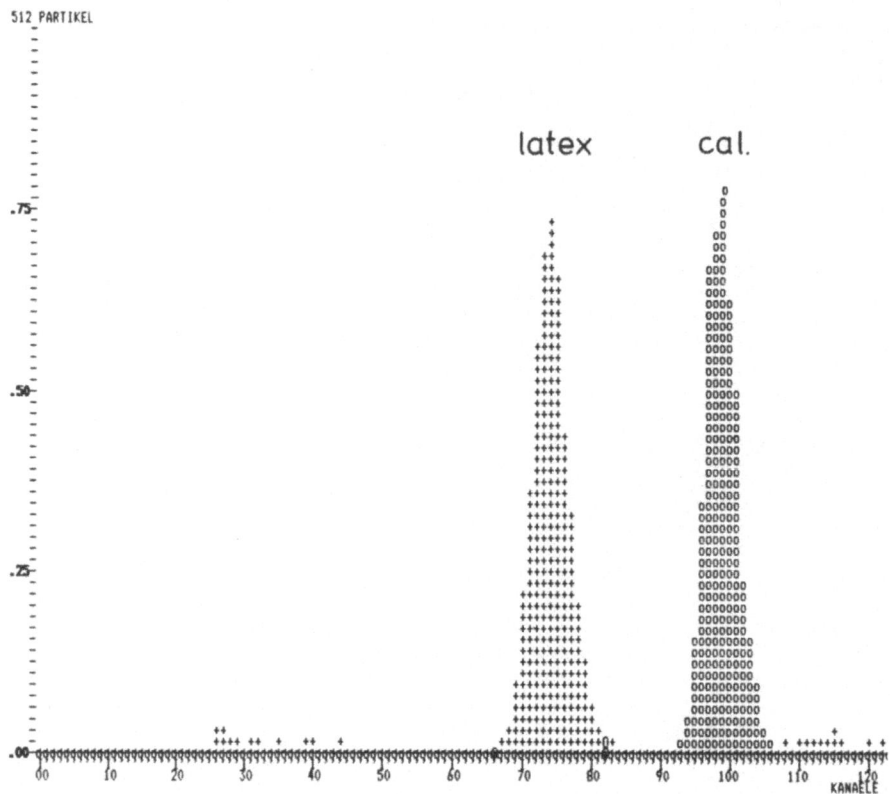

PARTIKEL: 4.31μ ltx
ELEKTROL.WID.:   60 Ωcm
MESSOEFFNUNG: 50 μm ∅
MESSSTROM: 50 μA
VERSTAERKUNG: 32,
EICHIMPULS: 0.4 mV
SAUGDRUCK: 0.1

CODENUMMER: 00 00 10
GESAMTSUMME: 002915
M1=066 - M2=082 = SUMME: 002390

STAT. WERTE ZW. M1 - M2:
MITTELWERT : 074.00 KAN.
STAND. ABW.: 02.65 KAN.
VAR. KOEFF.: 03.58 %

FIGURE 49. The calibrator pulse peak as indicator of noise in the transducer system. The calibrator pulses are *a priori* generated as uniform in height. Therefore, the distortion function disclosed by the calibrator peak is also superimposed on the particle distribution. Here high noise is superimposed. Both calibrator and particle distributions show high coefficients of variation compared with the low noise analysis of Fig. 50.

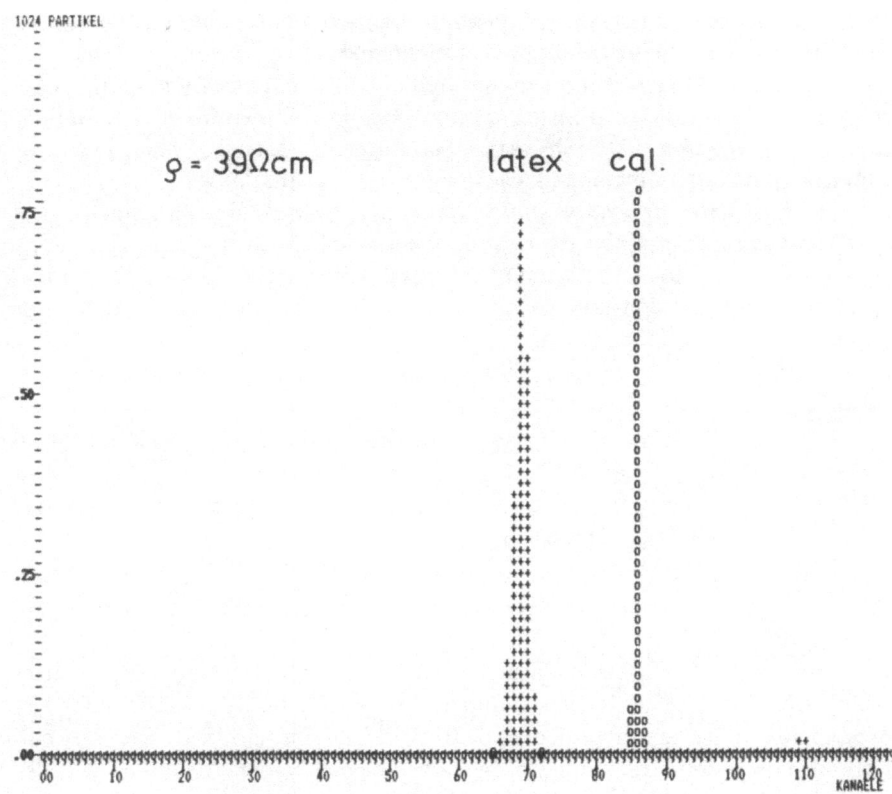

FIGURE 50. A low noise size analysis of the same 4.31-μm-diameter latex particles as in Fig. 49. Here the calibrator peak shows a narrow distribution, indicating that low noise distorts the analysis. This analysis is performed with an electrolyte of 39 Ω·cm. The 4-mV calibrator peak is found in channel 86. The mean of the latex peak is 68.94 (channels).

resent the noise function superimposed to the particle distribution. Figures 49 and 50 show particle distributions with high and low noise superimposed.

This method detects frequent low and high fluctuations affecting the baseline, e.g., electric noise from inside and outside the instrument, gas bubble motions, microphony, etc. Distortions produced by particles which move on different paths in an inhomogeneous field cannot be disclosed.

b. Resistivity, Blockage, and Temperature Control. The amplification of a current sensing amplifier depends on the ratio of the feedback resistor R_f to the resistance R found between the electrodes of the transducer. This feature, combined with the constant voltage pulses of the calibrator, form a valuable control organ in a sizing system.

If the orifice is partially blocked or the resistivity of the electrolyte changes, the amplification of the preamplifier changes according to the changed resistance ratio. The height of the calibrator pulses at the output of the amplifier changes according to the amplification as in equation (33). With increasing resistance R_o between the electrodes, the amplification and the height of the calibrator pulses decrease.

Figures 50 and 51 show how a calibrator peak changes if the resistivity of the electrolyte is changed. The position of the particle peak remains constant as expected.

The resistivity of physiological electrolytes used in flow cytometric instruments strongly depends on the temperature of the fluids. Many biological experiments must be performed at distinct incubation temperatures. Particularly for flow cytometric studies of biological or biochemical reactions of cells, a temperature control inside the transducer is very valuable.

The calibrator control system offers an elegant method of temperature control. Figure 52 shows how the electric resistivity of a physiological electrolyte changes with temperature.

From such curves we can derive the channel in which the calibrator peak must be found for a distinct temperature of the transducer if a starting point is defined at room temperature.

In general, one cannot be sure that a change of the control peak is not caused by a partial blockage of the orifice. This can be counterchecked by measuring uniform latex particles together with the cells. Normally, the latex peak does not change with change in resistivity (see Section III-B). Partial blockage, however, distorts the field configuration and thus also the shape and position of the latex distribution.

Figure 53 illustrates an experiment of temperature change with latex particles and a calibrator peak. Figure 53a shows the sequence of 32 distribution curves during warming up of a Metricell system from 10°C to about 38°C. The latex curve stays constant; the calibrator peak, however, changes with increasing temperature. Figures 53b and c show the distributions of 10°C and 38°C.

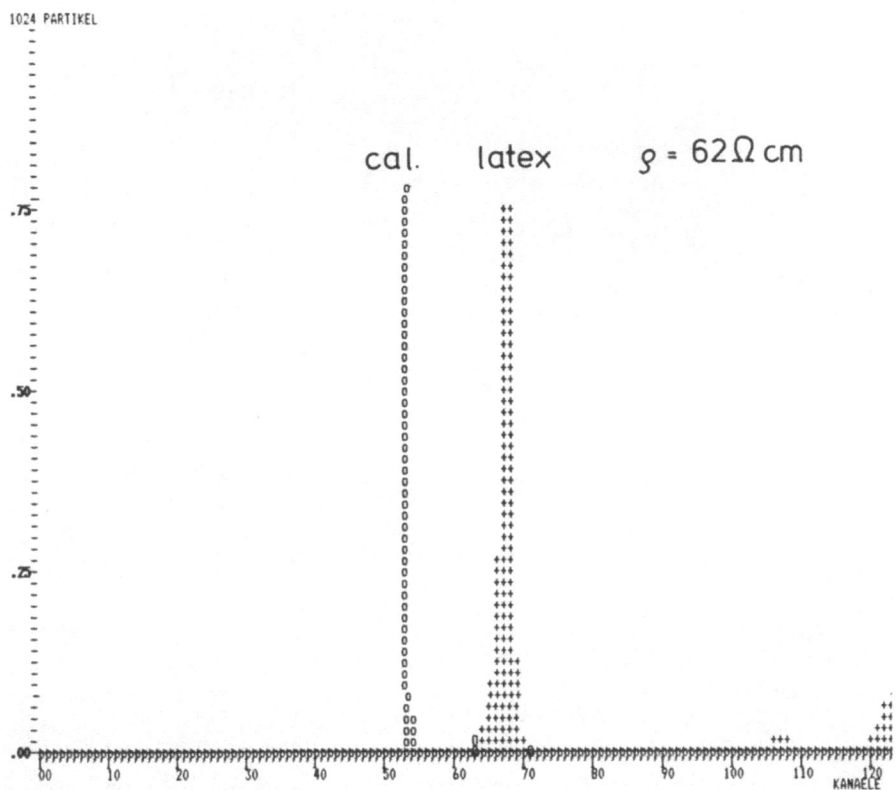

PARTIKEL: ltx 4.31μm∅
ELEKTROL.WID.: 62 Ωcm
MESSOEFFNUNG: 50 μm ∅
MESSSTROM: 0.8 mA
VERSTAERKUNG: 2
EICHIMPULS: 4mV
SAUGDRUCK: 0.1

CODENUMMER: 00 00 05
GESAMTSUMME: 002738
M1=063 - M2=071 = SUMME: 002162

STAT. WERTE ZW. M1 - M2:
MITTELWERT : 067.23 KAN.
STAND. ABW.: 01.14 KAN.
VAR. KOEFF.: 01.69 %

FIGURE 51. Analysis of 4.31-μm latex particles performed in an electrolyte 62-Ω·cm resistivity. The calibrator peak is now in channel 54. Compare Fig. 50: The inverted peak position ratio 54/86 = 0.62 equals the resistivity ratio 39/62 = 0.62. The latex peak, however, is kept constant on its channel position due to the compensating features of the current sensing amplifier.

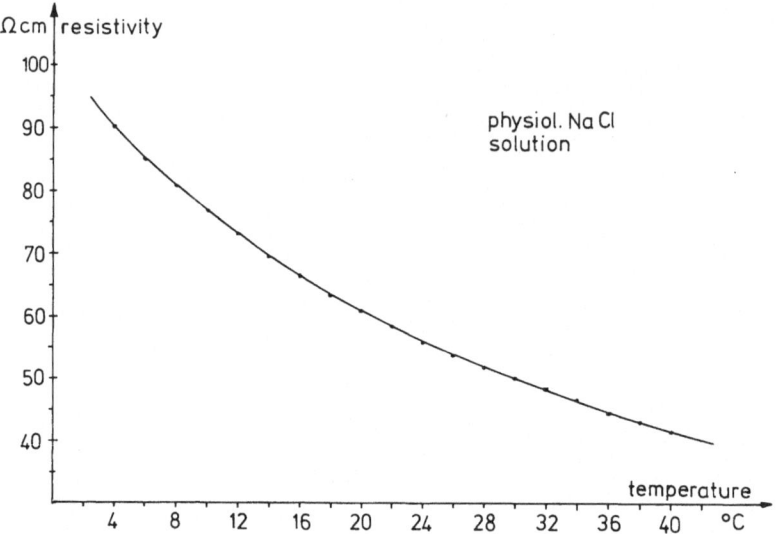

FIGURE 52. The specific resistivity of a physiological NaCl solution as a function of temperature.

B. Calibration with Short Orifices

Equation (36), which gave the basis for all our considerations in particle size calibration, is valid only if a homogeneous field exists inside the orifice in which the pulse maximum is generated. In orifices of length-to-diameter ratios less than 1.2, this condition is no longer true. In short orifices, due to the inhomogeneities of the electric field, pulse heights depend on the particle path (see Section II-B).

We have learned from the model experiments that particles at the axial path in the orifices that are most suited for practical use are more and more underevaluated in relation to the homogeneous field with a decreasing l/d ratio. Curve C of Fig. 54 shows the underevaluation of particles at the axial path as a function of different length-to-diameter ratios. In contrast, curve A shows how particles are measured in the homogenized field.

Curve C, which may be considered an error function at axial paths, can be used to correct size measurement being performed in short orifices. A correction factor, the so-called orifice factor f_k indicated at the right scale of Fig. 54, must be introduced into equation (36). It is calculated from the inverted underevaluation of the particles.

Equation (36) is modified to:

$$\Delta V = \frac{q^2(f_k)}{\rho_2 if} \Delta U \tag{48}$$

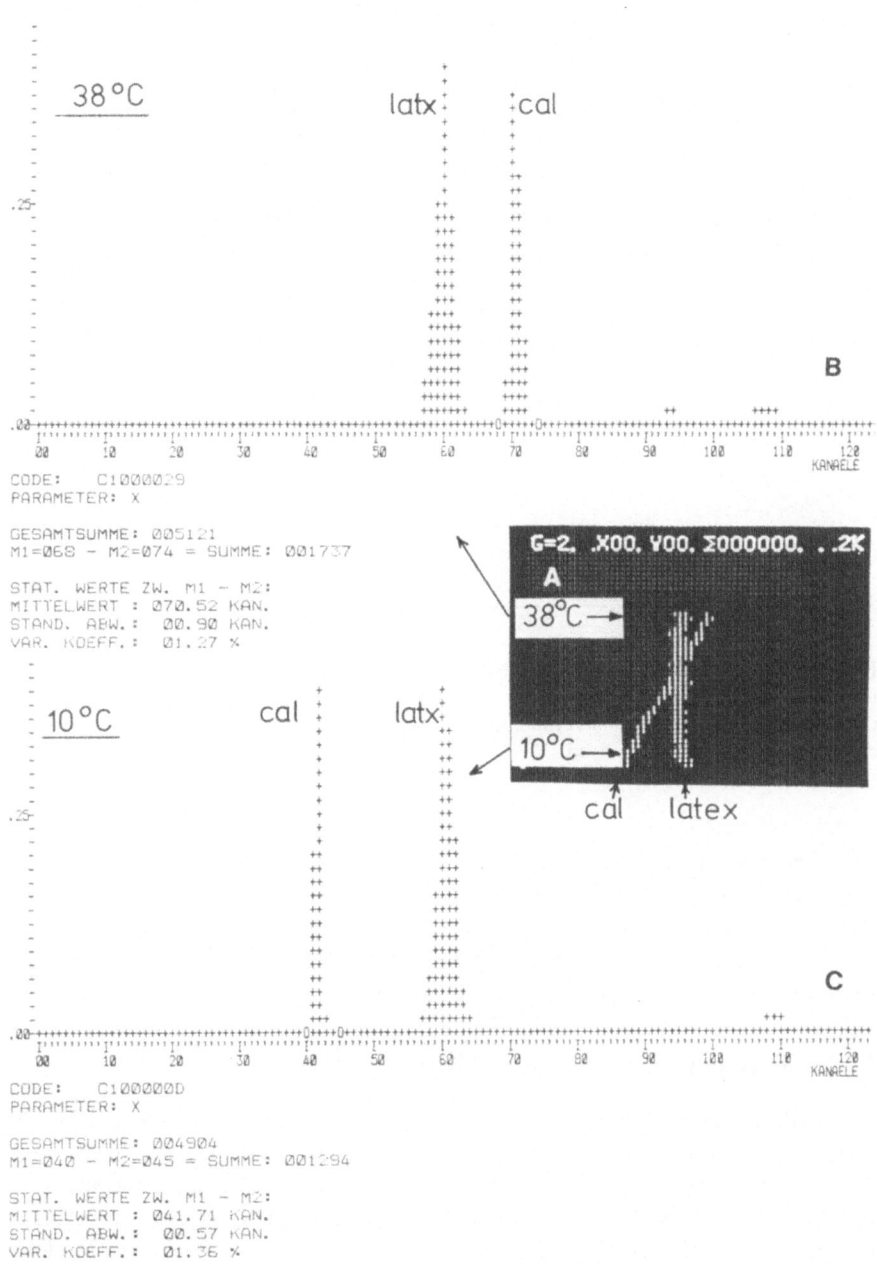

FIGURE 53. The Metricell calibrator peak as temperature control. (A) A sequence of 32 distribution curves of latex particles and calibrator peaks taken with a Metricell transducer and a CYTOMIC analyzer. The analysis starts in line 0 at a temperature of the transducer system of 10°C up to a temperature of 38°C in line 32. Each line represents a one-parameter distribution. The changing position of the calibrator peak is a measure of the actual temperature of the transducer. The position of the latex peak remains constant. (B) One-parameter curve of line 3 representing 12°C. (C) One-parameter curve of line 32 representing 38°C.

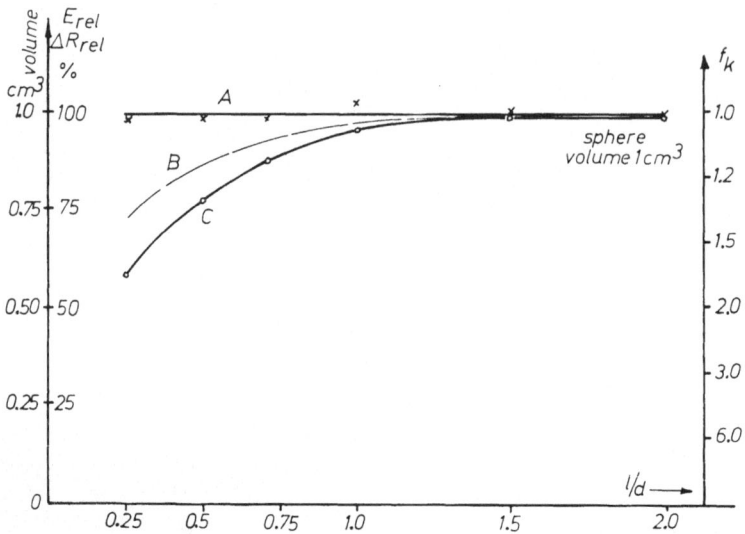

FIGURE 54. The evaluation of particles at the axial path of sizing orifices as a function of the length-to-diameter ratio. Curve A shows the evaluation in the homogeneous field and curve C in the natural field. From the underevaluation curve C, the compensation factor fk, which is indicated at the right scale, is derived.

There are mainly two reasons for not exclusively using long orifices. First, the coincidence probability increases with the increased l/d ratio, and second, the basic resistance R_o of an orifice increases with increasing length. High basic resistances complicate the optimal transducer–amplifier coupling and increase the basic electric noise in the system.

C. Calibration with Reference Particles

Calibration with particles of known size is normally performed under conditions that are identical with the conditions of the proper cell measurement. The cell measurement is related to the calibration measurement, and identical parameters such as the orifice cross section q, the resistivity ρ_2 of the electrolyte, or the current i and the orifice factor f_k can remain out of consideration.

The shape and conductivity factor f_s or f_e, however, must be carefully taken into account. If, for example, mammalian erythrocytes are calibrated with spherical latex particles, the shape factor ratio of the different particles ($1.05/1.5 = 0.7$) must be considered in the result (see equations 6–11).

Uncertainties lie in the absolute size specifications made by the manufacturers of artifically produced particles.

V. APPLICATIONS OF THE METRICELL CELL VOLUME ANALYZER

A. Erythrocyte Volume Distribution Curves of Young Rats

Figure 55 shows a computer-plotted time sequence of the volume distribution curves of erythrocytes of newborn rats (Valet *et al.,* 1972b; Ruhenstroth-Bauer *et al.,* 1974). During the first 40 days after birth, five different erythrocyte volume populations can be distinguished.

Populations I and II, existing at birth, are replaced by population III, which exists about 25 days. This is replaced by a second intermediate population IV and the final population V, which remains alone during the life of the rat. Populations I–III are electrophoretically slow; populations IV and V, however, are fast. Populations I and II probably are produced in the liver (Valet *et al.,* 1977). The production of the cells of the other populations in the hematopoietic organs is triggered by a diffusible factor (Hanser *et al.,* 1974, 1979).

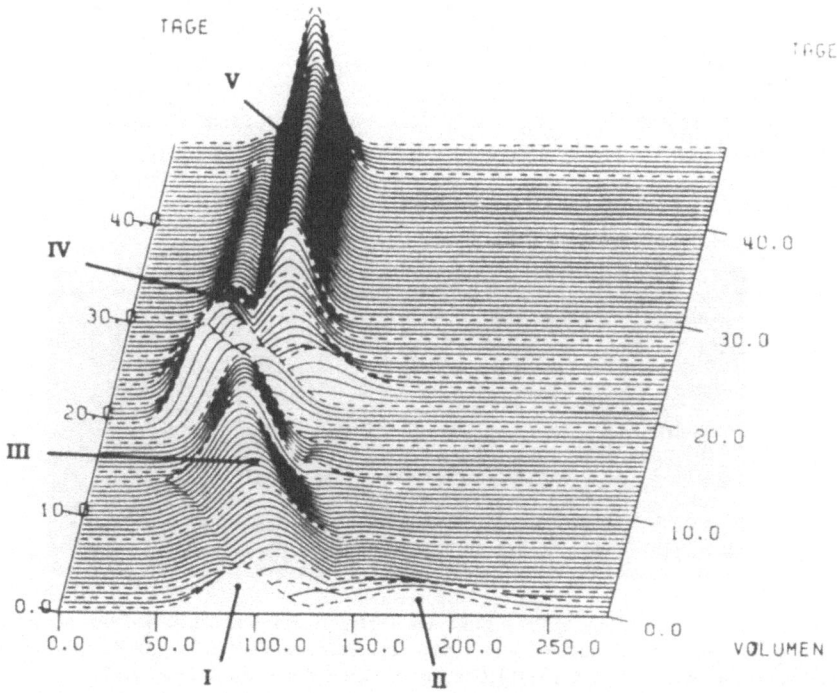

FIGURE 55. Time sequence of erythrocyte volume distributions of rats after birth. Orifice, 70-μm diameter; current, 0.4 mA. The broken lines mark the measured curves; the other curves are interpolated. During the first 40 days after birth, five different populations are found.

B. Red Blood Cell Aggregation in Humans

Boss *et al.* (1973, 1975, 1980) have shown that the red blood cells of patients with coronary artery disease risk factors show a significantly higher tendency to aggregate than the red blood cells of normal persons. With the Metricell instrument, singlets and multiplets of erythrocytes can be resolved easily. Coincidence is minimized by measuring with pulse rates in the range of up to 500 1/sec.

Figure 56a shows a red blood cell aggregation measurement of a normal person that indicates an erythrocyte aggregation index (EAI) of 14.36%. In contrast, Figure 56b, measured from the blood of a person with several coronary disease risk factors, indicates an EAI of 66.74%. S, D, and T mark the populations of singlets, doublets, and triplets of red blood cells that have been prepared according to the method described by Boss *et al.* (1980). The EAI is calculated as:

$$EAI = \frac{2D + 3T}{S + 2D + 3T} \cdot 100$$

The EAI values are automatically calculated by the new CYTOMIC analyzer described in Section VI-C-2. Investigations are in progress to find out whether this method can generally be used to detect the risk of heart infarction.

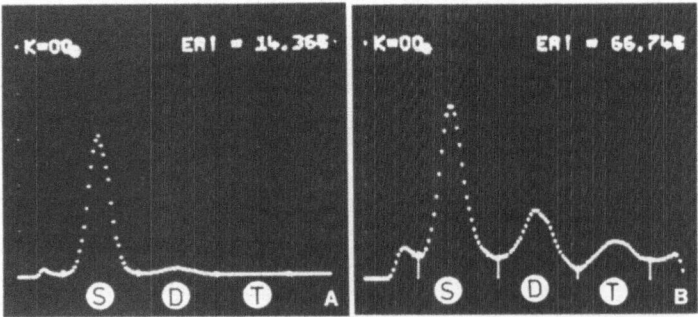

FIGURE 56. Determination of the erythrocyte aggregation. Orifice, 80-μm diameter; current, 0.6 mA. (S) Singlets; (D) doublets; (T) triplets. The evaluation is performed with the dedicated aggregation program of the CYTOMIC analyzer (see Section VI-C-2). (A) Blood of a normal person. Only a few doublets and no triplets are recognized in the histogram, and the erythrocyte aggregation index (EAI) is 14.36%. (B) Blood of a person with heart disease risk factors. Many doublets and triplets are found. The EAI is 66.74%.

C. Thrombocyte Volume Distributions Measured from Whole Blood

The investigation of thrombocytes in human blood is of interest in many aspects. Thrombocytes, however, are very sensitive to any treatment. Hence it is highly desirable to count and size thrombocytes without particular separation treatment directly from whole blood that is only diluted and immediately measured in the Metricell transducer.

Such measurements are problematic insofar that the number of thrombocytes in general is only 10% of the number of erythrocytes and their volume

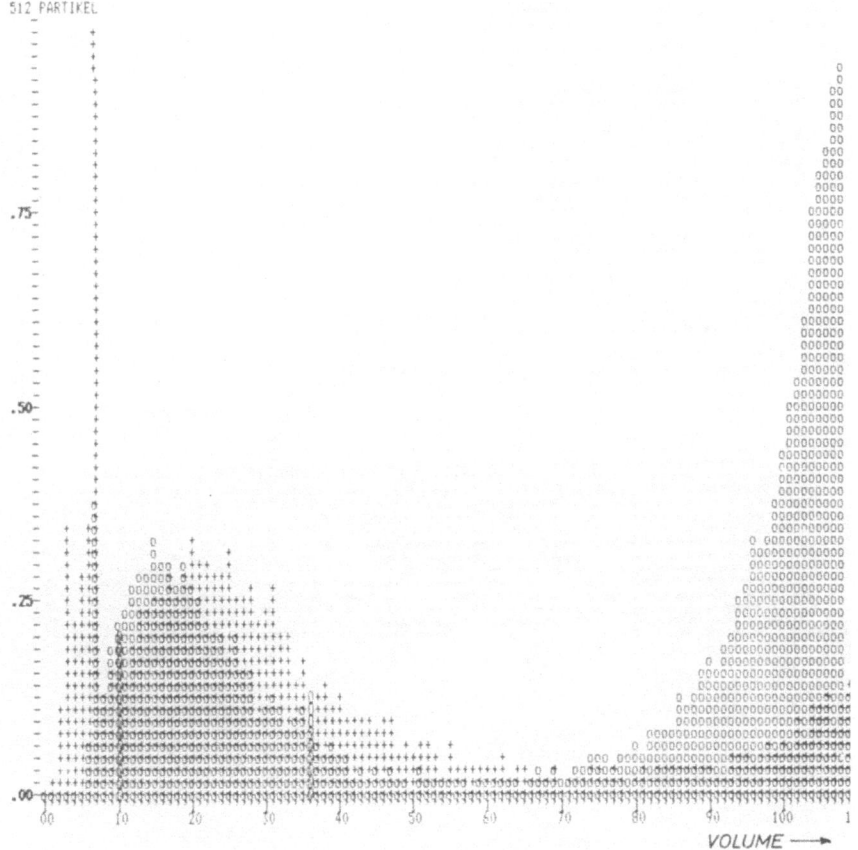

FIGURE 57. Thrombocyte size distributions from whole blood. Orifice diameter, 30 μm; current, 300 μA (0) and 400 μA (+). The slope at the right side belongs to the erythrocyte volume distribution. The compare print has been done with the CYTOMIC analyzer.

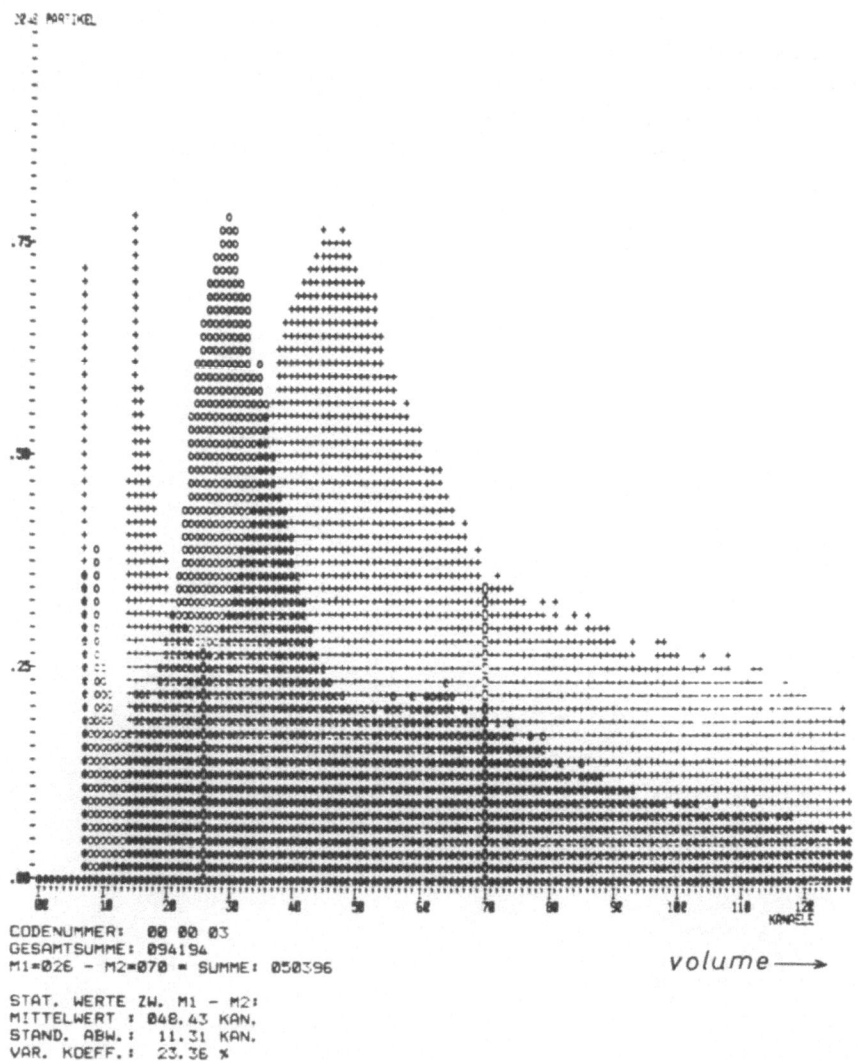

FIGURE 58. Volume distributions of pheochromocytoma cells with (+) and without (0) incubation in presence of nerve growth factor (NGF). The volume of the NGF-incubated cells is increased by 50–60%. Orifice, 80-μm diameter; current, 0.1 mA. The curves are printed by the compare print program of the CYTOMIC analyzer.

is in the 10% range of the erythrocyte volume. Here the recirculation problem becomes important (see Section II-D-5). Figure 57 shows a superposition of two size distribution curves of thrombocytes taken in a 30-μm orifice with currents of 300 (0) and 400 (+) μA. The slope of the curve at the right belongs to the erythrocyte distribution. The thrombocyte curves taken with the anti-recirculation flow on are clearly separated from the background noise.

D. Effect of Nerve Growth Factor on Pheochromocytoma Cells

Figure 58 demonstrates the effect of nerve growth factor (NGF) on pheochromocytoma cells. The experiment was performed in cooperation with R. Heumann (Department of Neurochemistry, Max-Planck-Institut für Psychiatrie). The hypertrophic effect of NGF on its target cells and the correlation with induction of fiber outgrowth can be investigated in an NGF-responsive pheochromocytoma cell line (PC12) which is grown in culture. The Metricell experiments show that after 2 days of incubation in the presence of NGF, the volume of the pheochromocytoma cells increases by 50–60%. At the same time, the number of cells with fiber outgrowth is increased by the same percentage (Heumann *et al.*, in preparation). The investigation of the relationship between cell volume and fiber outgrowth could be of importance for the characterization of neuronal regeneration.

VI. COMBINATION OF ELECTRICAL SIZING WITH MULTIPARAMETER OPTICAL FLOW ANALYSIS

A. The Fluorescence–Volume (FLUVO) Transducer

The performance and versatility of flow cytometers is strongly increased by combining electrical sizing with optical flow analysis, e.g., fluorescence or absorption measurements. The fluorescence technique is widely used in biology, particularly in studying cell proliferation by DNA analysis, or in immunology, where fluorescinated antibodies can be determined quantitatively. Cell volume as an additional parameter allows calculation of concentrations or densities of substances for each individual cell.

A multiparameter system that includes electrical sizing was described by Steinkamp *et al.* (1973). This instrument, which uses a laser for exciting the fluorescence, measures electrical volume and fluorescence at different points in the transducer, resulting in a time delay of volume and fluorescence pulses. The time delay complicates the pulse coordination and evaluation. The electrodes in the transducer are designed such that DNA fluorochromes are degraded by influence of the sizing current (Alabaster *et al.*, 1980).

Such drawbacks are avoided by a multiparameter instrument described by Kachel and Glossner (1976) and Kachel *et al.* (1977). This so-called FLUVO-Metricell device uses the electrical sizing instrumentation described above and the optical pulse fluorescence device described by Dittrich and Goehde (1969). With this epiillumination design, the exciting light of a relatively inexpensive high-pressure mercury lamp illuminates the sensitive zone of the orifice in axial direction.

Figure 59 shows a schematic graph of the FLUVO-Metricell instrument. The fluid system is driven by suction applied to (13), leaving the suspension prepared in the pipette tip (4) unpressurized in the open air. The central part of the transducer is the orifice (9), where the electrical and optical analysis of each cell is performed simultaneously. The electrode block which is connected with the current source CS and the preamplified V1 emits the current through the orifice to the grounded electrode (12).

The 100-W mercury lamp 18 serves as a light source for the optical system. The exciting filter (17) transmits the wavelength range necessary to excite the fluorescence of the dye, which is bound in the cells, and the dichroic mirror (14) reflects the excitation light toward the orifice but transmits the fluorescent light emitted by the cells. The objective (11) has a double function as in each epiilluminated microscope: it focuses the exciting light to the orifice and collects at the same time the emitted fluorescent light. The fluorescence pick up is very efficient since objectives with numerical apertures up to 1.3 are used.

Each cell, hydrodynamically focused onto a distinct path in the orifice by the injection tip (B), automatically must move through the plane of focus normally adjusted to the outlet of the orifice, and each cell produces its maximum fluorescence emission at this point. Signal fluctuations rising from small fluctuations of the particle path in radially illuminating systems do not exist with this design.

The emitted fluorescent light passes through the dichroic mirror (14) and is received by the photomultiplier tube (PMT) (19a) if (15) is a normal mirror. If (15) is a dichroic mirror, the fluorescent light of double-stained cells is split into two portions of different wavelengths which can be detected separately by the PMTs (19a) and (19b). In this way, three different parameters of each cell can be determined simultaneously with a rate of up to 3000 cells/sec.

The PMTs transform the fluorescent light pulses to electrical pulses which are amplified by the amplifiers VF1 and VF2 and processed together with the volume pulses in the control unit (CU) and evaluation unit (EV unit).

With the mirror (15) rotated by 90°, the optical system can be adjusted to the orifice by control of the operator over the ocular (20).

The fluid system works in a similar way as in the one-parameter Metricell system described in Section III-B. An additional cleaning flow (10) isolated by

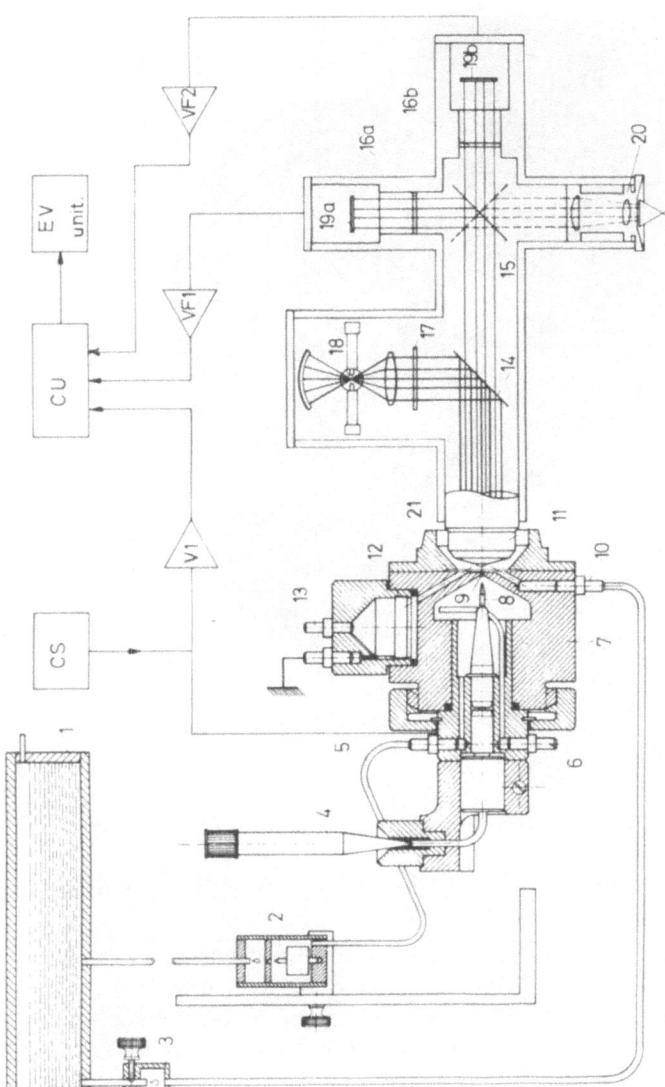

FIGURE 59. Schematic diagram of the FLUVO–Metricell three-parameter flow cytometer. (1) Tank for particle-free electrolyte; (2) regulating chamber movable in height which isolates the supply tank from the transducer and regulates the particle flow; (3) isolating adjustable restrictor valve; (4) pipette tip containing the particle suspension; (5) tube feeding the particle-free electrolyte; (6) bleeder valve; (7) transducer block; (8) particle injection tip; (9) orifice for electrical and optical cytometry; (10) cleaning channel; (11) objective for epillumination and fluorescence pickup; (12) ground electrodes; (13) suck connection; (14) dichroic mirror; (15) dichroic mirror discriminating the wavelengths for the two fluorescence parameters (must be rotated by 90° for adjusting orifice and optical path); (16a,b) fluorescence filters; (17) excitation filter; (18) 100-W mercury lamp; (19a,b) photomultiplier tubes; (20) adjustment ocular. (CS) Current source; (V1) volume amplifier; (VF1) fluorescence 1 amplifier; (VF2) fluorescence 2 amplifier; (CU) control unit for selecting priority modes; (EV) evaluation unit.

FIGURE 60. The FLUVO transducer module. It is coupled to the system by a bayonet-type fitting. For explanation, see Fig. 59.

the dropping chamber (3) moves the particles measured away from the sensitive zone.

Figure 60 shows the FLUVO transducer module, which is attached to the optical system by a bayonet-type fitting.

Figure 61 demonstrates the pulse coordination during a cell volume (upper trace) and fluorescence (lower trace) analysis. Resulting from the adjustment of the plane of focus of the fluorescence system to the outlet of the sizing orifice, the fluorescence pulses are delayed against the volume pulses by about 10 μsec.

B. Cell Volume–Cell Absorption Transducer

A multiparameter transducer particularly suited for cell size and absorption measurements in flow is shown in Figs. 62 and 63. This type of transducer may also be used for epiilluminated fluorescence measurements. Here, however, in contrast to the FLUVO chamber shown in Figs. 50 and 60, the optical axis is free for diaillumination. The particle injection tip (1) is mounted outside the optical axis and the objective (7) can work as illuminating condensor for the orifice (2). The objective (6) picks up the absorption light, which must be limited to the diameter of approximately one cell diameter by adjusting a diaphragm aperture in the intermediate image plane of the optical system.

Theoretically, with this chamber four parameters can be measured by simultaneously sizing cells, using the objective (6) as a two-parameter epiilluminating fluorescence device and the objective (7) as absorption detector for the exciting light of (6).

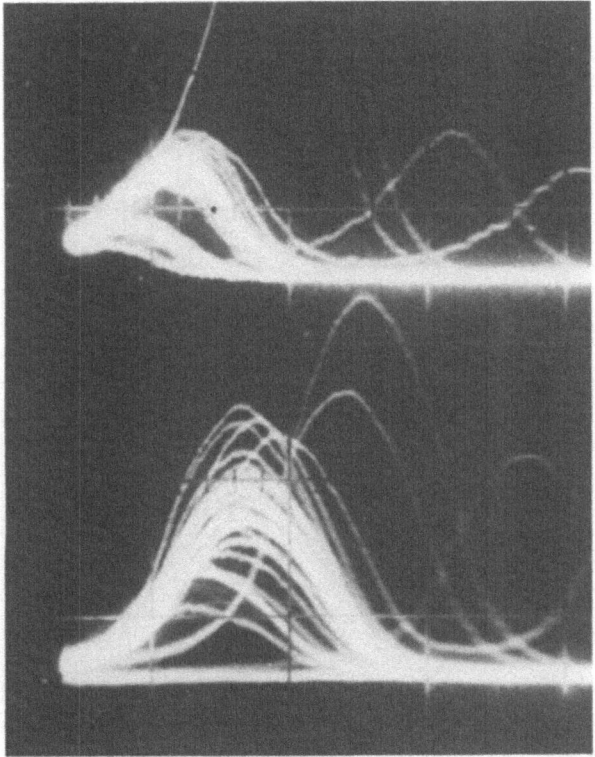

FIGURE 61. Oscilloscope pattern during two-parameter analysis. Upper trace: volume pulses; lower trace: fluorescence pulses. Time axis: 20 μsec/div. The optical plane of focus is adjusted to the outlet of the orifice. Therefore, the fluorescence pulse maxima lie in the falling slopes of the volume pulses.

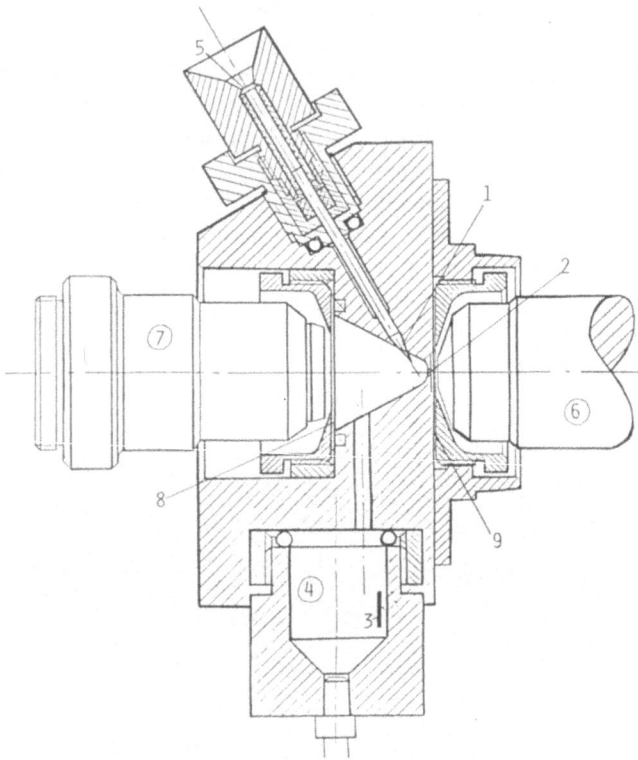

FIGURE 62. Cell volume–cell absorption transducer, which can also be used for fluorescence measurements. In contrast to the standard FLUVO transducer, the optical axis is free. (1) Particle injection tip; (2) orifice; (3) electrode; (4) electrode space; (5) fitting for particle injection; (6) objective which can be used for epifluorescence or absorption pickup; (7) objective for absorption illumination or absorption pickup.

C. Electronic Evaluation

1. State-of-the-Art System

The usefulness of a sizing system and of a multiparameter flow cytometer in generaly highly depends on the data evaluation capabilities available. Basic requirements are:

1. Capabilities for storing and displaying histograms.
2. Means for comparing histograms, for reading channel contents and integrals over histogram regions, and for calculating statistical numbers of cell populations.

FIGURE 63. Photograph of the transducer explained in Fig. 62.

3. Means for printing histograms and calculated results and for storing pulse heights and histograms on magnetic tape or on disk.

The usual way to evaluate data is to connect the PHA, which has collected the histograms, with a computer that performs the data manipulations and calculations.

The block diagram of Fig. 64 explains a typical multiparameter flow cytometric system where x, y, and z mark the modules of the x parameter (e.g., the cell volume), the y parameter [e.g., the fluorescence (1)], and the z parameter [e.g., the fluorescence (2)]. The user supplies the cells into the transducer, adjusts the amplification and trigger levels, and selects the parameters for the PHA analysis via the control unit, which also defines the priority

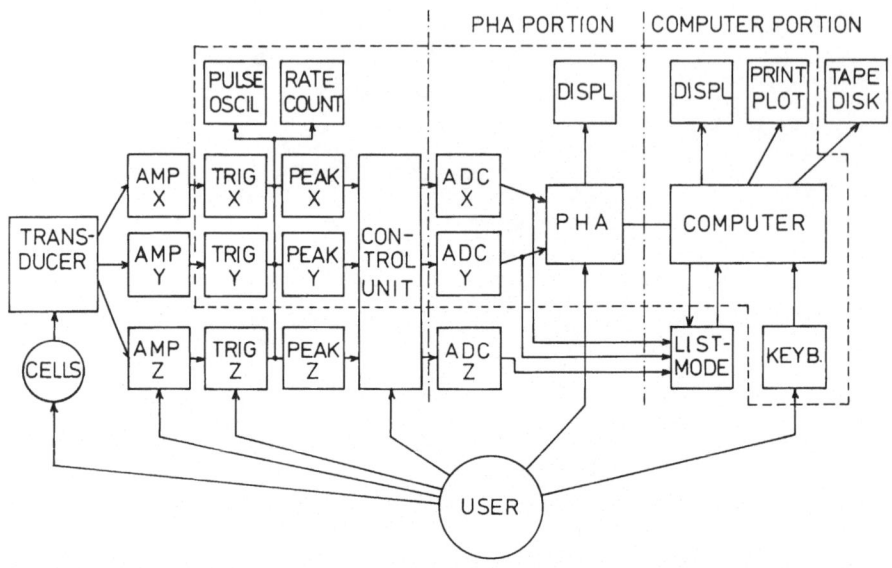

FIGURE 64. Block circuitry of a multiparameter evaluation unit. (AMP) Amplifier; (TRIG) Lower- and upper-level trigger adjust; (PEAK) detector for maximum pulse height; (ADC) analog to digital converter; (PHA) pulse height analyzer; (DISPL) display unit; (LIST MODE) unit which stores three values of each cell in the form of a list on magnetic tape. The modules within the broken line are included in the CYTOMIC microprocessor analyzer. X, Y, and Z identify the three parameters.

of one of the parameters (Kachel and Meier, 1980). The progress of the analysis is supervised on the pulse oscilloscope, the rate counter, and the PHA display.

For simultaneous three-parameter analysis, a computer-controlled list mode interface takes the three digital values of each cell and stores them on magnetic tape. From the list of values stored on tape, the computer can compose one-, two-, or three-parameter histograms according to the requirements of the actual experiment (Sharpless, 1979; Benker *et al.,* 1980).

2. Advanced Microprocessor Analyzer

Modern microprocessor technology allows small instruments to be built that include data acquisition, pulse control, pulse height analysis, and data processing in a single unit which is operated easily by calling dedicated functions. Such an instrument is described by Kachel *et al.* (1980b). The modules inside the broken line of Fig. 64 are included in this so-called "CYTOMIC" analyzer. The screen of this analyzer serves as a comprehensive information center. During size or combined analysis the original pulses, pulse rate meters, and the growing histogram are displayed.

In the display mode the one- and two-parameter histograms are shown on the screen. Marker points movable in the histograms by the operator limit regions of interest. Statistical numbers can be calculated and displayed.

With such preprogrammed evaluation units, the user is free from programming the computer to evaluate his measurements. Figure 65 shows the seven groups of functions available in the CYTOMIC analyzer. The data acquisition functions control the uptake of one- and two-parameter histograms. The display functions are concerned with the display of the histogram, with increasing or decreasing the channel groups and scale factors, with rotation of the two-parameter fields, and with comparison channel groups by superposition.

The process-managing functions move the marker points in the histo-

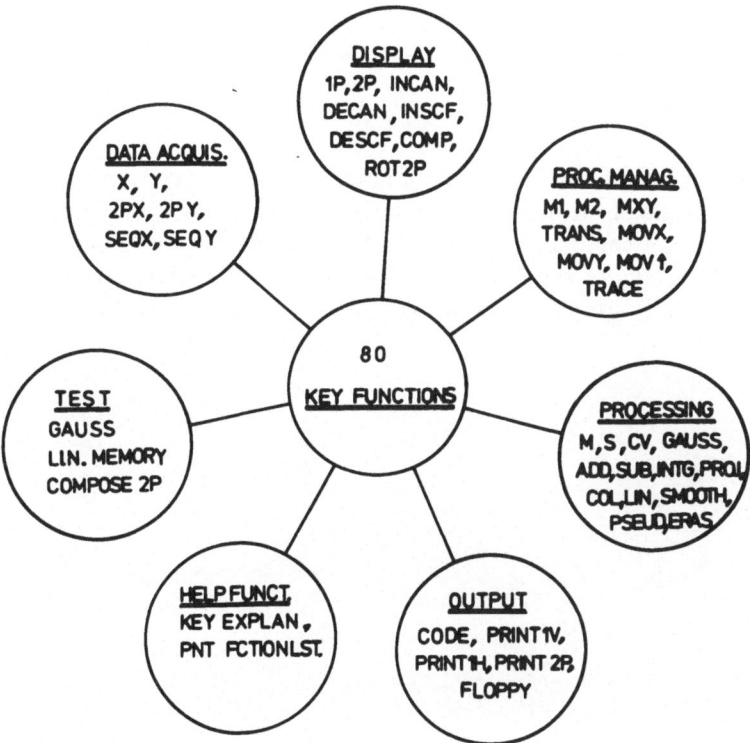

FIGURE 65. The seven groups of functions available in the CYTOMIC microprocessor analyzer. The groups are data acquisition, display, processing management, processing, output, help function, and test.

grams and shift the histograms in different directions. With the proper processing functions, statistical calculations are performed, channel group contents are added, subtracted, or smoothed, and integrals or projections are calculated, etc.

The output functions print the histograms and calculation results on an attached printer or transfer them to another data system. The help functions help the user who is not yet familiar with the keyboard to find the key for the function he wants to execute. The help function puts a graph of the keyboard on the screen, and the alphabetical list of the functions can be run through in the text line whereby the key belonging to each function is marked on the screen. Finally, a set of test functions allows testing of the hard- and software structures of the analyzer.

A photograph of the CYTOMIC analyzer with printer is shown in Fig. 66. Figure 67a–f shows examples of the most important one-parameter functions, and Fig. 68a–h gives examples of two-parameter functions available in the instrument.

D. Examples of Combined Electrical–Optical Cell Analysis

1. Cell Volumetry Gated by Fluorescence

Gating means that the control unit of the evaluation system is adjusted so that the pickup of the first cell parameter is allowed only if the second or third parameter or a combination of both is found within distinct limits. Figure 69 shows the results of such a gating experiment. DNA-stained (mithramycin) rat glioma cells are measured in the FLUVO Metricell cytometer. The right

FIGURE 66. The microprocessor-controlled two-parameter analyzer CYTOMIC. The standard version has a set of 80 one- and two-parameter functions built in which are specially developed for flow cytometric data acquisition and processing and which are called by the 18-element keyboard. The printer performs the graphic and alphanumeric output.

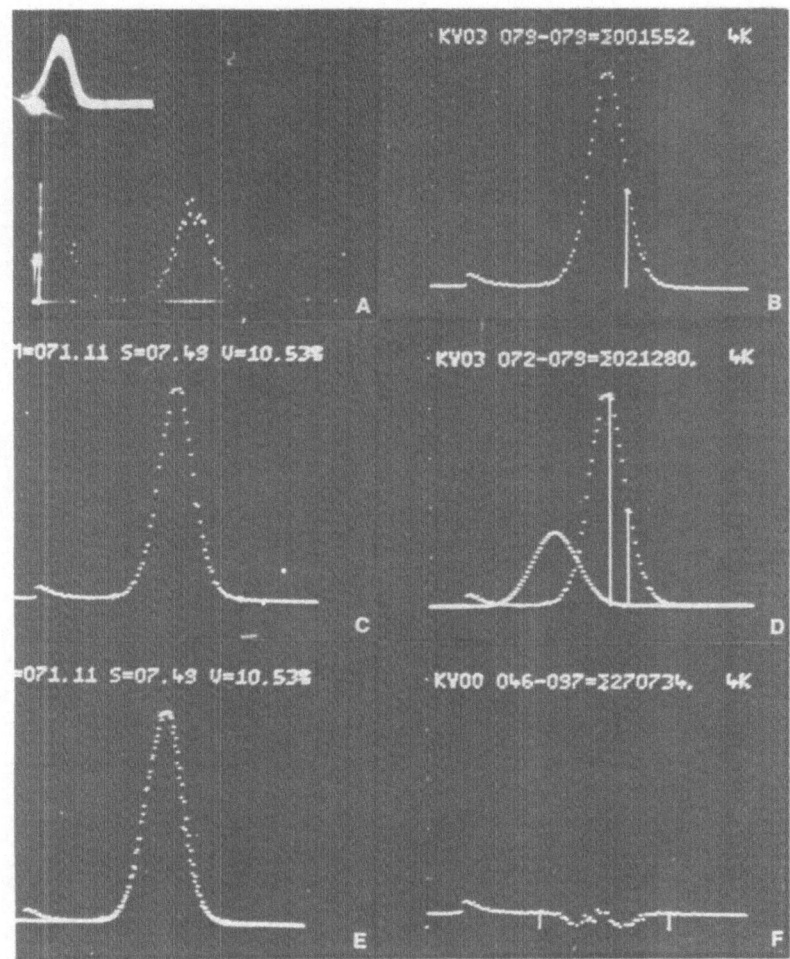

FIGURE 67. Examples of one-parameter application of the CYTOMIC analyzer. (A) Display during one-parameter sizing of chicken erythrocytes. Top left: volume pulses. Bottom right: the growing histogram. Bottom left: the analog rate meter. A small horizontal line marks the actual pulse rate on a scale of 500 particles/sec. (B) One-parameter display mode. KY 03 means that the content of the channel group 3 originates from the *y* parameter. The position of two marker points, which in this case are at the same position, are identified by 79–79. The number 1552 indicates the integral between the markers, in this case of the channel 79. 4K indicates the frequency scale factor of the histogram. (C) By calling the statistic function, mean value M, standard deviation S, and coefficient of variation V are calculated over the histogram between the marker points and displayed in the text line. (D) Two channel groups are compared by superimposition. (E) A gaussian distribution is calculated from M and S and superimposed onto the original curve. (F) The gaussian curve produced in (E) is subtracted from the original curve. The resulting histogram is shown.

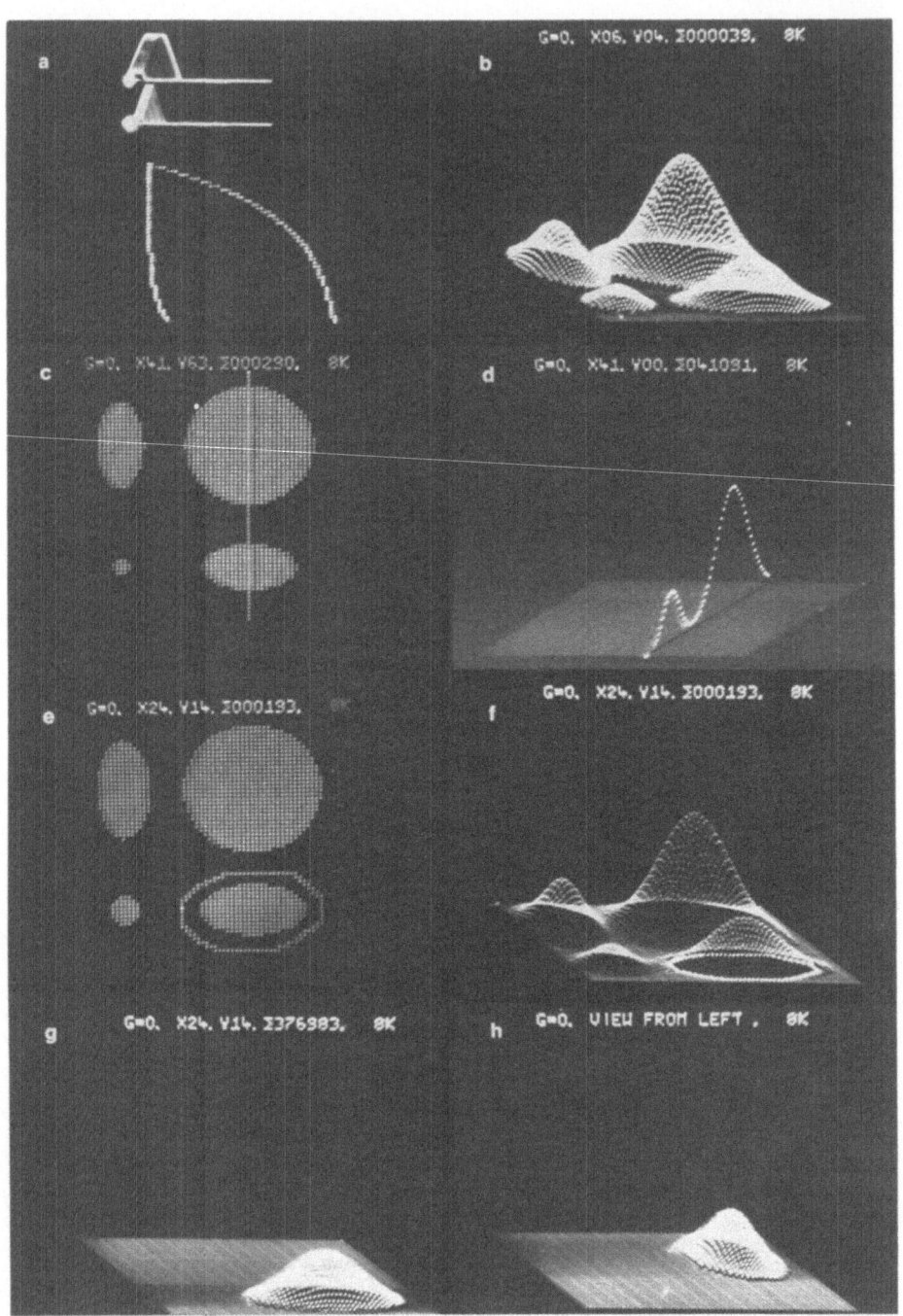

column shows the volume distributions of the cells and the left the DNA distributions. The curves of the first line are not influenced by one other. The arrow in the following lines indicates that the fluorescence measurement has gated the volume measurement. In the second line, only particles with a DNA signal had been allowed for volume analysis. Consequently, the nonfluorescent debris which distorts the ungated volume distribution of the first line is excluded from the volume analysis. The following lines show the volume distributions of the glioma cells which are in the G0/G1 phase selectively gated by the G0/G1 DNA fluorescence and the volume distribution curves and the S-phase and G2-phase volume distributions. With this technique, two-parameter information can be obtained with one-parameter evaluation equipment.

2. Two-Parameter Volume–Fluorescence Analysis

Figure 70 shows an experiment performed by Valet *et al.* (1979) to measure electrophoretic mobility of erythrocytes with the FLUVO Metricell analyzer. A mixture of 41% rat, 35% rabbit, and 24% human erythrocytes was stained with FITC-poly-L-ornithine. The fluorescinated polycations are quantitatively bound to the negative surface charge of the erythrocytes. Thus the fluorescence axis of Fig. 70a represents the total surface charge of the measured cells. From the simultaneously measured cell volume shown in the x axis of Fig. 70a, it is possible to calculate the cellular surface for each point in the histogram. Valet *et al.* (1979) have shown that the surface charge of the three erythrocyte populations measured by flow cytometry is well correlated with their electrophoretic mobility determined in a Zeiss microelectrophoresis system.

The curves of Fig. 70b represent the surface charge density distributions of the three erythrocyte populations.

FIGURE 68. Two-parameter examples of the CYTOMIC analyzer. (a) The screen of the analyzer during two-parameter data acquisition. Top: the display of the x and y pulses, which here originate from a pulse spectrum generator. The growing two-parameter display is shown as a top view scatter plot. (b) A two-parameter distribution with a section in height marked by an adjustable intensification. The two-parameter channel group (three groups are available) is identified by G = 0. X06 and Y04 are the coordinates of a marker point, and 39 is the content of the marked channel. (c) Top view onto the two-parameter field with a section in height. The column X41 is marked for integration by conducting the marker point along this line. (d) After execution of the integration function, column X41 remains alone on the screen. The integral in this case over the column X41 is 41091. (e) Top view onto the two-parameter field with a section in height and an integration trace around a peak. (f) The same as (e) without the intensified section in isometric view. (g) Isometric view onto the peak that was selected by the integration trace. The integral over the peak is 376983. (h) Figure (g) is rotated by 90° and is now seen from the left.

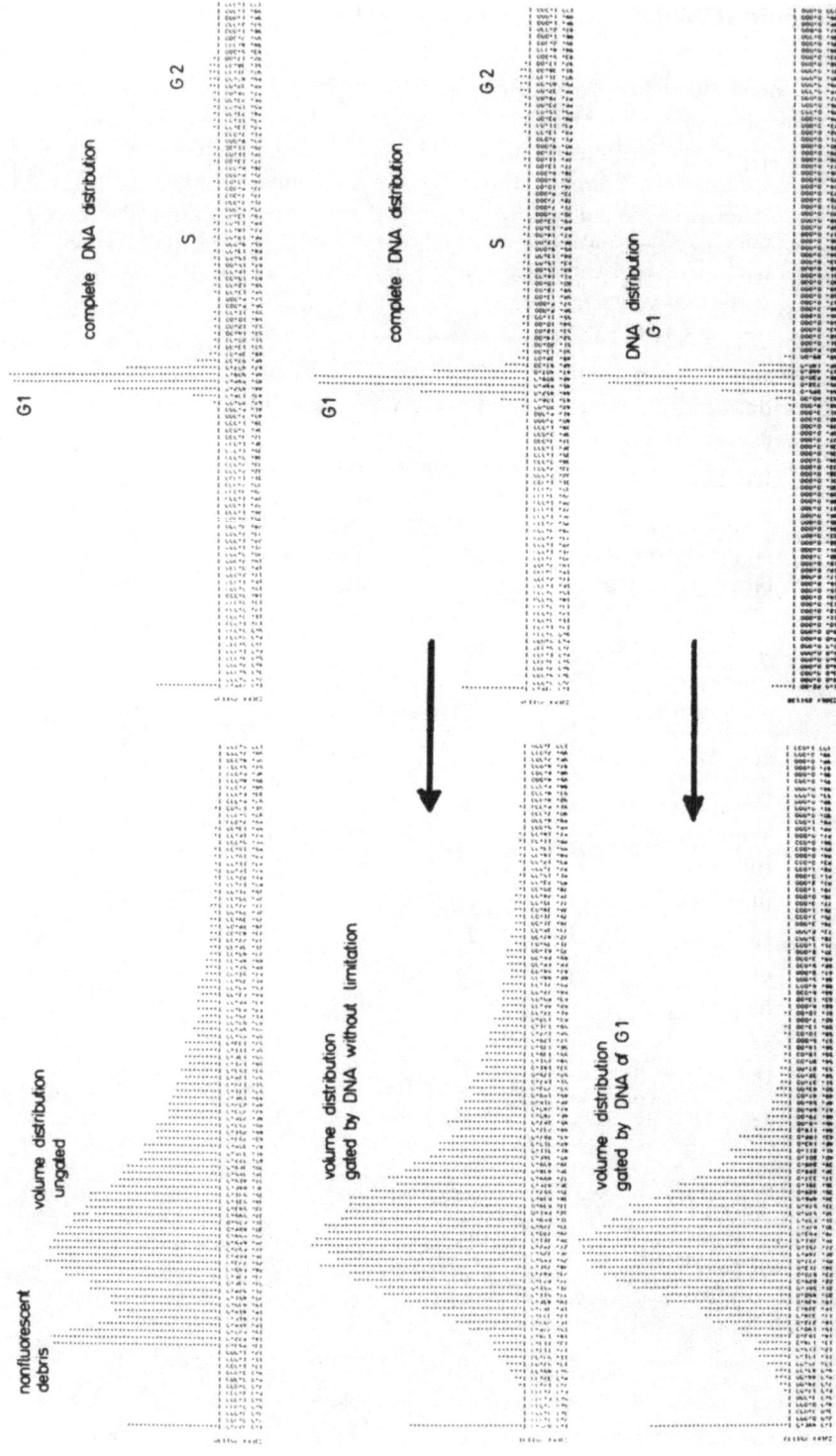

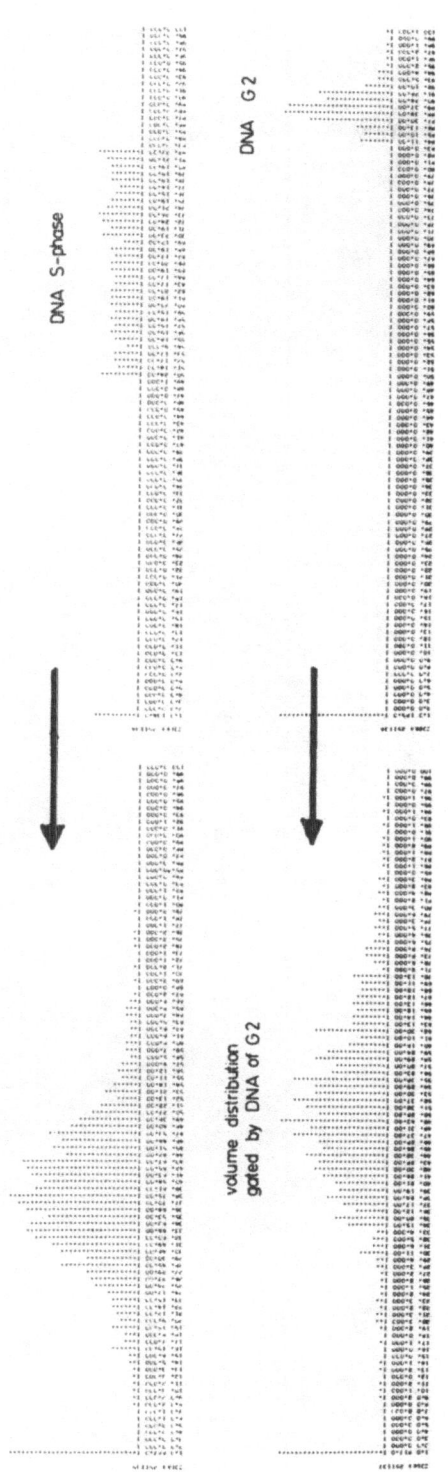

FIGURE 69. Cell volume measurements gated by the simultaneously measured DNA fluorescence. The rat glioma cells stained with mithramycin have been measured in the FLUVO–Metricell. Orifice diameter, 140 μm; current, 0.25 mA. Excitation of the DNA fluorescence at 330–490 nm, barrier filter 520 nm. The curves in the first line are measured without gating. In the following lines the fluorescence parameter has gated the cell volume measurement, i.e., only cells with a DNA fluorescence in the predetermined fluorescence window are allowed for the volume analysis. In this way, the nonfluorescent debris is completely removed from the volume distribution, and we can selectively determine the volume distributions of the cells that are in the different proliferation phases.

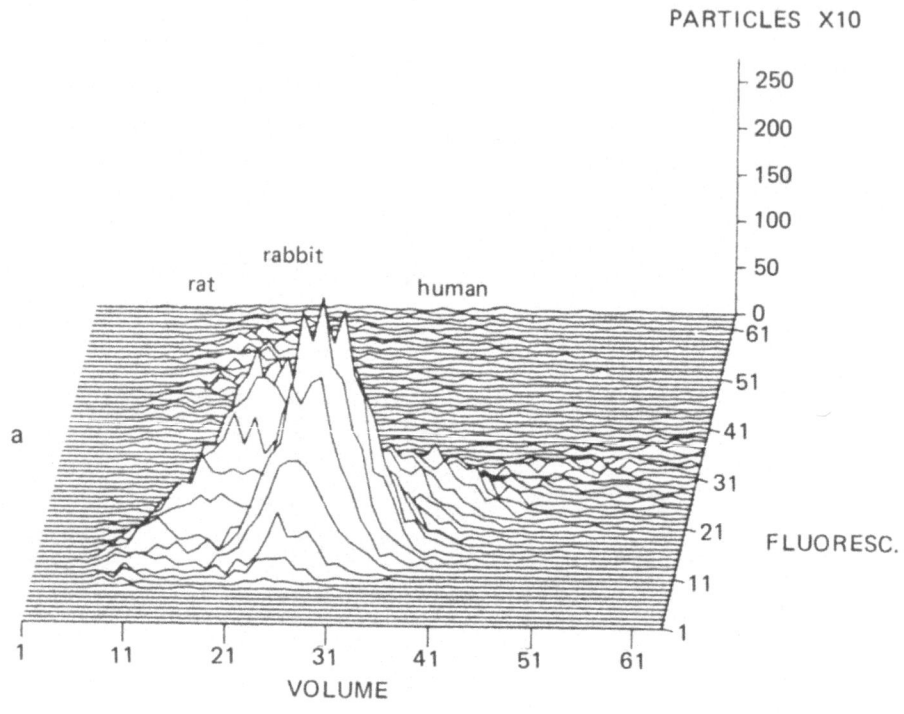

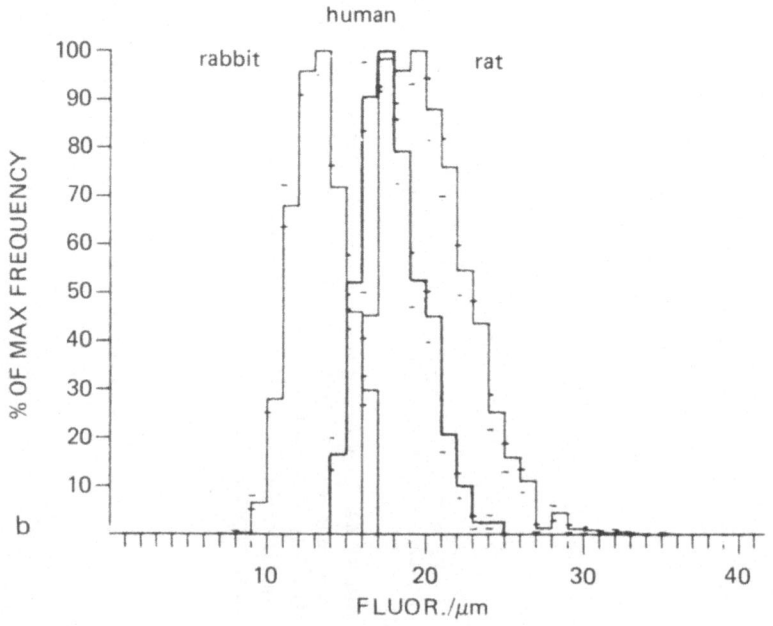

3. Two-Parameter Cell Volume–Cell Absorption Measurements

The two-parameter displays of Fig. 71 explain an experiment of complement lysis performed by Valet and Bauer (in preparation) with the cell volume–cell absorption Metricell instrument (see Figs. 62, 63). Antibody-coated sheep erythrocytes (sensitized) have been incubated at 30°C with guinea pig serum as complement source. The dilution was such that the lysis was completed within 5 min. The absorption light source was a 100-W halogen lamp without filter. Figure 71a shows the starting situation of the experiment. The unlysed erythrocytes are found in a peak with the peak values 20 (volume) and 42 (absorption). Figure 71b shows the situation after 4 min. Cell lysis is in progress. The lysing cells lose their hemoglobin and thus their main absorption; the peak of the lysed cells is in the absorption channel (8). The volume pulse height of the ghosts, however, is slightly increased by a real volume increase and by an increase of the shape factor of the cells, which are spherically transformed during lysis. An important result is that the cells are directly lysed by the complement without significant swelling, as is shown by the arrows. Figure 71c shows the final phase of the experiment: all cells are lysed and the peak channel of the ghost population is 32.5 (volume) and 9.5 (absorption).

4. Three-Parameter Cell Volume–Fluorescence Analysis

Figure 72 represents a three-parameter FLUVO–Metricell analysis of unfixed rat bone marrow cells that were incubated with fluorescein diacetate (FDA) and 4-methyl-umbelliferyl acetate (UMA) (Malin-Berdel and Valet, 1980). With a Fortran program developed by Valet (1980), cell clusters can be displayed as clouds in a three-dimensional space. Inside the clouds in this case, particle frequencies above 5% of the maximum channel content are found. Increasing the percentage shrinks the clouds; decreasing expands them. This type of display facilitates the recognition of interconnections between the parameters.

VII. SORTING ACTIVATED BY ELECTRICAL SIZING

The first electronic cell sorter built by Fulwyler (1965) was activated by the electrically measured cell volume. With the rise of the fluorescence-acti-

---

←

FIGURE 70. (a) Two-parameter volume–fluorescence measurement of a mixture of 41% rat, 35% rabbit, and 24% human erythrocytes (Valet *et al.,* 1979). The cells have been stained with FITC-poly-L-ornithine, which binds proportionally to the electrophoretic mobility of the cells. Orifice diameter, 85 μm; current, 0.5 mA. (b) Surface charge densities of the erythroycte populations whereby the cell surface is calculated from the cell volume (Valet *et al.,* 1979).

FIGURE 71. Experiment of complement lysis with the cell volume–cell absorption Metricell instrument. Sensitized antibody-coated sheep erythrocytes have been incubated in 30° with guinea pig serum as complement source (Valet and Bauer, in preparation). Orifice, 60-μm diameter; current, 0.7 mA. (A) Situation at the beginning of the experiment. The erythrocytes con-

```
 -I--------I---------I---------I---------I---------I--------I-----
  I                                                             I
  I                                                             I
  I                                                             I
  I                                                             I
 60                                                             *
  I                                                             I
  I       EA        (1x10 /ML) + C      3.5-4.5 MIN.            I
  I         SHEEP              GP                                I
  I                                                             I
  I                                                             I
  I                              1                              I
  I                       1 111111  1  1                        I
  I                       11111122111  111                      I
  I                   1 1222234222221211  1                     I
 50       22.5  \   111223333244333272111                      *
  I               \1112245453754433222111                      I
  I          20   12 3436678677546443212 11   1  UNLYZED ERYTHRO-I
  I             \ 1124 56899988877544332111                CYTES I
  I               122457 *****9985554331111                      I
  I          111213 69**  ******8775332 1111                     I
  I          11123679*** ********96544322121                     I
  I          11123559** *********65543321111                     I
  I          222468*** ********97654332111                       I
  I          2123388** ****8865332111111                        I
 40       122368********* ****876544212 1  1                    *
  I       222468******** ****99755532121011  1                  I
  I       22336********** ******9866443311111  11               I
  A I     1223467 9******** **9987555534222211                  I
  B I     1343467 8********* *9886654322311111                   I
  S I     23345589******** 7887555532321111                     I
  0 I   13343356 8 8899*5876655544322222111                     I
  R I   13323444 65686977676c4443222221112                      I
  P I   13332333556566575654443321301111    1                   I
  T I   1332 1122 3346555545443333222111  1  1                  I
130 I   143211112433345444343422222 12111  11                   *
  0 I   24211 11112332444444 3333232211112   1                  I
  N I   25311   1 11123233333334333211221111  1                 I
  I     1421 1  1  1 1222322372223322211111111                  I
  I     2621   1 112221122333322221211 111                      I
  I     1431       11112 2 12333322211221 1 1                    I
  I     2421    1 111122 122122322111111 1  1     ERYTHROCYTES   I
  I     2321    1 1111 2121112111 11  111              DURING     I
  I     3421    1 112 1122112111  11 1                Lysis       I
  I     2521      1 111222 2111  1 11                             I
 20    3621      11 111111111111  1                             *
  I     3531      111  111111 1                                  I
  I     4831      11  111111 111                   32            I
  I     6841       111111111111                                 I
  I     8 42      1111222222221111111   1           GHOSTS       I
  I     9 7311    11 1223444343332333122111 111                 I
  I     ***521111 11123456777 8*987987 66443 22211111           I
  I     ***9432212323577***********9976533322211111             I
  I     ****7543333578*********************8 7755434212 1 111    I
  I     ****8865568************************90 56332222 11111     I
 10    ****** 9889*************************86653332221111*      *
  I     ********99**************************98444232222 11       I
  I     ********98*************************98565432221211         I
  I     ******* 98************************86664432121111         I
  I     ******777779*******************9885463321211111 1        I
  I     ******875444789****************875433221111             I
  I     ****** 642221223557***********96554332211  1            I
  I     ***842111 11112354666 79867665743322 211                I
  I B   49521       11 11121111221  1                            *
 -I--------I---------I---------I---------I---------I---------I-----
  1        10        20   \    30        40        50        60
                            VOLUME
```

taining the normal content of hemoglobin show high absorption. (B) Situation during the experiment, after about 4 min. The cells lyse without swelling. The direction of the lysis is shown by the arrows. The lysed erythrocytes (ghosts) have lost their hemoglobin and show very low absorption. (C) At the end of the experiment, all cells are lysed and found in the low-absorption peak.

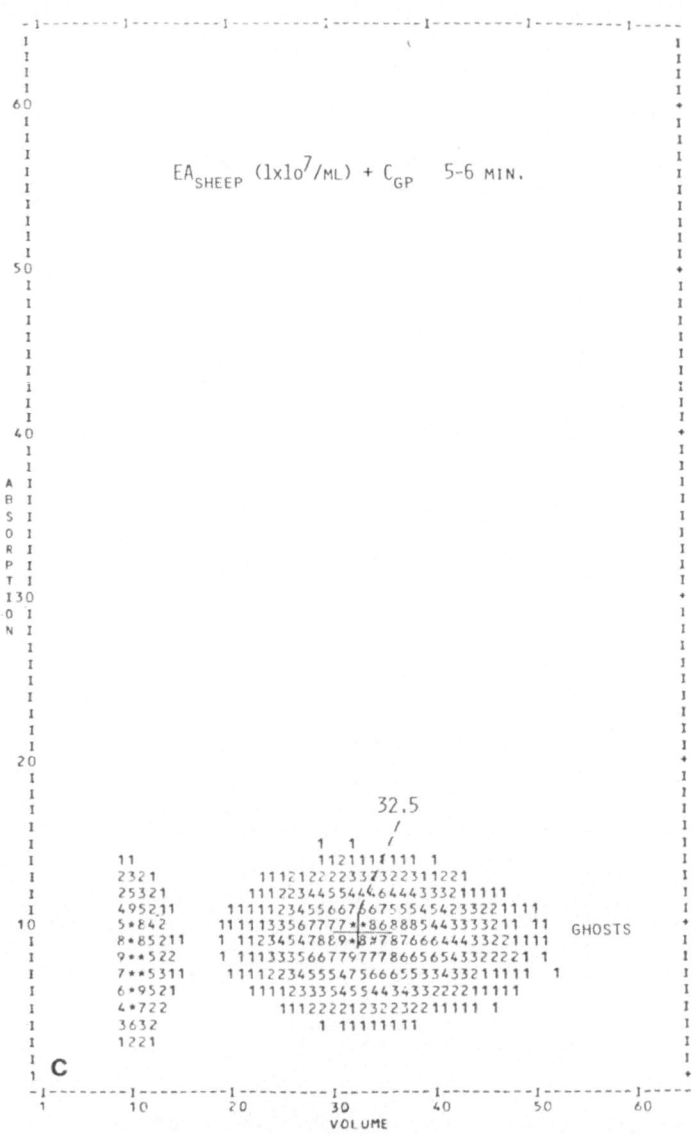

FIGURE 71. (Cont.)

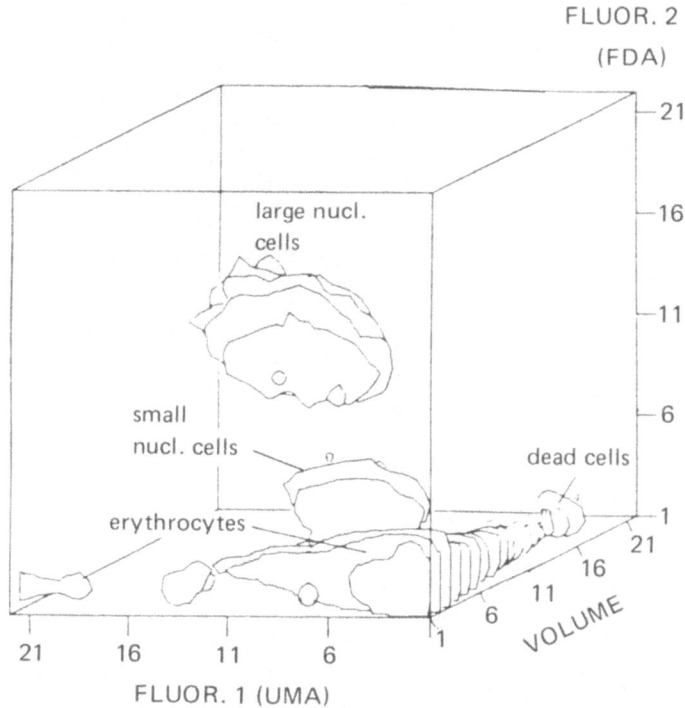

FIGURE 72. Graphic display of a three-parameter measurement of fluorescein diacetate (FDA), 4-methyl-umbelliferyl acetate (UMA) esterase activity, and cell volume of unfixed rat bone marrow cells after 10 min incubation at 25° (Malin-Berdel and Valet, 1980).

vated cell sorters (Bonner *et al.*, 1972), size-activated sorting stood back, although this method has some advantages over optically sizing methods. A multiparameter sorter including electrical size activation was described by Steinkamp *et al.* (1973). A cell-volume-activated sorting transducer developed by Menke *et al.* (1977) is shown in Fig. 73. The cell suspension ejected by the particle tube (2) is hydrodynamically focused to the volume sensing orifice (4). A second sheath flow (5) focuses the cells to the second orifice (6), from which the cells included into small droplets are ejected into the open air. A charging electrode activated by volume signals found in a preselected window charges the droplets that contain cells to be separated in the high-voltage field that is passed by the stream of droplets. Uniformity of the droplets is achieved by ultrasound excitation.

Figure 74a shows a volume histogram of human blood cells measured in

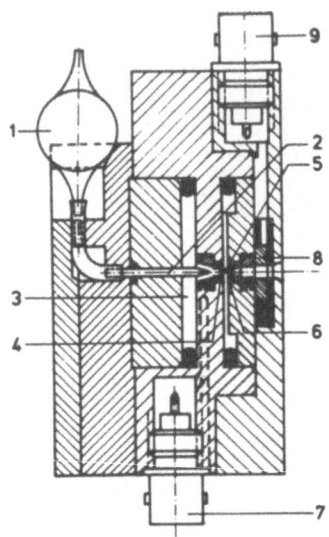

FIGURE 73. Schematic diagram of a volume-activated sorting transducer (Menke *et al.*, 1977). (1) Particle container; (2) particle tube; (3) space for the first sheath flow; (4) sizing orifice; (5) space for the second sheath flow; (6) orifice ejecting the stream that contains the cells into air; (7) connector for the piezo electric transducer (8), which excites the uniform droplet formation; (9) connector to the volume-sensing electrodes.

the Metricell volume analyzer. The result of sorting the granulocytes is shown in Fig. 74b, and the result of sorting the erythrocytes is shown in Fig. 74c.

Figure 75 shows the concept of a multiparameter sorting transducer that uses electrically measured cell size and epiillumination fluorescence for sort activation. Inexpensive mercury or xenon lamps can be used as light sources. The construction of such a sorter is in progress.

VIII. SIZE-TRIGGERED IMAGING IN FLOW

A. Instrumentation for Size-Triggered Imaging

Imaging in flow can be used for methodologically studying the behavior of cells in the cytometric transducer. Imaging in flow, however, may also be used for cytoanalytical studies. Flow cytometric instruments normally do not disclose morphological cell features. This disadvantage is overcome by combining flow analysis and imaging. Cells from a particular region of interest are imaged immediately during the flow analysis. In this way, rare cells can be

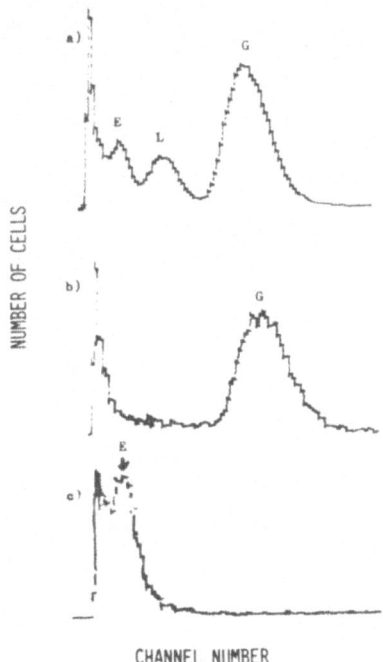

FIGURE 74. (a) Volume distribution curve of human blood cells measured with the Metricell before sorting. (E) Erythrocytes; (L) leukocytes; (G) granulocytes. (b) Volume distribution curve of the granulocytes after sorting out from the sample of (a). (c) Volume distribution curve of the erythrocytes after sorting out from the sample of (a).

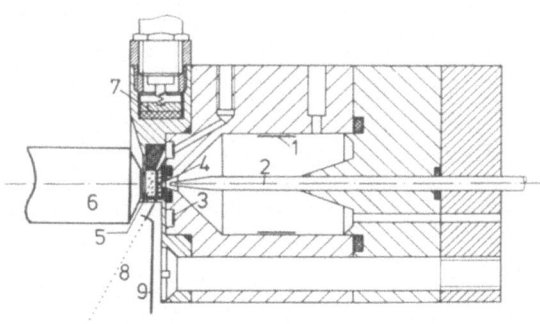

FIGURE 75. Multiparameter sorter transducer with epiillumination. A mercury lamp can be used for illumination. (1) Active electrode; (2) particle tube; (3) grounded electrode; (4) volume-sensing orifice; (5) optical window for epiillumination; (6) epiillumination objective; (7) piezo transducer; (8) droplet stream; (9) charging electrode.

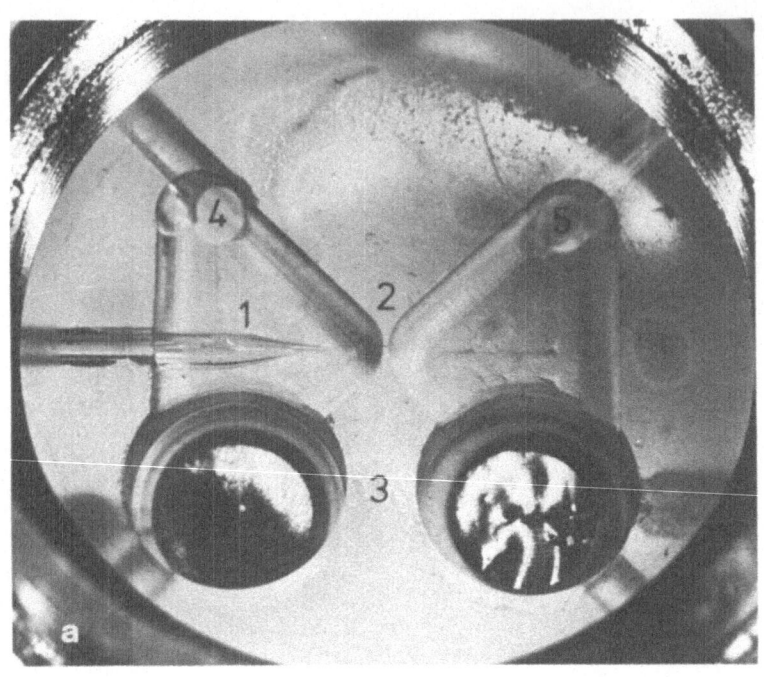

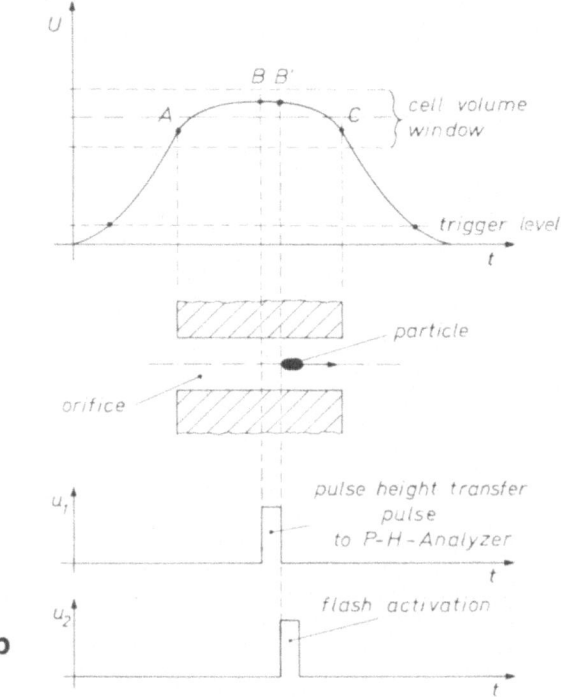

searched out from large populations since thousands of cells can pass the flow system each minute.

Electrical volume triggering is particularly suited in such an instrument since the nonoptical analyzing device is not in conflict with the optical imaging system. A volume-activated imaging transducer is shown in Fig. 76 (Kachel *et al.*, 1979, 1980a). Two excavations in a block of plexiglass that contain the circular electrodes are connected by the optically accessible sizing channel. If a cell of interest is detected by the volumetry circuit, a Nanolite flashlamp is triggered which illuminates the passing cell in the orifice channel for about 40 nsec, a time which is sufficient to generate an image of the cell on the retina of the human eye or on a photographic film. Such an instrument, therefore, can be used in two ways: first as a simple flow microscope, in which the observer sees only the cells selected by the cytometric system, and, second, as an image-storing device, in which a particle-controlled camera (Benker, 1978) stores the selected cell images (Fig. 77).

Figure 76b explains the trigger conditions for a cell passing through the sizing channel. The imaging decision first can be taken when the volume pulse has reached its maximum in point B. In the few microseconds which remain until the cell leaves the channel, the flash must be triggered if the volume pulse height is in the selected window.

The chamber of Fig. 76a has two disadvantages. First, the sizing orifice channel is limited at its top side by the coverglass covering the excavations of the plexi block. The thin layer of electrolyte between chamber and coverglass takes over more or less of the sizing current which then is lost for the sizing process. Since the thickness of the layer may change at random, the pulse heights measured are not stable in time. Second, the structured bottom of the sizing/imaging channel is found in the cell images as distorting background (Figs. 81, 82, 83a,b).

An imaging transducer that is improved in these points is shown in Fig. 78. Here the cells are sized in a definite cylindrical orifice, and the images are taken in the observation bore a short time after analysis when the cells have left the orifice. With this design, the cells are sized with stable pulse heights and imaged without distorting background (Fig. 83c).

FIGURE 76. (a) Transducer for analytical cell imaging. A coverglass covers the chamber at top. (1) Particle injection tip; (2) orifice channel; (3) electrodes; (4) sheath flow inlet; (5) suck connection. (b) Trigger conditions for the flash activation in the transducer of (a). (A) Point at which the particle enters the orifice; (B) pulse maximum containing the volume information; (B B′) pulse height transfer time; (C) particle leaving point. The time of 5–10 μsec between B′ and C is available for flash activation.

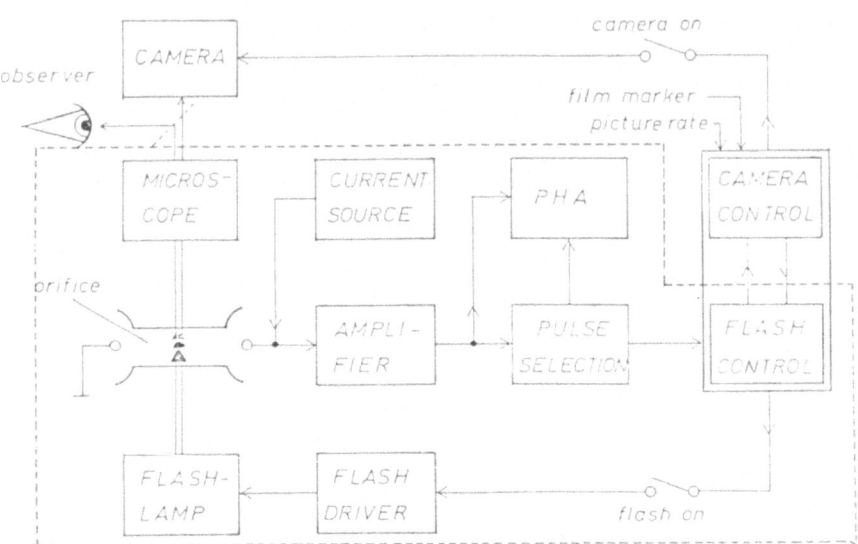

FIGURE 77. Schematic diagram of the imaging circuitry. (PHA) Pulse height analyzer. The elements inside the broken line represent a flow microscope in which the observer sees only the selected cells in the orifice channel.

B. Problems of Visualization Cell Structures

The purpose of analytical cell imaging in flow is to recognize morphological structures of selected cells. With normal diaillumination, the contrast of nuclear structures of native cells is not sufficient, and the flash intensities are not strong enough for using phase-contrast optics. Dyes for sufficient contrasting of nuclei of living cells in suspension are not known. Hueller (1980) has found that after formalin fixation a sufficient contrast of the nuclei of suspended cells is achieved with aluminum sulfate–carmine at pH 2.5.

A general problem is the definition of the morphological structure of a cell or a cell nucleus since both are three-dimensional bodies. The instrument, however, as any microscope, sees only the small section of the plane of focus. The position of the cell in the imaging point of the transducer determines whether the cell crosses the narrow plane of focus at all, and if it crosses, which section through the cell is seen.

A simple principle shown in Fig. 79 helps to overcome this "plane of focus" problem. If the hydrodynamically focused particle path is inclined by a small angle to the plane of focus, and if the cells are several times imaged at several positions in the field of view, different sections through the same cell are collected at the same photomicrograph. Such a device could be realized by

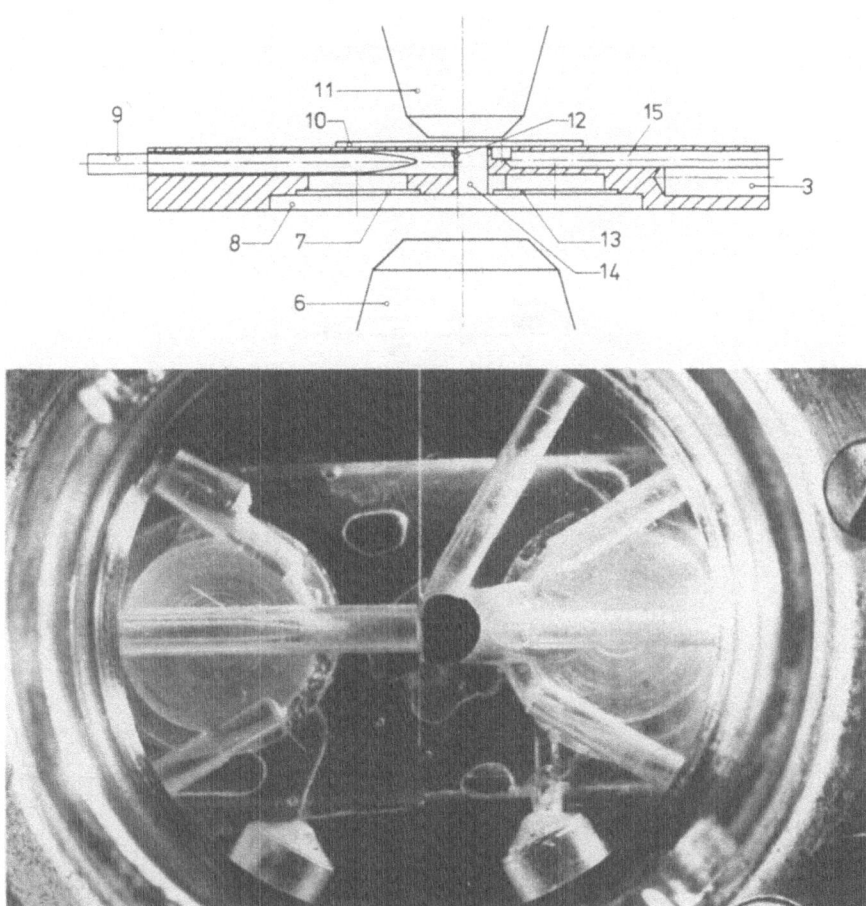

FIGURE 78. Improved imaging transducer. The cells are supplied through the particle tube (9) and focused through the orifice (12). Just outside the orifice, the cells are imaged in the observation bore (14). A second sheath flow passing along the electrodes and through the observation bore conducts the particle stream to the waste tube (15). (8, 10) Glass plates; (7) active electrode; (6) illumination; (11) objective taking the image.

a multiflash device or by constant illumination with an electronically controlled short-time shutter.

C. Examples of Cell Imaging in Flow

Figure 80 shows the complete imaging-in-flow system. The images of Figs. 81, 82, and 83a,b are taken with the transducer of Fig. 77a. The particles and

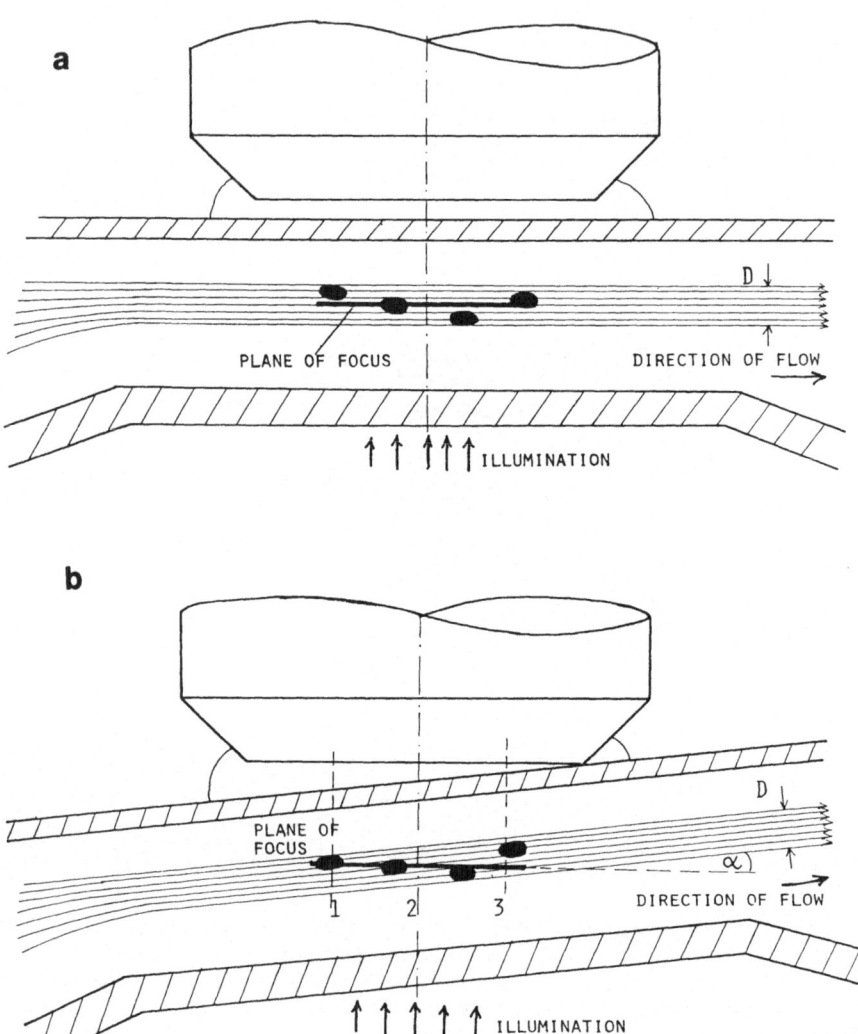

FIGURE 79. Simple principle to overcome the plane of focus problem. (a) The plane of focus and the direction of flow are parallel. The particles running at bottom or top of the tolerance range (D) are never focused. (b) The plane of focus is slightly inclined to the direction of flow. If cells are three times imaged at points 1–3, each cell is in focus at least at one of the imaging points, and with the angle small enough different sections of the same cell can be imaged on the same film.

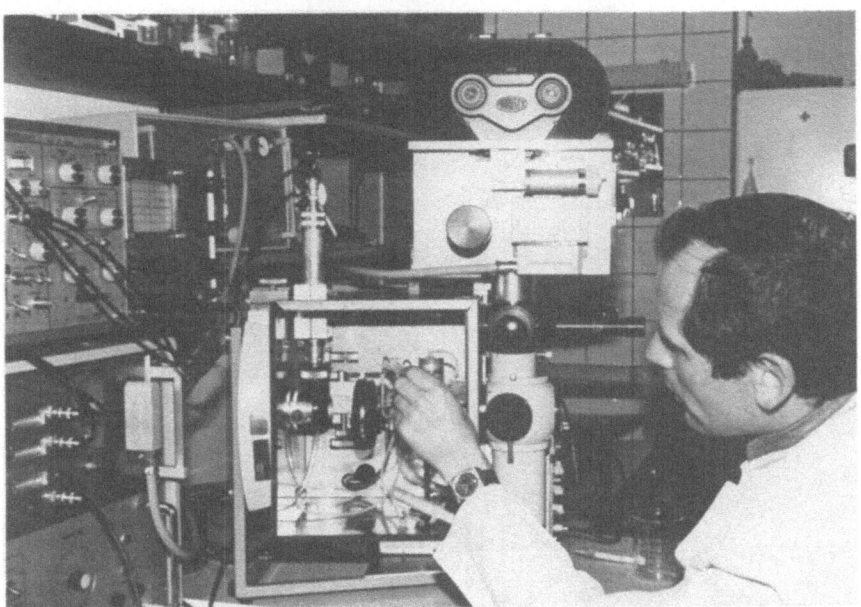

FIGURE 80. The complete imaging in flow instrumentation. At left, the pulse height analyzer with flash and camera control. In the middle, the shielded transducer system with flashlamp and the automatic camera.

cells are imaged directly in the orifice channel. Figure 81 shows a sequence of 20-μm-diameter latex particles and Fig. 82 an experiment with aggregates of human erythrocytes. In the volume distribution of Fig. 82a, the second peak was selected for imaging; the images on the film show doublets (Fig. 82b). If the events of the third peak are imaged, triplets are found on the film (Fig. 82c). The cells of Fig. 82a and b are white blood cells fixed in 1% glutaraldehyde and stained with hemalaun according to Mayer. This staining technique is not sufficient for clearly disclosing the structure of the nuclei. The rat bone marrow cells of Fig. 83b that are stained with aluminum sulfate–carmine and imaged in the transducer of Fig. 78 show better the nuclear structure of the cells. In addition, the background is completely neutral.

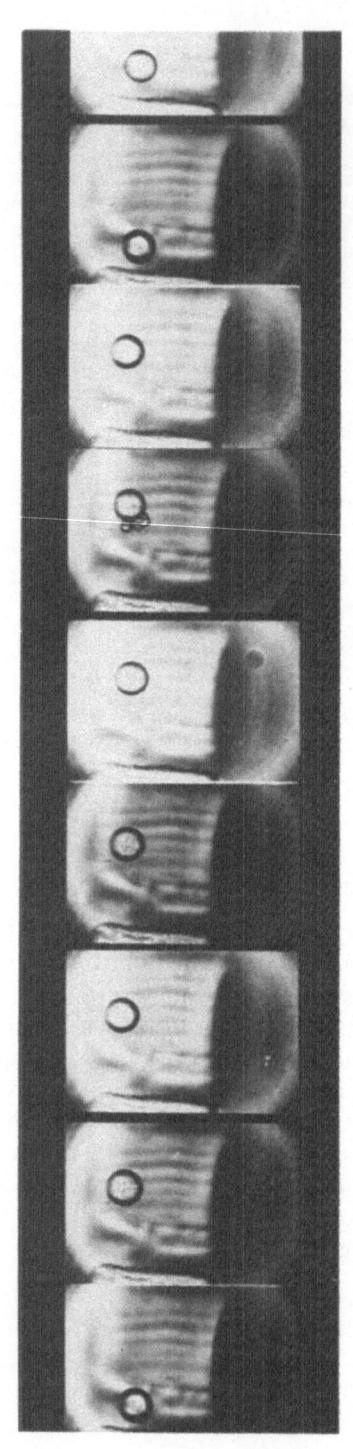

FIGURE 81. A sequence of 20-μm latex particles imaged in a channel of 100-μm width.

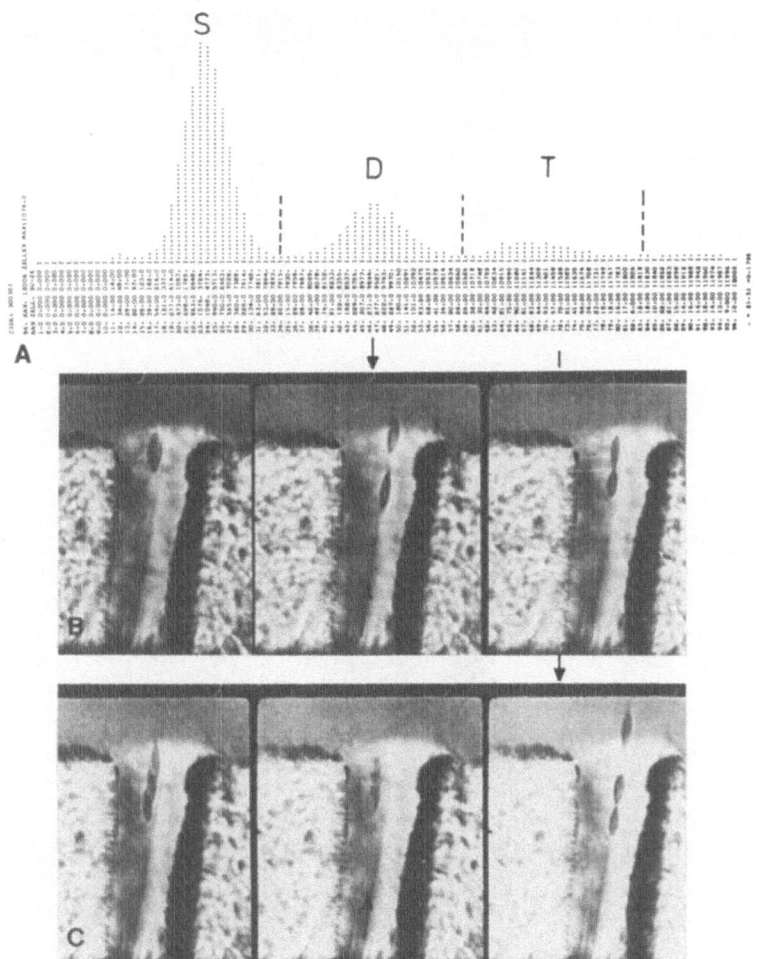

FIGURE 82. (A) Volume distribution of red blood cells from a patient with coronary heart disease. A great number of doublets (D) and triplets (T) are detected. (S) Singlets. (B) The cells of the doublet range are selectively triggered. The cell images show doublets as expected. The dimensions of the orifice are 40 × 60 × 120 μm (width × depth × length). (C) The cells of the triplet range are selectively triggered. The cell images show triplets.

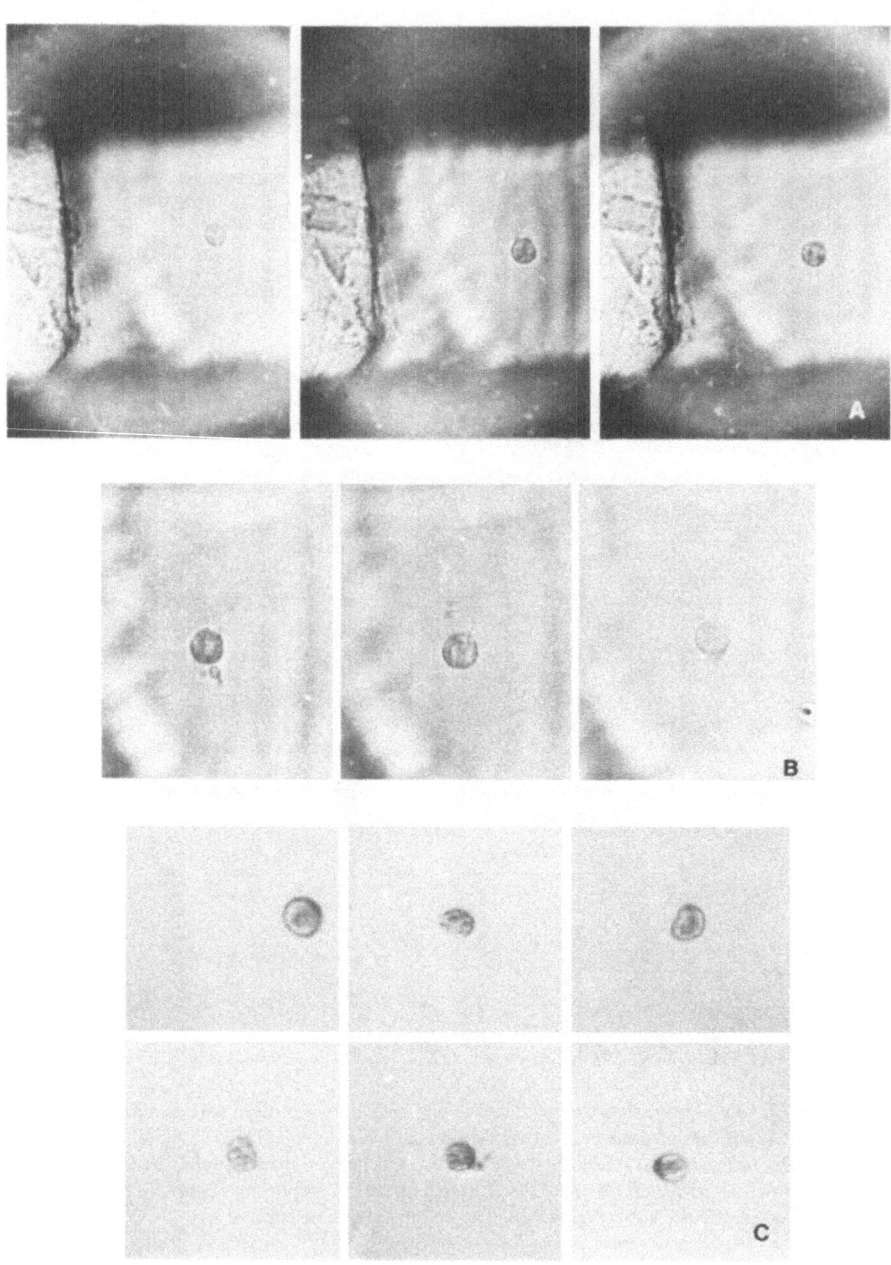

ACKNOWLEDGMENTS. The author wishes to thank Prof. Dr. G. Ruhenstroth-Bauer for his intensive advancing of the development of flow cytometric instrumentation; G. Benker, E. Glossner, R. Hüller, L. Haack, E. Kordwig, K. Lichtnau, H. Meier, K. Schedler, H. Schneider and W. Weiss for excellent technical contributions and assistance; and Dr. G. Valet and Dr. N. Boss for many helpful discussions, particularly on the aspects of biological and medical application of flow cytometric systems.

APPENDIX

Tables of Field Strength and Pulse Height Values in Orifices of Different Length-to-Diameter Ratios

The field strength and pulse height values are calculated by solving the Laplace equation for symmetry around the orifice axis by applying an iterative computer program.

For reasons of symmetry only the points of a quarter of the orifice region are tabulated.

The x scale in axial direction is normalized to the orifice diameter and defined such that the orifice inlet is found at $x/d = 0$. Negative x/d values are outside the orifice.

The radial distances from the orifice axis r are also normalized to the orifice diameter. The orifice axis is at $r/d = 0$ and the orifice wall at $r/d = 0.5$.

The field strength values E/E_m (in percent; E_m prints out as Em here) are normalized to the field strength in the homogeneous field, and the pulse height values U/U_m (in percent; U_m prints out as Um here) are normalized to the pulse height in the homogeneous field.

The homogeneous field is defined such that the current flowing through the orifice is uniformly distributed over the entire cross section of the orifice considered.

←_____

FIGURE 83. (A) Human white blood cells imaged in the $100 \times 100 \times 80$ μm orifice of the chamber of Fig. 76a. The cells are fixed in 1% glutaraldehyde and stained with hemalaun. (B) Other cells from the series shown in (A) with more magnification. The contrast of the nuclei is not sufficient and the background from the orifice channel is not neutral. (C) Bone marrow cells fixed with formalin and stained with aluminum sulfate-carmine at pH 2.5 imaged in the transducer of Fig. 78. The background is neutral and the nuclei are more contrasted.

LENGTH TO DIAMETER RATIO 0.5
----

orifice inlet at x/d = 0
orifice wall at r/d = 0.5

x/d	r/d = 0.0		r/d = 0.05		r/d = 0.10		r/d = 0.15		r/d = 0.20	
	E/Em	Ux/Um	E/Em	Ux/Um	E/Em	Ux/Um	E/Em	Ux/Um	E/Em	Ux/Um
	[%]	[%]	[%]	[%]	[%]	[%]	[%]	[%]	[%]	[%]
-1.50	5.09	0.26	5.08	0.26	5.07	0.26	5.05	0.25	5.02	0.25
-1.40	5.75	0.33	5.75	0.33	5.73	0.33	5.70	0.33	5.66	0.32
-1.30	6.55	0.43	6.54	0.43	6.52	0.43	6.49	0.42	6.44	0.41
-1.20	7.53	0.57	7.52	0.57	7.49	0.56	7.45	0.55	7.38	0.54
-1.10	8.73	0.76	8.72	0.76	8.68	0.75	8.62	0.74	8.54	0.73
-1.00	10.23	1.05	10.22	1.04	10.17	1.03	10.09	1.02	9.98	1.00
-0.90	12.13	1.47	12.11	1.47	12.05	1.45	11.94	1.43	11.80	1.39
-0.80	14.56	2.12	14.53	2.11	14.45	2.09	14.31	2.05	14.12	1.99
-0.70	17.70	3.13	17.66	3.12	17.55	3.08	17.37	3.02	17.11	2.93
-0.60	21.79	4.75	21.74	4.73	21.60	4.67	21.36	4.56	21.03	4.42
-0.50	27.15	7.37	27.09	7.34	26.91	7.24	26.62	7.09	26.21	6.87
-0.40	34.10	11.63	34.04	11.59	33.86	11.46	33.54	11.25	33.08	10.94
-0.30	42.92	18.42	42.87	18.38	42.74	18.27	42.49	18.06	42.11	17.73
-0.20	53.54	28.66	53.54	28.67	53.57	28.70	53.59	28.72	53.59	28.72
-0.10	65.14	42.44	65.25	42.57	65.58	43.01	66.13	43.73	66.94	44.81
0.00	76.06	57.86	76.27	58.17	76.93	59.18	78.07	60.94	79.80	63.68
0.10	84.30	71.06	84.56	71.51	85.37	72.89	86.76	75.28	88.82	78.89
0.20	88.60	78.49	88.86	78.96	89.69	80.44	91.04	82.88	92.96	86.42
0.30	88.57	78.44	88.81	78.88	89.73	80.51	91.02	82.85	92.98	86.46

x/d	r/d = 0.25		r/d = 0.3		r/d = 0.35		r/d = 0.40		r/d = 0.45	
	E/Em	Ux/Um	E/Em	Ux/Um	E/Em	Ux/Um	E/Em	Ux/Um	E/Em	Ux/Um
	[%]	[%]	[%]	[%]	[%]	[%]	[%]	[%]	[%]	[%]
-1.50	4.98	0.25	4.93	0.24	4.88	0.24	4.82	0.23	4.75	0.23
-1.40	5.62	0.32	5.56	0.31	5.49	0.30	5.41	0.29	5.33	0.28
-1.30	6.38	0.41	6.30	0.40	6.22	0.39	6.12	0.37	6.02	0.36
-1.20	7.30	0.53	7.21	0.52	7.10	0.50	6.97	0.49	6.84	0.47
-1.10	8.44	0.71	8.31	0.69	8.17	0.67	8.01	0.64	7.84	0.61
-1.00	9.85	0.97	9.68	0.94	9.50	0.90	9.29	0.86	9.06	0.82
-0.90	11.62	1.35	11.40	1.30	11.16	1.24	10.87	1.18	10.57	1.12
-0.80	13.88	1.93	13.58	1.85	13.25	1.76	12.88	1.66	12.47	1.55
-0.70	16.79	2.82	16.40	2.69	15.95	2.54	15.44	2.38	14.89	2.22
-0.60	20.60	4.25	20.09	4.04	19.48	3.79	18.79	3.53	18.03	3.25
-0.50	25.66	6.59	24.99	6.25	24.18	5.85	23.25	5.40	22.19	4.92
-0.40	32.44	10.53	31.63	10.00	30.60	9.37	29.36	8.62	27.90	7.79
-0.30	41.54	17.26	40.74	16.60	39.62	15.70	38.11	14.52	36.13	13.06
-0.20	53.49	28.62	53.21	28.32	52.57	27.64	51.32	26.34	49.07	24.08
-0.10	68.01	46.25	69.36	48.11	70.93	50.31	72.44	52.48	72.82	53.03
0.00	82.31	67.74	85.97	73.91	91.61	83.92	101.38	102.79	122.41	149.85
0.10	91.63	83.96	95.38	90.98	100.27	100.54	106.43	113.27	113.18	128.11
0.20	95.41	91.04	98.34	96.70	101.52	103.06	104.60	109.41	106.91	114.30
0.30	95.44	91.09	98.30	96.62	101.51	103.04	104.61	109.43	106.90	114.29

APPENDIX: FIELD AND PULSE HEIGHT, L/D = 0.5

x/d	r/d = 0.50 E/Em [%]	Ux/Um [%]	r/d = 0.55 E/Em [%]	Ux/Um [%]	r/d = 0.60 E/Em [%]	Ux/Um [%]	r/d = 0.65 E/Em [%]	Ux/Um [%]	r/d = 0.70 E/Em [%]	Ux/Um [%]
-1.50	4.68	0.22	4.60	0.21	4.52	0.20	4.43	0.20	4.34	0.19
-1.40	5.24	0.27	5.14	0.26	5.04	0.25	4.93	0.24	4.81	0.23
-1.30	5.90	0.35	5.78	0.33	5.65	0.32	5.51	0.30	5.37	0.29
-1.20	6.69	0.45	6.54	0.43	6.37	0.41	6.20	0.38	6.03	0.36
-1.10	7.65	0.58	7.45	0.55	7.24	0.52	7.02	0.49	6.80	0.46
-1.00	8.81	0.78	8.55	0.73	8.28	0.69	8.00	0.64	7.71	0.60
-0.90	10.24	1.05	9.90	0.98	9.54	0.91	9.18	0.84	8.81	0.78
-0.80	12.03	1.45	11.57	1.34	11.10	1.23	10.62	1.13	10.13	1.03
-0.70	14.29	2.04	13.67	1.87	13.03	1.70	12.36	1.53	11.73	1.38
-0.60	17.21	2.96	16.35	2.67	15.46	2.39	14.57	2.12	13.69	1.87
-0.50	21.04	4.43	19.82	3.93	18.57	3.45	17.32	3.00	16.10	2.59
-0.40	26.26	6.89	24.48	5.99	22.64	5.13	20.82	4.33	19.06	3.63
-0.30	33.72	11.37	30.96	9.59	28.10	7.89	25.30	6.40	22.69	5.15
-0.20	45.48	20.69	40.66	16.53	35.66	12.71	31.02	9.62	26.98	7.28
-0.10	69.09	47.73	57.01	32.50	46.20	21.34	37.74	14.25	31.34	9.82
0.00			105.72	111.77	54.06	29.23	41.57	17.28	33.44	11.18
0.10	115.38	133.13								
0.20	107.68	115.95								
0.30	107.67	115.94								

x/d	r/d = 0.75 E/Em [%]	Ux/Um [%]	r/d = 0.80 E/Em [%]	Ux/Um [%]	r/d = 0.85 E/Em [%]	Ux/Um [%]	r/d = 0.90 E/Em [%]	Ux/Um [%]	r/d = 0.95 E/Em [%]	Ux/Um [%]
-1.50	4.25	0.18	4.15	0.17	4.05	0.16	3.95	0.16	3.85	0.15
-1.40	4.70	0.22	4.58	0.21	4.47	0.20	4.34	0.19	4.23	0.18
-1.30	5.23	0.27	5.08	0.26	4.94	0.24	4.79	0.23	4.65	0.22
-1.20	5.85	0.34	5.67	0.32	5.49	0.30	5.31	0.28	5.13	0.26
-1.10	6.57	0.43	6.35	0.40	6.12	0.37	5.90	0.35	5.68	0.32
-1.00	7.43	0.55	7.14	0.51	6.86	0.47	6.58	0.43	6.31	0.40
-0.90	8.44	0.71	8.07	0.65	7.72	0.60	7.36	0.54	7.02	0.49
-0.80	9.65	0.93	9.17	0.84	8.71	0.76	8.27	0.68	7.84	0.61
-0.70	11.10	1.23	10.47	1.10	9.88	0.98	9.30	0.87	8.76	0.77
-0.60	12.83	1.65	12.01	1.44	11.23	1.26	10.49	1.10	9.80	0.96
-0.50	14.92	2.23	13.81	1.91	12.78	1.63	11.82	1.40	10.94	1.20
-0.40	17.41	3.03	15.90	2.53	14.52	2.11	13.27	1.76	12.15	1.48
-0.30	20.33	4.13	18.23	3.33	16.39	2.69	14.78	2.19	13.38	1.79
-0.20	23.54	5.54	20.65	4.27	18.23	3.32	16.20	2.62	14.48	2.10
-0.10	26.46	7.00	22.68	5.14	19.68	3.87	17.26	2.98	15.28	2.34
0.00	27.71	7.68	23.49	5.52	20.25	4.10	17.67	3.12	15.56	2.42

x/d	r/d = 1.00 E/Em [%]	Ux/Um [%]
-1.50	3.75	0.14
-1.40	4.10	0.17
-1.30	4.50	0.20
-1.20	4.95	0.25
-1.10	5.46	0.30
-1.00	6.04	0.36
-0.90	6.69	0.45
-0.80	7.43	0.55
-0.70	8.25	0.68
-0.60	9.15	0.84
-0.50	10.13	1.03
-0.40	11.15	1.24
-0.30	12.15	1.48
-0.20	13.02	1.70
-0.10	13.64	1.86
0.00	13.85	1.92

APPENDIX: FIELD AND PULSE HEIGHT, L/D = 0.8

LENGTH TO DIAMETER RATIO L/D = 0.8

orifice inlet at x/d = 0
orifice wall at r/d = 0.5

x/d	r/d = 0.00 E/Em [%]	Ux/Um [%]	r/d = 0.05 E/Em [%]	Ux/Um [%]	r/d = 0.10 E/Em [%]	Ux/Um [%]	r/d = 0.15 E/Em [%]	Ux/Um [%]	r/d = 0.20 E/Em [%]	Ux/Um [%]
-1.50	5.12	0.26	5.12	0.26	5.10	0.26	5.08	0.26	5.05	0.26
-1.40	5.78	0.33	5.78	0.33	5.76	0.33	5.74	0.33	5.70	0.32
-1.30	6.58	0.43	6.62	0.44	6.56	0.43	6.52	0.43	6.47	0.42
-1.20	7.56	0.57	7.55	0.57	7.52	0.57	7.48	0.56	7.42	0.55
-1.10	8.77	0.77	8.76	0.77	8.72	0.76	8.66	0.75	8.58	0.74
-1.00	10.27	1.05	10.25	1.05	10.20	1.04	10.13	1.03	10.02	1.00
-0.90	12.17	1.48	12.15	1.48	12.08	1.46	11.98	1.44	11.84	1.40
-0.80	14.60	2.13	14.57	2.12	14.49	2.10	14.35	2.06	14.16	2.00
-0.70	17.74	3.15	17.70	3.13	17.59	3.09	17.41	3.03	17.15	2.94
-0.60	21.85	4.77	21.80	4.75	21.65	4.69	21.41	4.59	21.08	4.44
-0.50	27.21	7.41	27.16	7.38	26.98	7.28	26.69	7.12	26.27	6.90
-0.40	34.20	11.69	34.14	11.65	33.95	11.53	33.63	11.31	33.16	11.00
-0.30	43.07	18.55	43.03	18.51	42.89	18.39	42.63	18.17	42.23	17.84
-0.20	53.80	28.94	53.80	28.94	53.82	28.96	53.81	28.96	53.78	28.92
-0.10	65.63	43.07	65.72	43.20	66.02	43.59	66.52	44.25	67.27	45.25
0.00	77.01	59.30	77.19	59.58	77.78	60.49	78.80	62.09	80.39	64.63
0.10	86.16	74.23	86.37	74.60	87.04	75.76	88.20	77.79	89.95	80.90
0.20	92.24	85.08	92.42	85.41	92.98	86.45	93.91	88.19	95.23	90.69
0.30	95.48	91.17	95.62	91.43	96.04	92.25	96.71	93.53	97.63	95.32
0.40	96.48	93.07	96.60	93.31	96.99	94.07	97.54	95.13	98.30	96.63

x/d	r/d = 0.25 E/Em [%]	Ux/Um [%]	r/d = 0.30 E/Em [%]	Ux/Um [%]	r/d = 0.35 E/Em [%]	Ux/Um [%]	r/d = 0.40 E/Em [%]	Ux/Um [%]	r/d = 0.45 E/Em [%]	Ux/Um [%]
-1.50	5.02	0.25	4.97	0.25	4.92	0.24	4.85	0.24	4.79	0.23
-1.40	5.65	0.32	5.59	0.31	5.53	0.31	5.45	0.30	5.37	0.29
-1.30	6.41	0.41	6.34	0.40	6.25	0.39	6.15	0.38	6.05	0.37
-1.20	7.34	0.54	7.24	0.52	7.13	0.51	7.01	0.49	6.87	0.47
-1.10	8.47	0.72	8.35	0.70	8.21	0.67	8.04	0.65	7.87	0.62
-1.00	9.88	0.98	9.72	0.94	9.53	0.91	9.32	0.87	9.09	0.83
-0.90	11.66	1.36	11.44	1.31	11.19	1.25	10.91	1.19	10.61	1.12
-0.80	13.91	1.94	13.62	1.86	13.29	1.77	12.91	1.67	12.51	1.56
-0.70	16.83	2.83	16.44	2.70	15.99	2.56	15.48	2.40	14.93	2.23
-0.60	20.65	4.27	20.13	4.05	19.52	3.81	18.83	3.55	18.07	3.26
-0.50	25.72	6.62	25.05	6.27	24.23	5.87	23.29	5.43	22.24	4.94
-0.40	32.52	10.58	31.70	10.05	30.67	9.41	29.42	8.65	27.95	7.81
-0.30	41.65	17.35	40.83	16.67	39.69	15.76	38.17	14.57	36.19	13.10
-0.20	53.65	28.78	53.34	28.45	52.67	27.74	51.39	26.41	49.12	24.13
-0.10	68.26	46.59	69.54	48.35	71.04	50.46	72.50	52.56	72.84	53.05
0.00	82.73	68.44	86.22	74.34	91.68	84.06	101.31	102.64	122.21	149.35
0.10	92.38	85.35	95.74	91.65	100.21	100.41	105.99	112.35	112.48	126.53
0.20	96.93	93.95	98.98	97.98	101.25	102.51	103.51	107.13	105.24	110.75
0.30	98.73	97.47	99.93	99.86	101.11	102.24	102.15	104.34	102.84	105.75
0.40	99.18	98.37	100.12	100.24	101.00	102.02	101.76	103.54	102.26	104.57

APPENDIX: FIELD AND PULSE HEIGHT, L/D = 0.8

x/d	r/d = 0.50 E/Em [%]	Ux/Um [%]	r/d = 0.55 E/Em [%]	Ux/Um [%]	r/d = 0.60 E/Em [%]	Ux/Um [%]	r/d = 0.65 E/Em [%]	Ux/Um [%]	r/d = 0.70 E/Em [%]	Ux/Um [%]
-1.50	4.71	0.22	4.64	0.21	4.55	0.21	4.47	0.20	4.37	0.19
-1.40	5.27	0.28	5.18	0.27	5.07	0.26	4.96	0.25	4.85	0.24
-1.30	5.93	0.35	5.81	0.34	5.68	0.32	5.55	0.31	5.41	0.29
-1.20	6.72	0.45	6.57	0.43	6.40	0.41	6.24	0.39	6.06	0.37
-1.10	7.68	0.59	7.48	0.56	7.27	0.53	7.05	0.50	6.83	0.47
-1.00	8.84	0.78	8.58	0.74	8.31	0.69	8.03	0.65	7.75	0.60
-0.90	10.28	1.06	9.94	0.99	9.58	0.92	9.71	0.85	8.84	0.78
-0.80	12.07	1.46	11.61	1.35	11.13	1.24	10.65	1.13	10.16	1.03
-0.70	14.33	2.05	13.71	1.88	13.07	1.71	12.42	1.54	11.77	1.38
-0.60	17.25	2.97	16.39	2.68	15.50	2.40	14.61	2.13	13.72	1.88
-0.50	21.08	4.45	19.86	3.95	18.61	3.46	17.36	3.01	16.13	2.60
-0.40	26.30	6.92	24.52	6.01	22.68	5.14	20.85	4.35	19.10	3.65
-0.30	33.76	11.40	31.01	9.61	28.14	7.92	25.34	6.42	22.73	5.17
-0.20	45.53	20.73	40.70	16.56	35.69	12.74	31.06	9.65	27.01	7.30
-0.10	69.10	47.75	57.03	32.53	46.23	21.37	37.78	14.27	31.37	9.84
0.00			105.69	111.70	54.08	29.25	41.60	17.31	33.47	11.20
0.10	114.58	131.29								
0.20	105.81	111.97								
0.30	103.07	106.23								
0.40	102.44	104.93								

x/d	r/d = 0.75 E/Em [%]	Ux/Um [%]	r/d = 0.80 E/Em [%]	Ux/Um [%]	r/d = 0.85 E/Em [%]	Ux/Um [%]	r/d = 0.90 E/Em [%]	Ux/Um [%]	r/d = 0.95 E/Em [%]	Ux/Um [%]
-1.50	4.28	0.18	4.18	0.17	4.09	0.17	3.99	0.16	3.89	0.15
-1.40	4.74	0.22	4.62	0.21	4.50	0.20	4.38	0.19	4.26	0.18
-1.30	5.27	0.28	5.12	0.26	4.98	0.25	4.83	0.23	4.68	0.22
-1.20	5.88	0.35	5.70	0.33	5.52	0.31	5.34	0.29	5.16	0.27
-1.10	6.61	0.44	6.38	0.41	6.16	0.38	5.93	0.35	5.71	0.33
-1.00	7.46	0.56	7.18	0.51	6.89	0.48	6.61	0.44	6.34	0.40
-0.90	8.48	0.72	8.11	0.66	7.75	0.60	7.40	0.55	7.06	0.50
-0.80	9.68	0.94	9.21	0.85	8.75	0.77	8.30	0.69	7.87	0.62
-0.70	11.13	1.24	10.51	1.10	9.91	0.98	9.34	0.87	8.80	0.77
-0.60	12.87	1.66	12.04	1.45	11.26	1.27	10.52	1.11	9.83	0.97
-0.50	14.96	2.24	13.84	1.92	12.81	1.64	11.85	1.40	10.97	1.20
-0.40	17.45	3.04	15.93	2.54	14.55	2.12	13.31	1.77	12.19	1.49
-0.30	20.37	4.15	18.27	3.34	16.43	2.70	14.82	2.20	13.41	1.80
-0.20	23.57	5.56	20.69	4.28	18.27	3.34	16.23	2.63	14.52	2.11
-0.10	26.50	7.02	22.71	5.16	19.71	3.89	17.30	2.99	15.31	2.35
0.00	27.74	7.70	23.52	5.53	20.28	4.11	17.70	3.13	15.59	2.43

x/d	r/d = 1.00 E/Em [%]	Ux/Um [%]
-1.50	3.79	0.14
-1.40	4.14	0.17
-1.30	4.54	0.21
-1.20	4.99	0.25
-1.10	5.49	0.30
-1.00	6.07	0.37
-0.90	6.72	0.45
-0.80	7.46	0.56
-0.70	8.28	0.69
-0.60	9.19	0.84
-0.50	10.16	1.03
-0.40	11.18	1.25
-0.30	12.18	1.48
-0.20	13.06	1.70
-0.10	13.67	1.87
0.00	13.89	1.93

```
                    LENGTH TO DIAMETER RATIO L/D = 1.0
                    --------------------------------------

orifice inlet at x/d = 0
orifice wall at  r/d = 0.5
```

x/d	r/d = 0.00 E/Em [%]	Ux/Um [%]	r/d = 0.05 E/Em [%]	Ux/Um [%]	r/d = 0.10 E/Em [%]	Ux/Um [%]	r/d = 0.15 E/Em [%]	Ux/Um [%]	r/d = 0.20 E/Em [%]	Ux/Um [%]
-1.50	5.12	0.26	5.11	0.26	5.10	0.26	5.08	0.26	5.04	0.25
-1.40	5.77	0.33	5.77	0.33	5.75	0.33	5.73	0.33	5.69	0.32
-1.30	6.57	0.43	6.56	0.43	6.54	0.43	6.51	0.42	6.46	0.42
-1.20	7.54	0.57	7.53	0.57	7.50	0.56	7.46	0.56	7.40	0.55
-1.10	8.74	0.76	8.73	0.76	8.69	0.76	8.63	0.75	8.55	0.73
-1.00	10.23	1.05	10.22	1.04	10.17	1.03	10.09	1.02	9.98	1.00
-0.90	12.12	1.47	12.10	1.46	12.04	1.45	11.94	1.42	11.79	1.39
-0.80	14.54	2.11	14.51	2.11	14.43	2.08	14.29	2.04	14.10	1.99
-0.70	17.66	3.12	17.63	3.11	17.52	3.07	17.34	3.01	17.08	2.92
-0.60	21.75	4.73	21.70	4.71	21.56	4.65	21.32	4.55	20.99	4.41
-0.50	27.10	7.34	27.04	7.31	26.86	7.22	26.57	7.06	26.15	6.84
-0.40	34.04	11.59	33.99	11.55	33.80	11.42	33.48	11.21	33.01	10.90
-0.30	42.88	18.38	42.83	18.35	42.69	18.23	42.44	18.01	42.04	17.68
-0.20	53.56	28.69	53.56	28.69	53.58	28.70	53.57	28.70	53.54	28.66
-0.10	65.35	42.71	65.44	42.83	65.74	43.21	66.23	43.86	66.96	44.84
0.00	76.72	58.86	76.89	59.12	77.46	60.00	78.48	61.58	80.05	64.07
0.10	85.91	73.80	86.11	74.15	86.77	75.29	87.89	77.25	89.60	80.28
0.20	92.15	84.91	92.31	85.22	92.83	86.18	93.70	87.80	94.96	90.17
0.30	95.77	91.71	95.87	91.92	96.23	92.60	96.79	93.68	97.56	95.18
0.40	97.54	95.14	97.61	95.28	97.85	95.75	98.21	96.44	98.69	97.40
0.50	98.07	96.19	98.15	96.33	98.33	96.68	98.60	97.21	99.01	98.03

x/d	r/d = 0.25 E/Em [%]	Ux/Um [%]	r/d = 0.30 E/Em [%]	Ux/Um [%]	r/d = 0.35 E/Em [%]	Ux/Um [%]	r/d = 0.40 E/Em [%]	Ux/Um [%]	r/d = 0.45 E/Em [%]	Ux/Um [%]
-1.50	5.01	0.25	4.96	0.25	4.91	0.24	4.85	0.23	4.78	0.23
-1.40	5.64	0.32	5.58	0.31	5.52	0.30	5.44	0.30	5.36	0.29
-1.30	6.40	0.41	6.32	0.40	6.24	0.39	6.14	0.38	6.04	0.36
-1.20	7.32	0.54	7.22	0.52	7.11	0.51	6.99	0.49	6.86	0.47
-1.10	8.42	0.71	8.32	0.69	8.18	0.67	8.02	0.64	7.85	0.62
-1.00	9.85	0.97	9.69	0.94	9.50	0.90	9.29	0.86	9.06	0.82
-0.90	11.61	1.35	11.40	1.30	11.15	1.24	10.87	1.18	10.57	1.12
-0.80	13.86	1.92	13.57	1.84	13.24	1.75	12.86	1.65	12.46	1.55
-0.70	16.76	2.81	16.37	2.68	15.92	2.54	15.42	2.38	14.87	2.21
-0.60	20.56	4.23	20.04	4.02	19.44	3.78	18.75	3.52	17.99	3.24
-0.50	25.61	6.56	24.93	6.22	24.13	5.82	23.19	5.38	22.14	4.90
-0.40	32.38	10.48	31.56	9.96	30.53	9.32	29.29	8.58	27.83	7.74
-0.30	41.46	17.19	40.64	16.52	39.51	15.61	37.99	14.44	36.02	12.98
-0.20	53.40	28.52	53.09	28.19	52.42	27.48	51.15	26.17	48.89	23.91
0.10	67.95	46.17	69.21	47.90	70.70	49.98	72.15	52.05	72.49	52.55
0.00	82.36	67.84	85.81	73.64	91.24	83.25	100.81	101.63	121.63	147.95
0.10	92.00	84.63	95.29	90.80	99.70	99.41	105.44	111.17	111.87	125.15
0.20	96.57	93.26	98.53	97.08	100.71	101.42	102.90	105.87	104.57	109.34
0.30	98.47	96.97	99.51	99.02	100.52	101.04	101.41	102.84	102.01	104.06
0.40	99.23	98.46	99.80	99.60	100.33	100.65	100.77	101.54	101.03	102.08
0.50	99.41	98.83	99.88	99.75	100.25	100.51	100.59	101.18	100.77	101.54

APPENDIX: FIELD AND PULSE HIGHT, L/D = 1.0

x/d	r/d = 0.50		r/d = 0.55		r/d = 0.60		r/d = 0.65		r/d = 0.70	
	E/Em	Ux/Um	E/Em	Ux/Um	E/Em	Ux/Um	E/Em	Ux/Um	E/Em	Ux/Um
	[%]	[%]	[%]	[%]	[%]	[%]	[%]	[%]	[%]	[%]
-1.50	4.71	0.22	4.63	0.21	4.55	0.21	4.46	0.20	4.37	0.19
-1.40	5.26	0.28	5.17	0.27	5.06	0.26	4.96	0.25	4.84	0.23
-1.30	5.92	0.35	5.80	0.34	5.67	0.32	5.54	0.31	5.40	0.29
-1.20	6.71	0.45	6.56	0.43	6.39	0.41	6.22	0.39	6.05	0.37
-1.10	7.66	0.59	7.46	0.56	7.25	0.53	7.04	0.50	6.81	0.46
-1.00	8.82	0.78	8.56	0.73	8.29	0.69	8.01	0.64	7.73	0.60
-0.90	10.24	1.05	9.90	0.98	9.51	0.90	9.19	0.84	8.82	0.78
-0.80	12.02	1.45	11.57	1.34	11.09	1.23	10.61	1.13	10.13	1.03
-0.70	14.27	2.04	13.65	1.86	12.98	1.69	12.37	1.53	11.72	1.37
-0.60	17.17	2.95	16.32	2.66	15.44	2.38	14.55	2.12	13.67	1.87
-0.50	20.99	4.41	19.78	3.91	18.53	3.43	17.29	2.99	16.06	2.58
-0.40	26.18	6.86	24.41	5.96	22.58	5.10	20.76	4.31	19.01	3.62
-0.30	33.61	11.30	30.87	9.53	28.01	7.85	25.22	6.36	22.63	5.12
-0.20	45.31	20.53	40.51	16.41	35.53	12.62	30.92	9.56	26.89	7.23
-0.10	68.77	47.29	56.76	32.22	46.01	21.17	37.60	14.14	31.23	9.76
0.00			105.17	110.60	53.82	28.97	41.41	17.15	33.32	11.10
0.10	113.95	129.84								
0.20	105.13	110.53								
0.30	102.21	104.48								
0.40	101.14	102.30								
0.50	100.83	101.67								

x/d	r/d = 0.75		r/d = 0.80		r/d = 0.85		r/d = 0.90		r/d = 0.95	
	E/Em	Ux/Um	E/Em	Ux/Um	E/Em	Ux/Um	E/Em	Ux/Um	E/Em	Ux/Um
	[%]	[%]	[%]	[%]	[%]	[%]	[%]	[%]	[%]	[%]
-1.50	4.28	0.18	4.18	0.17	4.09	0.17	3.99	0.16	3.89	0.15
-1.40	4.73	0.22	4.61	0.21	4.50	0.20	4.38	0.19	4.26	0.18
-1.30	5.26	0.28	5.11	0.26	4.97	0.25	4.82	0.23	4.68	0.22
-1.20	5.87	0.34	5.69	0.32	5.51	0.30	5.33	0.28	5.16	0.27
-1.10	6.59	0.43	6.37	0.41	6.14	0.38	5.92	0.35	5.70	0.33
-1.00	7.44	0.55	7.16	0.51	6.88	0.47	6.60	0.44	6.33	0.40
-0.90	8.45	0.71	8.08	0.65	7.73	0.60	7.38	0.54	7.04	0.50
-0.80	9.65	0.93	9.18	0.84	8.72	0.76	8.27	0.68	7.85	0.62
-0.70	11.09	1.23	10.47	1.10	9.88	0.98	9.31	0.87	8.77	0.77
-0.60	12.82	1.64	12.00	1.44	11.22	1.26	10.48	1.10	9.80	0.96
-0.50	14.90	2.22	13.79	1.90	12.76	1.63	11.81	1.39	10.93	1.20
-0.40	17.38	3.02	15.87	2.52	14.50	2.10	13.25	1.76	12.14	1.47
-0.30	20.28	4.11	18.19	3.31	16.36	2.68	14.76	2.18	13.36	1.78
-0.20	23.47	5.51	20.60	4.24	18.19	3.31	16.17	2.61	14.46	2.09
-0.10	26.38	6.96	22.61	5.11	19.63	3.85	17.23	2.97	15.25	2.33
0.00	27.61	7.62	23.41	5.48	20.21	4.08	17.63	3.11	15.51	2.41

x/d	r/d = 1.00	
	E/Em	Ux/Um
	[%]	[%]
-1.50	3.79	0.14
-1.40	4.14	0.17
-1.30	4.53	0.21
-1.20	4.98	0.25
-1.10	5.49	0.30
-1.00	6.06	0.37
-0.90	6.71	0.45
-0.80	7.44	0.55
-0.70	8.26	0.68
-0.60	9.16	0.84
-0.50	10.13	1.03
-0.40	11.14	1.24
-0.30	12.14	1.47
-0.20	13.01	1.69
-0.10	13.62	1.85
0.00	13.83	1.91

APPENDIX: FIELD AND PULSE HEIGHT, L/D = 1.2

LENGTH TO DIAMETER RATIO L/D = 1.2
----
orifice inlet at x/d = 0
orifice wall at r/d = 0.5

x/d	r/d = 0.00 E/Em [%]	Ux/Um [%]	r/d = 0.05 E/Em [%]	Ux/Um [%]	r/d = 0.10 E/Em [%]	Ux/Um [%]	r/d = 0.15 E/Em [%]	Ux/Um [%]	r/d = 0.20 E/Em [%]	Ux/Um [%]
-1.50	5.16	0.27	5.16	0.27	5.14	0.26	5.12	0.26	5.09	0.26
-1.40	5.82	0.34	5.82	0.34	5.80	0.34	5.77	0.33	5.73	0.33
-1.30	6.62	0.44	6.62	0.44	6.59	0.43	6.56	0.43	6.51	0.42
-1.20	7.59	0.58	7.59	0.58	7.56	0.57	7.51	0.56	7.45	0.55
-1.10	8.80	0.77	8.79	0.75	8.75	0.77	8.69	0.76	8.61	0.74
-1.00	10.30	1.06	10.29	1.06	10.24	1.05	10.16	1.03	10.05	1.01
-0.90	12.20	1.49	12.18	1.48	12.12	1.47	12.01	1.44	11.87	1.41
-0.80	14.63	2.14	14.61	2.13	14.52	2.11	14.38	2.07	14.19	2.01
-0.70	17.77	3.16	17.74	3.15	17.63	3.11	17.44	3.04	17.19	2.95
-0.60	21.88	4.79	21.83	4.77	21.69	4.70	21.45	4.60	21.11	4.46
-0.50	27.25	7.43	27.20	7.40	27.02	7.30	26.73	7.14	26.31	6.92
-0.40	34.24	11.72	34.18	11.68	33.99	11.56	33.67	11.34	33.20	11.02
-0.30	43.12	18.60	43.07	18.55	42.94	18.44	42.68	18.22	42.28	17.88
-0.20	53.85	29.00	53.86	29.01	53.88	29.03	53.88	29.03	53.84	28.99
-0.10	65.72	43.19	65.81	43.31	66.11	43.70	66.60	44.36	67.34	45.35
0.00	77.15	59.52	77.33	59.80	77.91	60.70	78.92	62.29	80.50	64.80
0.10	86.42	74.68	86.62	75.03	87.27	76.16	88.40	78.15	90.11	81.20
0.20	92.73	85.98	92.89	86.28	93.41	87.25	94.27	88.87	95.52	91.24
0.30	96.45	93.03	96.55	93.22	96.89	93.88	97.43	94.92	98.17	96.36
0.40	98.42	96.86	98.47	96.97	98.68	97.37	98.99	97.98	99.40	98.80
0.50	99.33	98.67	99.37	98.75	99.50	99.00	99.67	99.34	99.92	99.84
0.60	99.60	99.21	99.64	99.29	99.72	99.45	99.87	99.74	100.06	100.13

x/d	r/d = 0.25 E/Em [%]	Ux/Um [%]	r/d = 0.30 E/Em [%]	Ux/Um [%]	r/d = 0.35 E/Em [%]	Ux/Um [%]	r/d = 0.40 E/Em [%]	Ux/Um [%]	r/d = 0.45 E/Em [%]	Ux/Um [%]
-1.50	5.05	0.26	5.00	0.25	4.95	0.25	4.89	0.24	4.83	0.23
-1.40	5.69	0.32	5.63	0.32	5.56	0.31	5.48	0.30	5.40	0.29
-1.30	6.45	0.42	6.37	0.41	6.29	0.40	6.19	0.38	6.09	0.37
-1.20	7.37	0.54	7.28	0.53	7.17	0.51	7.04	0.50	6.91	0.48
-1.10	8.51	0.72	8.38	0.70	8.24	0.68	8.08	0.65	7.91	0.63
-1.00	9.92	0.98	9.75	0.95	9.57	0.92	9.35	0.87	9.13	0.83
-0.90	11.69	1.37	11.47	1.32	11.22	1.26	10.94	1.20	10.64	1.13
-0.80	13.95	1.95	13.65	1.86	13.32	1.77	12.95	1.68	12.54	1.57
-0.70	16.87	2.84	16.47	2.71	16.02	2.57	15.51	2.41	14.96	2.24
-0.60	20.69	4.28	20.17	4.07	19.56	3.82	18.86	3.56	18.10	3.28
-0.50	25.76	6.64	24.67	6.09	24.27	5.89	23.33	5.44	22.27	4.96
-0.40	32.56	10.60	31.74	10.07	30.71	9.43	29.46	8.68	27.99	7.83
-0.30	41.70	17.39	40.87	16.71	39.74	15.79	38.21	14.60	36.23	13.13
-0.20	53.71	28.84	53.39	28.50	52.72	27.79	51.44	26.46	49.17	24.18
-0.10	68.33	46.69	69.60	48.44	71.10	50.55	72.55	52.64	72.90	53.14
0.00	82.82	68.58	86.29	74.46	91.74	84.17	101.36	102.75	122.32	149.61
0.10	92.51	85.58	95.81	91.80	100.24	100.49	106.00	112.36	112.47	126.49
0.20	97.12	94.32	99.07	98.16	101.25	102.52	103.43	106.99	105.11	110.49
0.30	99.06	98.13	100.05	100.10	101.04	102.09	101.92	103.87	102.50	105.06
0.40	99.85	99.71	100.36	100.72	100.81	101.63	101.20	102.42	101.44	102.89
0.50	100.17	100.34	100.45	100.84	100.69	101.38	100.88	101.78	101.00	102.01
0.60	100.26	100.52	100.47	100.94	100.65	101.30	100.81	101.63	100.87	101.74

APPENDIX: FIELD AND PULSE HEIGHT ... 1.2

x/d	r/d = 0.50 E/Em [%]	Ux/Um [%]	r/d = 0.55 E/Em [%]	Ux/Um [%]	r/d = 0.60 E/Em [%]	Ux/Um [%]	r/d = 0.65 E/Em [%]	Ux/Um [%]	r/d = 0.70 E/Em [%]	Ux/Um [%]
-1.50	4.75	0.23	4.67	0.22	4.59	0.21	4.53	0.21	4.41	0.19
-1.40	5.31	0.28	5.21	0.27	5.11	0.26	5.00	0.25	4.89	0.24
-1.30	5.97	0.36	5.85	0.34	5.72	0.33	5.58	0.31	5.44	0.30
-1.20	6.76	0.46	6.61	0.44	6.44	0.41	6.27	0.39	6.10	0.37
-1.10	7.73	0.60	7.52	0.57	7.30	0.53	7.09	0.50	6.87	0.47
-1.00	8.88	0.79	8.62	0.74	8.35	0.70	8.07	0.65	7.78	0.61
-0.90	10.31	1.06	9.97	0.99	9.61	0.92	9.25	0.86	8.88	0.79
-0.80	12.10	1.46	11.64	1.36	11.17	1.25	10.69	1.14	10.20	1.04
-0.70	14.36	2.06	13.74	1.89	13.10	1.72	12.45	1.55	11.80	1.39
-0.60	17.73	0.03	16.42	2.70	15.53	2.41	14.64	2.14	13.76	1.89
-0.50	21.12	4.46	19.90	3.96	18.65	3.48	17.39	3.03	16.16	2.61
-0.40	26.34	6.94	24.55	6.03	22.71	5.16	20.89	4.36	19.13	3.66
-0.30	33.80	11.43	31.05	9.64	28.17	7.94	25.38	6.44	22.76	5.18
-0.20	45.57	20.77	40.74	16.60	35.73	12.77	31.10	9.67	27.05	7.32
-0.10	69.15	47.82	57.08	32.58	46.27	21.41	37.82	14.30	31.41	9.87
0.00			105.76	111.84	54.13	29.30	41.65	17.35	33.51	11.23
0.10	114.55	131.23								
0.20	105.67	111.67								
0.30	102.70	105.46								
0.40	101.53	103.08								
0.50	101.04	102.09								
0.60	100.90	101.81								

x/d	r/d = 0.75 E/Em [%]	Ux/Um [%]	r/d = 0.80 E/Em [%]	Ux/Um [%]	r/d = 0.85 E/Em [%]	Ux/Um [%]	r/d = 0.90 E/Em [%]	Ux/Um [%]	r/d = 0.95 E/Em [%]	Ux/Um [%]
-1.50	4.32	0.19	4.22	0.18	4.13	0.17	4.02	0.16	3.93	0.15
-1.40	4.77	0.23	4.65	0.22	4.54	0.21	4.42	0.20	4.30	0.18
-1.30	5.30	0.28	5.16	0.27	5.01	0.25	4.86	0.24	4.72	0.22
-1.20	5.92	0.35	5.74	0.33	5.56	0.31	5.38	0.29	5.20	0.27
-1.10	6.64	0.44	6.42	0.41	6.19	0.38	5.97	0.36	5.75	0.33
-1.00	7.50	0.56	7.21	0.52	6.93	0.48	6.65	0.44	6.38	0.41
-0.90	8.51	0.72	8.14	0.66	7.78	0.61	7.43	0.55	7.09	0.50
-0.80	9.72	0.94	9.24	0.85	8.78	0.77	8.33	0.69	7.91	0.63
-0.70	11.17	1.25	10.54	1.11	9.95	0.99	9.37	0.88	8.83	0.78
-0.60	12.90	1.66	12.08	1.46	11.29	1.28	10.55	1.11	9.87	0.97
-0.50	14.99	2.25	13.88	1.93	12.85	1.65	11.88	1.41	11.01	1.21
-0.40	17.49	3.06	15.97	2.55	14.59	2.13	13.34	1.78	12.22	1.49
-0.30	20.40	4.16	18.30	3.35	16.46	2.71	14.85	2.21	13.45	1.81
-0.20	23.61	5.57	20.72	4.29	18.30	3.35	16.27	2.65	14.55	2.12
-0.10	26.53	7.04	22.75	5.17	19.76	3.90	17.33	3.00	15.35	2.36
0.00	27.77	7.71	23.55	5.55	20.33	4.13	17.74	3.15	15.61	2.44

x/d	r/d = 1.00 E/Em [%]	Ux/Um [%]
-1.50	3.82	0.15
-1.40	4.18	0.17
-1.30	4.57	0.21
-1.20	4.88	0.23
-1.10	5.53	0.31
-1.00	6.11	0.37
-0.90	6.76	0.46
-0.80	7.49	0.56
-0.70	8.32	0.69
-0.60	9.22	0.85
-0.50	10.20	1.04
-0.40	11.22	1.26
-0.30	12.22	1.49
-0.20	13.09	1.71
-0.10	13.70	1.88
0.00	13.92	1.94

LENGTH TO DIAMETER RATIO L/D = 1.5

orifice inlet at x/d = 0
orifice wall at r/d = 0.5

x/d	r/d = 0.00 E/Em [%]	Ux/Um [%]	r/d = 0.05 E/Em [%]	Ux/Um [%]	r/d = 0.10 E/Em [%]	Ux/Um [%]	r/d = 0.15 E/Em [%]	Ux/Um [%]	r/d = 0.20 E/Em [%]	Ux/Um [%]
-1.50	5.17	0.27	5.16	0.27	5.14	0.26	5.12	0.26	5.09	0.26
-1.40	5.82	0.34	5.82	0.34	5.80	0.34	5.78	0.33	5.73	0.33
-1.30	6.61	0.44	6.60	0.44	6.58	0.43	6.55	0.43	6.50	0.42
-1.20	7.59	0.58	7.58	0.57	7.55	0.57	7.51	0.56	7.44	0.55
-1.10	8.77	0.77	8.77	0.77	8.73	0.76	8.67	0.75	8.59	0.74
-1.00	10.27	1.06	10.26	1.05	10.21	1.04	10.13	1.03	10.02	1.00
-0.90	12.16	1.48	12.15	1.48	12.08	1.46	11.98	1.43	11.83	1.40
-0.80	14.57	2.12	14.54	2.12	14.46	2.09	14.33	2.05	14.14	2.00
-0.70	17.71	3.14	17.67	3.12	17.56	3.08	17.38	3.02	17.12	2.93
-0.60	21.79	4.75	21.74	4.73	21.59	4.66	21.36	4.56	21.02	4.42
-0.50	27.13	7.36	27.07	7.33	26.90	7.24	26.61	7.08	26.19	6.86
-0.40	34.08	11.61	34.02	11.57	33.84	11.45	33.52	11.23	33.05	10.92
-0.30	42.92	18.42	42.87	18.38	42.73	18.26	42.49	18.04	42.08	17.71
-0.20	53.60	28.73	53.60	28.73	53.62	28.75	53.58	28.75	53.58	28.71
-0.10	65.40	42.77	65.49	42.89	65.79	43.28	66.28	43.93	67.02	44.91
0.00	76.77	58.94	76.95	59.22	77.53	60.11	78.54	61.68	80.11	64.17
0.10	86.00	73.96	86.21	74.31	86.85	75.43	87.97	77.38	89.67	80.41
0.20	92.29	85.18	92.45	85.47	92.96	86.41	93.81	88.01	95.05	90.35
0.30	96.01	92.19	94.56	89.41	96.44	93.01	96.97	94.03	97.69	95.44
0.40	98.00	96.05	98.06	96.16	98.25	96.54	98.54	97.11	98.95	97.90
0.50	99.01	98.03	99.05	98.11	99.15	98.31	99.29	98.59	99.50	98.99
0.60	99.51	99.03	99.51	99.03	99.57	99.14	99.64	99.27	99.75	99.50
0.70	99.68	99.37	99.68	99.37	99.75	99.50	99.78	99.55	99.85	99.70

x/d	r/d = 0.25 E/Em [%]	Ux/Um [%]	r/d = 0.30 E/Em [%]	Ux/Um [%]	r/d = 0.35 E/Em [%]	Ux/Um [%]	r/d = 0.40 E/Em [%]	Ux/Um [%]	r/d = 0.45 E/Em [%]	Ux/Um [%]
-1.50	5.06	0.26	5.01	0.25	4.96	0.25	4.89	0.24	4.83	0.23
-1.40	5.69	0.32	5.62	0.32	5.56	0.31	5.48	0.30	5.40	0.29
-1.30	6.44	0.42	6.37	0.41	6.28	0.39	6.19	0.38	6.08	0.37
-1.20	7.36	0.54	7.27	0.53	7.16	0.51	7.03	0.49	6.90	0.48
-1.10	8.49	0.72	8.36	0.70	8.22	0.68	8.06	0.65	7.89	0.62
-1.00	9.89	0.98	9.73	0.95	9.54	0.91	9.33	0.87	9.11	0.83
-0.90	11.65	1.36	11.43	1.31	11.19	1.25	10.91	1.19	10.61	1.13
-0.80	13.90	1.93	13.61	1.85	13.27	1.76	12.90	1.66	12.50	1.56
-0.70	16.80	2.82	16.41	2.69	15.96	2.55	15.45	2.39	14.91	2.22
-0.60	20.60	4.24	20.08	4.03	19.47	3.79	18.79	3.53	18.03	3.25
-0.50	25.65	6.58	24.97	6.24	24.13	5.82	23.23	5.40	22.18	4.92
-0.40	32.41	10.51	31.59	9.98	30.57	9.34	29.32	8.60	27.86	7.76
-0.30	41.50	17.22	40.69	16.55	39.55	15.64	38.04	14.47	36.07	13.01
-0.20	53.45	28.57	53.14	28.24	52.47	27.54	51.20	26.21	48.95	23.96
-0.10	67.99	46.23	69.26	47.97	70.75	50.05	72.20	52.13	72.55	52.63
0.00	82.42	67.93	85.87	73.74	91.30	83.36	100.89	101.78	121.75	148.22
0.10	92.06	84.74	95.34	90.90	99.75	99.50	105.47	111.24	111.91	125.24
0.20	96.64	93.39	98.59	97.20	100.75	101.50	102.92	105.93	104.59	109.39
0.30	98.57	97.16	99.55	99.11	100.52	101.05	101.38	102.78	101.95	103.94
0.40	99.37	98.75	99.85	99.70	100.29	100.58	100.66	101.33	100.90	101.80
0.50	99.71	99.42	99.94	99.89	100.14	100.28	100.32	100.64	100.40	100.80
0.60	99.86	99.72	99.98	99.96	100.07	100.15	100.17	100.34	100.21	100.41
0.70	99.92	99.83	99.99	99.98	100.05	100.09	100.10	100.23	100.12	100.24

x/d	r/d = 0.50 E/Em [%]	Ux/Um [%]	r/d = 0.55 E/Em [%]	Ux/Um [%]	r/d = 0.60 E/Em [%]	Ux/Um [%]	r/d = 0.65 E/Em [%]	Ux/Um [%]	r/d = 0.70 E/Em [%]	Ux/Um [%]
-1.50	4.76	0.23	4.68	0.22	4.59	0.21	4.51	0.20	4.42	0.20
-1.40	5.31	0.28	5.22	0.27	5.11	0.26	5.01	0.25	4.89	0.24
-1.30	5.97	0.36	5.85	0.34	5.72	0.33	5.58	0.31	5.44	0.30
-1.20	6.75	0.46	6.60	0.44	6.44	0.41	6.27	0.39	6.09	0.37
-1.10	7.70	0.59	7.50	0.56	7.29	0.53	7.08	0.50	6.86	0.47
-1.00	8.86	0.78	8.60	0.74	8.33	0.69	8.05	0.65	7.77	0.60
-0.90	10.28	1.06	9.94	0.99	9.59	0.92	9.23	0.85	8.86	0.78
-0.80	12.06	1.46	11.61	1.35	11.13	1.24	10.65	1.13	10.17	1.03
-0.70	14.31	2.05	13.70	1.88	13.05	1.70	12.41	1.54	11.76	1.38
-0.60	17.21	2.96	16.35	2.67	15.47	2.39	14.59	2.13	13.71	1.88
-0.50	21.03	4.42	19.82	3.93	18.57	3.45	17.33	3.00	16.11	2.59
-0.40	26.22	6.87	24.45	5.98	22.62	5.11	20.80	4.33	19.06	3.63
-0.30	33.65	11.33	30.91	9.55	28.05	7.87	25.26	6.38	22.67	5.14
-0.20	45.36	20.57	40.56	16.45	35.58	12.66	30.96	9.59	26.93	7.25
-0.10	68.82	47.36	56.81	32.27	46.06	21.21	37.65	14.18	31.28	9.78
0.00			105.25	110.78	53.86	29.01	41.47	17.20	33.38	11.14
0.10	113.95	129.85								
0.20	105.18	110.63								
0.30	102.13	104.30								
0.40	101.00	102.01								
0.50	100.43	100.86								
0.60	100.24	100.49								
0.70	100.14	100.28								

x/d	r/d = 0.75 E/Em [%]	Ux/Um [%]	r/d = 0.80 E/Em [%]	Ux/Um [%]	r/d = 0.85 E/Em [%]	Ux/Um [%]	r/d = 0.90 E/Em [%]	Ux/Um [%]	r/d = 0.95 E/Em [%]	Ux/Um [%]
-1.50	4.33	0.19	4.23	0.18	4.13	0.17	4.03	0.16	3.94	0.16
-1.40	4.78	0.23	4.66	0.22	4.55	0.21	4.42	0.20	4.31	0.19
-1.30	5.31	0.28	5.16	0.27	5.02	0.25	4.87	0.24	4.73	0.22
-1.20	5.92	0.35	5.74	0.33	5.56	0.31	5.38	0.29	5.20	0.27
-1.10	6.64	0.44	6.41	0.41	6.19	0.38	5.96	0.36	5.75	0.33
-1.00	7.49	0.56	7.20	0.52	6.92	0.48	6.64	0.44	6.37	0.41
-0.90	8.49	0.72	8.13	0.66	7.77	0.60	7.42	0.55	7.08	0.50
-0.80	9.69	0.94	9.22	0.85	8.76	0.77	8.32	0.69	7.89	0.62
-0.70	11.13	1.24	10.51	1.11	9.92	0.98	9.35	0.87	8.81	0.78
-0.60	12.86	1.65	12.04	1.45	11.26	1.27	10.53	1.11	9.84	0.97
-0.50	14.93	2.23	13.83	1.91	12.80	1.64	11.85	1.40	10.98	1.20
-0.40	17.42	3.03	15.91	2.53	14.54	2.11	13.29	1.77	12.18	1.48
-0.30	20.32	4.13	18.23	3.32	16.40	2.69	14.80	2.19	13.40	1.80
-0.20	23.51	5.53	20.64	4.26	18.23	3.32	16.21	2.63	14.50	2.10
-0.10	26.42	6.98	22.65	5.13	19.68	3.87	17.27	2.98	15.29	2.34
0.00	27.64	7.64	23.44	5.50	20.26	4.10	17.68	3.13	15.54	2.42

x/d	r/d = 1.0 E/Em [%]	Ux/Um [%]
-1.50	3.83	0.15
-1.40	4.19	0.18
-1.30	4.58	0.21
-1.20	5.03	0.25
-1.10	5.53	0.31
-1.00	6.10	0.37
-0.90	6.75	0.46
-0.80	7.48	0.56
-0.70	8.30	0.69
-0.60	9.20	0.85
-0.50	10.17	1.04
-0.40	11.19	1.25
-0.30	12.18	1.48
-0.20	13.05	1.70
-0.10	13.66	1.86
0.00	13.86	1.92

REFERENCES

Adams, R. B. (1968) Particle stream position effects in electrical sizing, *Biophys. Soc. Abstr.* **8**:A112.

Adams, R. B., and Gregg, E. C. (1972) Pulse shapes from particles traversing Coulter orifice fields, *Phys. Med. Biol.* **17**:830.

Adams, R. B., Voelker, W. H., and Gregg, E. C. (1967) Electrical counting and sizing of mammalian cells in suspension. An experimental evaluation, *Phys. Med. Biol.* **12**:79.

Adell, R., Skaklak, R., and Branemark, P. J. (1970) A preliminary study of rheology of granulocytes, *Blut* **21**:91.

Ahrens, O., Albrecht, U., and Rajewsky, M. F. (1980) Microprocessor based data acquisition system for flow cytometers, in: *Flow Cytometry IV* (O. Laerum, T. Lindmo, and E. Thoroud, eds.), Universiteteforlaget Bergen, Bergen, Norway, pp. 112-115.

Alabaster, O., Glaubiger, D. L., Hamilton, V. T., Bentley, S. A., Shackney, S. E., Skramstad, K. S., and Chen, R. F. (1980) Electrolytic degradation of DNA fluorochromes during flow cytometric measurement of electronic cell volume, *J. Histochem, Cytochem.* **28**:330.

Allan, R. S., and Mason, S. G. (1962) Particle behaviour in shear and electric fields. I. Deformation and burst of fluid drops. *Proc. R. Soc.* **A,267**:45.

Anderson, J. L. and Quinn, J. A. (1971) The relationship between particle size and signal in Coulter-type counters, *Rev. Sci. Instr.* **42**:1257.

Benker, G. (1978) Entwurf und Aufbau einer elektronisch gesteuerten Bildaufzeichnungseinheit

Chase, R. L., and Poulo, L. R. (1967) A high precision dc restorer, *IEEE Trans. Nucl, Sci.* **167**:83.

Chaussy, L., Baethmann, A., and Lubitz, W. (1981) Electrical sizing of nerve and glia cells in the study of cell volume regulation, in: *Cerebral Microcirculation and Metabolism* (J. Cervaos-Navarro and E. Fritschka, eds.), Raven Press, New York.

Coulter Electronics (1971) Deutsche Patentschrift 2,153,123. Corresponds to U.S. Patents 84,440; 101,352; 113,165; 113,920.

Coulter Electronics (1972) Deutsche Patentschrift 2,216,826. Corresponds to U.S. Patents 132,771; 142,531.

Coulter, W. H. (1953) Means for counting particles suspended in a fluid, U.S. Patent No. 2,656,508.

Coulter, W. H. (1956) High speed automatic blood cell counter and cell size analyzer, *Proc. Natl. El. Conf., Chicago* **12**:1034.

Coulter, W. H. (1966) Manual to Coulter counter.

Coulter, W. H., Hogg, W. R., Moran, J. P., and Claps, W. A. (1959) Particle analyzing device, U.S. Patent No. 3,259,842.

Cutts, H. J. (1972) Balanced salt solutions, in: *Cell Separation Methods in Hematology* (H. J. Cutts, ed.), Academic Press, New York, pp. 169–174.

Davies, R. (1978) Recent progress in rapid-response and on-line methods for particle size analysis, *Am. Lab.* **10**(4):97.

Davies, R., Karuhn, R., and Graf, J. (1975) Studies on the Coulter counter, part II. Investigations into the effect of flow direction and angle of entry of a particle on both particle volume and pulse shape, *Powder Technol.* **12**;157.

Deaver, J. R. (1978) Modeling limits to cell size, *Am. Biol. Teacher* **40**;502.

DeBlois, R. W., and Bean, C. P. (1970) Counting and sizing of submicron particles by the resistive pulse technique, *Rev. Sci. Instr.* **41**:909.

De Blois, R. W., Mayyasi, S. A., Schildlovsly, G., Wesley, R., and Wolff, J. S. (1974) Virus counting and analysis by the resistive pulse (Coulter counter) technique, *Proc. Am. Assoc. Cancer Res.* **15**:104.

Dittrich, W., and Goehde, W. (1969) Impulsfluorometrie bei Einzelzellen in Suspensionen, *Z. Naturforsch.* **24b**:360.

Doljanski, F., Zajicek, G., and Naaman J. (1966) The size distribution of normal human red blood cells, *Life Sci.* **5**: 2095.

Eck, B. (1961) *Technische Stroemungslehre,* Springer, Berlin.

Ferris, C. D. (1963) Four-electrode electronic bridge for electrolyte impedance determinations, *Rev. Sci. Instr.* **34**:109.

Fricke, H. (1924) A mathematical treatment of the electric conductivity and capacity of disperse systems, *Phys. Rev.* **24**:575.

Fricke, H. (1953) The Maxwell-Wagner dispersion in a suspension of ellipsoids, *J. Phys. Chem.* **57**:934.

Fulwyler, M. J. (1965) Electronic separation of biological cells by volume, *Science* **150**:910.

Gebicki, J. M., and Hunter, F. W. (1964) Determination of swelling and disintegration of mitochondria with an electronic particle counter, *J. Biol. Chem.* **293**:631.

Geigy, A. G. (Basel) (1975) *Wissenschaftliche Tabellen,* Thieme Verlag, Stuttgart.

Goldsmith, H. L. (1971) Deformation of human red blood cells in tube flow, *Biorheology* **7**:233.

Golibersuch, D. C. (1973) Observation of aspherical particle rotation in Poiseuille flow via the resistance pulse technique, *Biophys. J.* **13**:265.

Grant, J. L., Britton, M. C., and Kurtz, T. E. (1960) Measurement of red blood cell volume with the electronic cell counter, *Am. J. Clin. Pathol.* **33**:138.

Gray, J. W., and Dean, P. W. (1980) Display and analysis of flow cytometric data, *Ann. Rev. Biophys. Bioeng.* **9**:509.

Gregg, E. C., and Steidley, K. D. (1965) Electrical counting and sizing of mammalian cells in suspension, *Biophys. J.* **5**:393.

Grover, N. B., Naaman, J., Ben Sasson, S., and Doljanski, F. (1969a) Electrical sizing of particles in suspension. I. Theory, *Biophys. J.* **9**:1398.

Grover, N. B., Naaman, J., Ben Sasson, S., Doljanski, F., and Nadev, E. (1969b) Electrical sizing of particles in suspension. II. Experiments with rigid spheres, *Biophys. J.* **9**:1415.

Grover, N. B., Naaman, J., Ben Sasson, S., and Doljanski, F. (1972) Electrical sizing of particles in suspension. III. Rigid spheroids and blood cells, *Biophys. J.* **12**:1099.

Groves, M. R. (1980) Application of the electrical sizing principle of Coulter to a new multiparameter system, *IEEE Trans. Biomed. Eng.* **27**:364.

Gutmann, J. (1966) Elektronische Verfahren zur Ermittlung statistischer Masszahlen einiger medizinisch wichtiger Daten, *Elektromedizin* **11**:62.

Haigh, G. T. (1973) Current normalizer for particle size analysis apparatus, U.S. Patent No. 3,745,455.

Hanser, H., Valet, G., Boss, N., and Ruhenstroth-Bauer, G. (1974) Origin and regulation of the different erythrocyte volume populations in the newborn rat, XV Congress of the International Society of Hematology, Jerulalem, p. 318.

Hanser, G., Valet, G., and Ruhenstroth-Bauer, G. (1979) Multi gene switches differentiation in the erythropoietic development of the young rat induced by diffusible substances, *Hoppe Seyler's Z. Physiol. Chem.* **360**:277.

Harvey, R. J. (1968) Measurement of cell volumes by electric sensing zone instruments, *Meth. Cell Physiol.* **3**:1.

Harvey, R. J., and Marr, A. G. (1966) Measurement of size distributions of bacterial cells, *J. Bacteriol.* **92**:805.

Haynes, J. L. (1980) High resolution particle analysis—Its application to platelet counting and suggestions for further application in blood cell analysis, *Blood Cells* **6**:201.

Haynes, J. L., and Shoor, B. A. (1978) Particle density measuring system, U.S. Patent No. 4, 110, 604.

Helleman, P. W. (1972) *The Coulter Electronic Particle Counter*, (the Netherlands) De Bilt, Holland.

Howard, R. B., and Pesch, L. A. (1968) Respiratory activity of intact isolated parenchymal cells from rat liver, *J. Biol. Chem.* **243**:3105.

Hueller, R. (1980) Untersuchungen an einem Durchflusszytometer mit selektiver Zellabbildung, Thesis, Fachhochschule München.

Hurley, J. (1970) Sizing particles with a Coulter counter, *Biophys. J.* **10**:74.

Jeltsch, E., and Zimmermann, U. (1979) Particles in a homogeneous electrical field: A model for the electrical breakdown of living cells in a Coulter counter, *J. Electroanal. Chem.* **104**:349.

Kachel, V. (1970) Measuring chamber for cell volume according to Coulter, *XIII Cong. Hematol. Muenchen Abstr.* pp. 392–392.

Kachel, V. (1972) Methods of analysis and correction of instrumental errors in the electronic method of Coulter for particle sizing, Doctoral Thesis, Technische Universität Berlin.

Kachel, V. (1973) The improvement of resolution in Coulter particle sizing by an electronic method, *Blut* **27**:270.

Kachel, V. (1974a) Methodology and results of optical investigations of form-factors during determination of cell volumes according to Coulter, *Microsc. Acta* **75**:419.

Kachel, V. (1974b) Schaltung zur Eichung einer Anordnung zur Partikelvolumenmessung, German Patent No. 24, 28, 082. 8.

Kachel, V. (1975) An improved device according to Coulter to measure volumes of cells and particles equipped with a particle independent calibrating system, *Biomed. Tech.* **20** (Suppl. Vol May) :191.

Kachel, V. (1976) Basic principles of electrical sizing of cells and particles and their realization in the new instrument METRICELL, *J. Histochem. Cytochem.* **24**:211.

Kachel, V. (1979) Electrical resistance pulse sizing (Coulter sizing), in: *Flow Cytometry and Sorting* (M. Melamed, P. Mullaney, M. Mendelsohn, eds.), John Wiley and Sons, New York, pp. 61–104.

Kachel, V., and Glossner, E. (1976) Vorrichtung zur Durchfuehrung von mindestens zwei Messungen von Eigenschaften in einer Partikelsuspension suspendierter Partikel, German Patent No. 26, 56, 624. 1.

Kachel, V., and Glossner, E. (1977) Vorrichtung zur Messung bestimmter Eigenschaften in einer Partikelsuspension suspendierter Partikel, German Patent No. 27, 50, 447, 8.

Kachel, V., and Meier, H. (1980) Control unit for on-line handling of three-parameter flow cytometric data, in: *Flow Cytometry IV* (O. Laerum, T. Lindmo, and E. Thoroud, eds.), Universitetsforlaget Bergen, pp. 120–124.

Kachel, V., and Menke, E. (1979) Hydrodynamic properties of flow cytometric instruments, in: *Flow Cytometry and Sorting* (M. Melamed, P. Mullaney, M. Mendelsohn, eds.), John Wiley and Sons, New York, pp. 41–59.

Kachel, V., Metzger, H., and Ruhenstroth-Bauer, G. (1970) The influence of the particle path on the volume distribution curves according to the Coulter method, *Z. Ges. Exp. Med.* **153**:331.

Kachel, V., Glossner, E., and Kordwig, E. (1974) Vorrichtung zur Messung bestimmter Eigenschaften in einer Fluessigkeit suspendierter Partikel, German Patent No. 24, 620, 63. 1

Kachel, V., Glossner, E., Kordwig, E., and Ruhenstroth-Bauer, G. (1977) FLUVO-METRICELL, a combined cell volume and cell fluorescence analyzer, *J. Histochem. Cytochem.* **25**:804.

Kachel, V., Benker, G., Lichtnau, K., Valet, G., and Glossner, E. (1979) Fast imaging in flow: A means of combining flow-cytometry and image analysis, *J. Histochem. Cytochem.* **27**:335.

Kachel, V., Benker, G., Weiss, W., Glossner, E., Valet, G., and Ahrens, O. (1980a) Problems of fast imaging in flow, in: *Flow Cytometry IV* (O. Laerum, T. Lindmo, and E. Thoroud, eds.), Universitetsforlaget Bergen, Bergen, Norway, pp. 49–55.

Kachel, V., Schneider, H., and Schedler, K. (1980b) A new flow cytometric pulse height analyzer offering microprocessor controlled data acquisition and statistical analysis, *Cytometry* **1**:175.

Karuhn, R., Davies, R., Kaye, B. H., and Clinch, M. J. (1975) Studies on Coulter counter. Part 1: Investigations into the effect of orifice geometry and flow direction on the measurement of particle volume, *Powder Technol.* **11**:157.

Kay, D. B., Cambier, J. L., and Wheeless, L. L. (1979) Imaging in flow, *J. Histochem. Cytochem.* **27**:329.

Koller, A. (1970) POT 123 Numerische Berechnung von Potentialfeldern Data Praxis, Siemens Publication, Bereich Datenverarbeitung, Munich.

Kubek, D. J., and Shuler, M. L. (1978) Electronic measurement of plant cell number and size in suspension culture, *J. Exp. Botany* **29**:511.

Kubitschek, H. E. (1958) Electronic counting and sizing of bacteria, *Nature (London)* **182**:234.

Kubitschek, H. E. (1960) Electronic measurement of particle size, *Research (London)* **13**:128.

Langhaar, H. L. (1942) Steady flow in the transition length of a straight tube, *J. Appl. Mech.* **9**:A55.

Leif, R. C., and Thomas, R. A. (1973) Electronic cell-volume analysis by use of AMAC I transducer, *Clin. Chem.* **19**:853.

Lewis, H. D., and Goldman, A. (1965) Proper analysis of Coulter counter data, *Rev. Sci. Instr.* **36**:868.

Lusbaugh, C. C., Maddy, J. A., and Basman, N. J. (1962a) Electronic measurement of cellular volumes. I. Calibration of the apparatus, *Blood* **20**:233.

Lusbaugh, C. C., Baseman, J. J., and Glascock, B. (1962b) Electronic measurement of cellular volumes. II. Frequency distribution of erythrocyte volumes, *Blood* **20**:241.

Malin-Berdel, J., and Valet G. (1980) Flow cytometric determination of esterase and phosphatase activities and kinetics in hematopoietic cells with fluorogenic substrates, *J. Histochem, Cytochem.* **1**:222.

Mattern, C. F. T., Brackett, F. S., and Olson, B. J. (1957) Determination of number and size of particles by electrical gating: Blood cells, *J. Appl. Physiol.* **10**:56.

Maxwell, J. C. (1883) *Lehrbuch der Elektrizitaet und des Magnetismus*, Springer-Verlag, Berlin.

Mel, H. C., and Yee, J. P. (1975) Erythrocyte size and deformability studies by resistive pulse spectroscopy, *Blood Cells* **1**:391.

Menke, E., Kordwig, E., Stuhlmueller, P., Kachel, V., and Ruhenstroth-Bauer, G. (1977) A volume activated cell sorter, *J. Histochem. Cytochem.* **25**:796.

Mercer, W. B. (1966) Calibration of Coulter counters for particles 1µ diameter, *Rev. Sci. Instr.* **37**:1515.

Merrill, J. T., Veizades, N., Hulett, H. R., Wolf, P. L., and Herzenberg, L. A. (1971) An improved cell volume analyzer, *Rev. Sci. Instr.* **42**:1157.

Metzger, H., Kachel, V., and Ruhenstroth-Bauer, G. (1971) The influence of particle size, form and consistence on the right skewness of Coulter volume distribution curves, *Blut* **23**:143.

Metzger, H., Valet, G., Kachel, V., and Ruhenstroth-Bauer, G. (1972) The calibration by electronic means of Coulter counters for determination of absolute particle volumes, *Blut* **25**:179.

Miller, G. G., and Wuest, L. J. (1972) Volume analysis of human red cells. II. The nature of the residue, *Ser. Hematol.* **5**(2):128.

Miller, R. G., Wuest, L. J., and Cowan, D. H. (1972) Volume analysis of human red cells. I. The general procedures, *Ser. Hematol.* **5**, (2):105.

Nash, G. B., Tathan, P. E. R., Powell, T., Twist, V. W., Speller, R. D., and Loverock, L. T. (1979) Size measurements on isolated rat heart cells using Coulter analysis and light scatter flow cytometry, *Biochim. Biophys. Acta* **587**:99.

Nevius, D. B. (1963) Osmotic error in electronic determinations of red cell volume, *Am. J. Clin. Pathol.* **39**:38.

Newbould, F. H. (1974) Electronic counting of somatic cells in farm bulk tank milk, *J. Milk Food* **37**:504.

Otto, F. (1970) Granulocytenisolierung aus dem Blut des Menschen und der Tiere, *Blut* **21**:290.

Otto, F., and Schmid, D. O. (1970) Lymphocytenisolierung aud dem Blut des Menschen und der Tiere, *Blut* **21**:118.

Patzelt, R. (1968) Improved base-line stabilization for pulse amplifiers, *Nucl. Instr. Meth.* **59**:283.

Paulus, J. M. (1975) Platelet size in man, *Blood* **46**:321.

Prantl, L. (1965) *Fuehrer durch die Stroemungslehre*, Verlag F. Vieweg und Sohn, Braunschweig.

Price-Jones, C. (1910) The variations in the sizes of red blood cells, *Br. Med. J.* **2**:1418.

Princen, L. H. (1966) Improved determination of calibration and coincidence correction constants for coulter counters, *Rev. Sci. Instr.* **37**:1416.

Princen, L. H., and Kwolek, W. F. (1965) Coincidence corrections for particle size determinations with the Coulter counter, *Rev. Sci. Instr.* **36**:646.

Rackham, S. J., and Sherlock, R. A. (1979) A pulse height analyzer for displaying Coulter counter particle size distributions, *IEEE Trans. Biomed. Eng.* **26**:436.

Robinson, L. B. (1961) Reduction of baseline shift in pulse amplitude measurements, *Rev. Sci. Instr.* **32**:1057.

Ross, D. W. (1978) The significance of leukemic cell volume distribution, *Nouv. Rev. Franc. Hematol.* **20**:297.

Ruhenstroth-Bauer, G., Valet, G., Kachel, V., and Boss, N. (1974) The electrical volume determination of blood cells in erythropoiesis, smokers, patients with myocardial infarction and leukemia and of liver cell nuclei, *Naturwissenschaften* **61**:260.

Rumscheidt, F. D., and Mason, S. G. (1961) Particle motion in sheared suspensions. XII. Deformation and burst of fluid drops in shear and hyperbolic flow, *J. Colloid Sci.* **16**:238.

Sadikov, I. N. (1967) Motion of a viscous fluid in the initial section of a flat channel, *Inzh. Fiz. Zh.* **12**:219.

Salzman, G. C., Mullaney, P. F., and Coulter, J. R. (1973) A Coulter volume spectrometer employing a potential sensing technique, *Biophys. Soc. Abstr.* **17**:302a.

Scheer, U., and Schellong, G. (1979) The prognostic value of measuring cell size in acute childhood leukemia, *Klin. Paediat.* **191**:127.

Schmid-Schoenbein, H., and Wells, R. (1969) Fluid drop like transition of erythrocytes under shear, *Science* **165**:288.

Schmid-Schoenbein, H., Wells, R. E., and Goldstone, J. (1971) Fluid drop like behaviour of erythrocytes. Disturbance in pathology and its quantification, *Biorheology* **7**:227.

Schultz, J., and Nitsche, H. J. (1972) Nachweis des transzellulaeren Ionenflusses bei der Volumenbestimmung von nativen Humanerythrozyten, Telefunken Publication N1/EP/V 1698.

Schulz, J., and Thom, R. (1973) Electrical sizing and counting of platelets in whole blood, *Med. Biol. Eng.* **1973**:447.

Shank, B. B., Adams, R. B., Steidley, K. D., and Murphy, L. R. (1969) A physical explanation of the bimodal distribution obtained by electronic sizing of erythrocytes, *J. Lab. Clin. Med.* **74**:630.

Sharpless, T. K. (1979) Cytometric data processing, in: *Flow Cytometry and Sorting* (M. Melamed, P. Mullaney and M. Mendelsohn, eds.), John Wiley and Sons, New York, p. 367.

Shuler, M. L., Aris, R., and Tsuchiya, H. M. (1972) Hydrodynamic focusing and electronic cell sizing techniques, *Appl. Microbiol.* **24**:384.

Smith, A. M. O. (1960) Remarks on transition in a round tube, *J. Fluid Mech.* **7**:565.

Smither, R. (1975) Use of a Coulter counter to detect discrete changes in cell numbers and volume during growth of Escherichia coli, *J. Appl. Bacteriol.* **39**:157.

Smythe, W. R. (1961) Flow around a sphere in a circular tube, *Phys. Fluids* **4**:756.

Smythe, W. R. (1964) Flow around spheroids in a circular tube, *Phys. Fluids* **7**:633.

Spielman, L., and Goren, S. L. (1968) Improving resolution in Coulter counting by hydrodynamic focusing, *J. Colloid Interface Sci.* **26**:175.

Steen, H. B., and Nielsen, V. (1979) Lymphocyte blastogenesis studied by volume spectroscopy, *Scand. J. Immunol.* **10**:135.

Steinkamp, J. A., Fulwyler, M. J., Coulter, J. R., Hiebert, R. D., Horney J. L., and Mullaney, P. F. (1973) A new multiparameter separator for microscopic particles and biological cells, *Rev. Sci. Instr.* **44**:1301.

Strackee, J. (1966) Coincidence loss in blood counters, *Med. Biol. Eng.* **4**:97.

Tatsumi, T. (1952) Stability of the laminar inlet-flow prior to the formation of Poiseuille regime. I., *J. Phys. Soc. Jpn.* **7**:489.

Thom, R. (1968) Zur Rechtsschiefe der Erythrozyten Volumenverteilungskurven, Coulter Symposion Bad Nenndorf pp. 33–36.

Thom, R. (1972a) Method and result by improved electronic cell sizing, in: *Modern Concepts in Hematology* (G. Izak, ed.), Academic Press, New York, pp. 191–200.

Thom, R. (1972b) Vergleichende Untersuchungen zur elektronischen Zellvolumenanalyse, Telefunken Publikation N1/EP/V 1698.

Thom, R., and Kachel, V. (1971) Fortschritte 'fuer die elektrcr elektronische Groessenbestim-
mung von Blutkoerperchen, *Blut* **21**:48.

Thom, R., Hampe, A., and Sauerbrey, G. (1969) Die elektronische Volumenbestimmung von
Blutkoerperchen und ihre Fehlerquellen, *Z. Gesamte Exp. Med.* **151**:331.

Thomas, R. A., Cameron, B. F., and Leif, R. C. (1974) Computer based electronic cell volume
analysis with the AMAC II transducer, *J. Histochem, Cytochem.* **22**:626.

Tietze, U., and Schenk, C. (1971) *Halbleiter Schaltungstechnik,* 2nd ed., Springer, Berlin, Hei-
delberg, New York.

Valet, G. (1980) Graphical representation of three-parameter flow cytometer histograms by a
newly developed FORTRAN IV computer program, in: *Flow Cytometry IV* (O. Laerum,
T. Lindmo, and E. Thoroud, eds.), Universitetsforlaget Bergen, Bergen, Norway, pp. 125–
129.

Valet, G. and Opferkuch, W. (1975) Mechanism of complement-induced cell lysis demonstrating
a three step mechanism of EAC1-8 cell lysis by C9 and of a non-osmotic swelling of erythrocytes, *J. Immunol.* **115**:1028.

Valet, G., Metzger, H., Kachel, V., and Ruhenstroth-Bauer, G. (1972a) The volume distribution
curves of rat erythrocytes after whole body X-irradiation, *Blut* **24**:274.

Valet, G., Megzger, H., Kachel, V., and Ruhenstroth-Bauer, G. (1972b) The demonstration of
several erythrocyte populations in the rat, *Blut* **24**:42.

Valet, G., Hanser, H., Metzger, H., and Ruhenstroth-Bauer, G. (1974) Several erythrocyte populations in the blood of the newborn rat, mouse, guinea pig and in the human fetus, XVth
Congress of the International Society of Hematology Jerusalem, p. 317.

Valet, G., Schindler, R., Hanser, H., and Ruhenstroth-Bauer, G. (1975a) Several erythrocyte
populations of different mean volume in the young sheep and rat with different electrophoretic mobilities, 3rd Meeting of the European–African Division of the International Society
of Hematology, London, Abstr. 18, (2).

Valet, G., Silz, S., Metzger, H., and Ruhenstroth-Bauer, G. (1975b) Electrical sizing of liver
cell nuclei by the particle beam method. Mean volume, volume distribution and electrical
resistance, *Acta Hepato-Gastroenterol.* **22**:274.

Valet, G., Hofmann, H., and Ruhenstroth-Bauer, G. (1976) The computer analysis of volume
distribution curves: Demonstrating of two erythrocyte populations of different size in the
young guinea pig and analysis of the mechanism of immune lysis of cells by antibody and
complement, *J. Histochem. Cytochem.* **24**:231.

Valet, G., Hanser, G., and Ruhenstroth-Bauer, G. (1977) Ein neues Konzept der Haematopoese
im Saeugetierorganismus waehrend der Nachgeburtsphase: Nachweis mehrerer Volumen-
populationen der Erythrozyten bei Ratten, Maeusen, Meerschweinchen, Kaninchen,
Schafen und Ziegen, *Blut* **34**:413.

Valet, G., Fischer, B., Sundergeld, A., Hanser, G., Kachel, V., and Ruhenstroth-Bauer, G.
(1979) Simultaneous flow cytometric DNA and volume measurements of bone marrow cells
as sensitive indicator of abnormal proliferation patterns in rat leukemias, *J. Histochem.
Cytochem.* **27**:398.

van Dilla, M. A., Fulwyler, M. J., and Boone, J. U. (1967) Volume distribution and separation
of normal human leukocytes, *Proc. Soc. Exp. Biol. Med.* **125**:367.

Velick, S., and Gorin, M. (1940) The electrical conductance of suspensions of ellipsoids and its
relation to the study of avian erythrocytes, *J. Gen. Physiol.* **23**:753.

von Behrens, W., and Edmondson, S. (1976) Comparison of techniques improving the resolution
of standard Coulter cell sizing systems, *J. Histochem. Cytochem.* **24**:247.

Wales, M., and Wilson, J. N. (1961) Theory of coincidence in Coulter counters, *Rev. Sci. Instr.*
32:1132.

Wales, M., and Wilson, J. N. (1962) Coincidence in Coulter counters, *Rev. Sci. Instr.* **33**:575.

Waterman, C. S., Atkinson, E. E., Wilkins, B., Fischer, C. L., and Kimzey, S. I. (1975) Improved measurement of erythrocyte volume distribution by aperture-counter signal analysis, *Clin. Chem.* **21**:1201.

Weber, A., and Mueller, E. (1978) Investigations on factors influencing the accuracy of bull spermatozoa-counts with an electronic particle counter (Coulter counter), *Zuchthygiene* **13**:97.

Weed, R. J., and Bowdler, A. J. (1967) The influence of hemoglobin concentration on the distribution pattern of the volumes of human erythrocytes, *Blood* **29**:297.

Wendt, G. (1958) *Elektrische Felder und Wellen,* in: *Handbuch der Physik,* Volume XVI, Springer-Verlag, Berlin, pp. 148–164.

Wilkins, B., Fraudolig, J. E., and Fischer, C. L. (1970) An interpretation of red cell volume distributions measured by pulse height analysis, *J. Assoc. Adv. Med. Instr.* **4**:99.

Winter, H., and Sheard, R. P. (1965) The skewness of volume distribution curves of erythrocytes, *Austr. J. Exp. Biol. Med. Sci.* **43**:687.

Zimmermann, U., Pilwat, G., and Riemann, F. (1974) Dielectric breakdown of cell membranes, *Biophys. J.* **14**:881.

Zimmermann, U., Schulz, J., and Pilwat, G. (1975) Transcellular ion flow in *Escherichia coli* B and electrical sizing of bacteria, *Biophys. J.* **13**:1005.

Index